Second Edition

Integrated Product and

and

Process Design and Development

The Product Realization Process

Edward B. Magrab
Satyandra K. Gupta
F. Patrick McCluskey
Peter A. Sandborn

CRC Press
Taylor & Francis Group
Boca Raton London New York

CRC Press is an imprint of the
Taylor & Francis Group, an **informa** business

CRC Press
Taylor & Francis Group
6000 Broken Sound Parkway NW, Suite 300
Boca Raton, FL 33487-2742

First issued in paperback 2019

ISBN-13: 978-1-4200-7060-6 (hbk)
ISBN-13: 978-0-367-38537-8 (pbk)

Library of Congress Cataloging-in-Publication Data

Integrated product and process design and development : the product realization process / Edward B. Magrab ... [et al.]. -- 2nd ed.
      p. cm.
   Includes bibliographical references and index.
   ISBN 978-1-4200-7060-6 (alk. paper)
   1. New products. 2. Production engineering. 3. Design, Industrial. 4. Quality control. I. Magrab, Edward B. II. Magrab, Edward B. Integrated product and process design and development. III. Title.

TS170.M34 2009
658.5'75--dc22

2009003653

Visit the Taylor & Francis Web site at
http://www.taylorandfrancis.com

and the CRC Press Web site at
http://www.crcpress.com

Dedication

To

June Coleman Magrab

Contents

Preface — Second Edition

Since the first edition of this book appeared more than a decade ago, the product realization process has undergone a number of significant changes due, in large part, to globally competitive corporations that are producing innovative, visually appealing, quality products within shorter and shorter development times.

This second edition reflects these advances while still meeting the goal of the first edition: to present a thorough treatment of the modern tools used in the integrated product realization process. The book presents a coherent and detailed introduction to the creation of high quality products by using an integrated approach to the product realization process. It emphasizes the role of the customer and how one translates customer needs into product requirements and specifications. It provides methods that can be used to perform product cost analyses and gives numerous suggestions on how to generate and evaluate product concepts that will satisfy the customers' needs. It then introduces several important product development steps that are usually considered simultaneously: materials and manufacturing processes selection and assembly procedures. It then considers the impact that life-cycle goals, environmental aspects, and safety requirements have on the product's outcome. Lastly, the design of experiments and the six sigma philosophy are briefly introduced as one means of attaining quality.

The book provides numerous figures and tables to illustrate the various ideas, concepts, and methods presented, and two book-long examples provide the reader with a realistic sense of how a product's creation progresses through its various stages. It will be found that the book contains a large amount of specific information that normally appears in many separate sources.

To capture the newer aspects of the product realization process, the author was fortunate to have had three of his colleagues help him enhance the original material. Dr. Satyandra K. Gupta read the entire manuscript, made numerous suggestions for improvements, and added new material on in-mold assembly, layered manufacturing, and bio-inspired concept generation. Dr. Peter Sandborn completely rewrote Chapter 3, "Product Cost Analysis." This chapter now explains how one computes manufacturing cost, costs of ownership, and life-cycle costs of products and systems, and how these costs can influence a design team's decision-making process. Dr. F. Patrick McCluskey extensively revised Chapter 8, "Material Selection," and added new sections on such modern materials as engineered plastics, ceramics, composites, and smart materials. In addition, the first chapter has been rewritten to reflect the advances that have been made during the last decade and to place the product realization process in its new context. The section on concept generation has been expanded to include bio-inspired concept generation and TRIZ.

The book can be used as a single, comprehensive source on the integrated product realization method. The material has been used successfully in the Department of Mechanical Engineering at the University of Maryland at the senior level for a decade. Since many companies are now expecting newly graduated engineers to have the capabilities, approaches, and skills associated with the approach presented in this book, it should prove useful to both beginning and experienced engineers who may need to learn more about the modern approach to the product realization process. The integrated product realization method has applicability in the development of mechanical and electromechanical products; aircraft systems and subsystems; electronic packaging and fabrication; building design and construction; and in the development and procurement of military hardware.

Edward B. Magrab
Satyandra K. Gupta
F. Patrick McCluskey
Peter A. Sandborn
College Park, Maryland

Preface—First Edition

The product realization process during the last decade has undergone a number of very important changes, many of them brought about by the increasing international competition based on quality, cost, and time-to-market. The material in this book presents the development of the integrated product and process design and development (IP^2D^2) team method, which has been successfully used to conceptualize, design, and rapidly produce competitively-priced quality products. The IP^2D^2 descriptor was selected to indicate, in the broadest sense, the overlapping, interacting, and iterative nature of all of the aspects that impact the product realization process. The method is a continuous process whereby a product's cost, performance and features, value, and time-to-market lead to a company's increased profitability and market share.

The new paradigm for the IP^2D^2 team approach is to consider a very broad set of requirements, objectives, and constraints in a more or less overlapping manner prior to the start of the detailed design process. This approach to the product creation process is one in which the evaluation and selection of the final candidate solution are made from a comprehensive list of alternatives that initially appear to satisfy a set of functional requirements and their constraints. Hence, the goal of the book is to create an attitude toward design that encourages creativity and innovation, while considering as an integral, and equally important part of the product development process, the more or less simultaneous consideration of customer requirements and satisfaction, quality, reliability, manufacturing methods and material selection, assembly, cost, the environment, scheduling, and so on. The book also demonstrates the need for the members of an IP^2D^2 team to represent many different types of knowledge and company constituencies; from business, marketing, purchasing, and service to design, materials, manufacturing, and production.

The book details the means of implementing an integrated approach to the product realization process, and contains a large amount of specific information that is normally widely scattered throughout many sources. It emphasizes customer satisfaction and its relationship to the product's definition, and presents and illustrates proven methods that have been used successfully to create products. The book give numerous figures and tables to illustrate the various ideas, concepts, and methods presented, and includes two book-long examples to provide the reader with a realistic sense of how a product's creation progresses through its various stages. It is felt that these two examples will greatly enhance the understanding of the various stages of the IP^2D^2 process. However, to gain the most benefit from the process described in this book, one should participate in the process.

There is a catch-22 situation in trying to convey the integrated nature of the new product realization process. The IP^2D^2 method is more or less a simultaneous and iterative one; however, when one introduces the method, it must be done sequentially. Therefore, when introducing the method, the way it is learned and the way it is applied in practice after it has been learned will differ in this regard. That is, the steps that are learned in a sequential manner will be applied in an overlapping and iterative manner, and with differing time scales. The method described here contains all the components as presently applied; however, different organizations tend to apply them to differing degrees depending on their products and on their policies.

The material in this book is arranged in the following manner. The first three chapters introduce the IP^2D^2 method in context with its evolution to its present form, define quality and show how it now is one of the driving forces in product development, outline the goals and methods that have been successfully used to realize a product; explain what the IP^2D^2 method is and the order in which its tasks are usually implemented; and, lastly, identify the factors that influence a product's cost.

Chapters 4 to 6 give specific procedures that an IP^2D^2 team can use to obtain customer needs, convert these needs into a multilevel set of functional requirements for the product, and generate and evaluate numerous candidate solutions and embodiments to arrive at a product that satisfies the customer.

Chapters 7 to 10 present the most important aspects of design for X, that is, a design process that produces products that maximize the individual desirable product characteristics—the Xs. Chapters 7 to 9 cover assembly methods, materials selection, and manufacturing processes, three very important aspects of the product development cycle that affect the product's cost, time-to-market, producibility, plant productivity, and product reliability. Chapter 10 presents numerous specific suggestions on how the IP^2D^2 team can satisfy several manufacturing, marketing, social, life-cycle, and environmental requirements, which sometimes place conflicting constraints on the product.

The last chapter, Chapter 11, introduces a very powerful statistical technique that can be used to improve a product and the processes that make it.

The book can be used as a single, comprehensive source on the IP^2D^2 method. The material has been used successfully in the Department of Mechanical Engineering at the University of Maryland at the junior and senior levels and at the graduate level. Since many companies are now expecting newly graduated engineers to have the capabilities, approaches, and skills associated with the approach presented in this book, it should prove useful to both beginning and experienced engineers who may need to learn more about the modern approach to the product realization process. The IP^2D^2 method has applicability in the development of mechanical and electromechanical products, aircraft systems and subsystems, electronic packaging, building design and construction, and in the development and procurement of military hardware.

The author was very fortunate during the generation of the final manuscript to have many of his colleagues from the Mechanical Engineering Department at the University of Maryland at College Park provide considerable input that led to many improvements. Drs. George Dieter and Shapour Azarm read the entire manuscript and provided numerous suggestions and insights. Dr. MarjorieAnn Natishan was very helpful with the material appearing in Chapters 8 and 9. Most of the material in these two chapters was taken from a portion of the master's thesis of Arun Kunchithapatham, who integrated, under the author's direction, this material into a computer tool called the Design Advisor. Drs. Ioannis Minis and Guang Ming Zhang read Chapter 11, and Dr. Minis provided its example #4. In addition, Dr. Minis also made substantial contributions to Section 10.10. Dr. Zhang also provided a large amount of feedback from the use of the final manuscript in his fall 1996 junior course, Product Engineering and Manufacturing. Melvin Dedicatoria did the vast majority of the drawings.

The two book-long problems, the drywall taping system and the steel frame joining tool, are a synthesis of the final results of semester-long projects submitted by the students from the author's fall 1994 and fall 1996 graduate class, Design for Manufacture, respectively. The data used in examples #1 and #3 in Chapter 11 were obtained from the reports submitted by the students in the author's graduate course Advanced Engineering Statistics. The material in Table 6.5 is a synthesis of the results submitted by the students in a two-semester senior course, Integrated Product and Process Development, taught by Dr. David Holloway during 1995–96.

Support to produce many aspects of this book was provided by a very generous grant from the Westinghouse Foundation, and by an ARPA/NSF Technology Reinvestment Project award titled "Preparing Engineers for Manufacturing in the 21st Century," of which the author was director.

Authors

Edward B. Magrab is emeritus professor, Department of Mechanical Engineering, College of Engineering, University of Maryland at College Park and former director of the Manufacturing Program in the College's Engineering Research Center. He has done research in the integration of design and manufacturing. Prior to joining the University of Maryland he held several supervisory positions in the Center for Manufacturing Engineering at the National Institute of Standards of Technology (NIST) over a 12-year period, including head of the Robot Metrology Group and manager of the vertical machining work station in their Automated Manufacturing Research Facility. Dr. Magrab went to NIST after spending 9 years on the faculty in the Department of Mechanical Engineering at Catholic University of America in Washington, DC. Dr. Magrab has written seven books and numerous journal articles, and holds one patent. He is a life fellow of the American Society of Mechanical Engineers and a registered professional engineer in Maryland.

Satyandra K. Gupta is a professor in the Mechanical Engineering Department and the Institute for Systems Research at the University of Maryland. He is interested in developing computational foundations for next-generation computer-aided design and manufacturing systems. His research projects include generative process planning for machining, automated manufacturability analysis, automated generation of redesign suggestions, generative process planning for sheet metal bending, automated tool design for sheet metal bending, assembly planning and simulation, extraction of lumped parameter simulation models for microelectromechanical systems, distributed design and manufacturing for solid freeform fabrication, 3D shape search, reverse engineering, and automated design of multistage and multipiece molds. He has authored or coauthored more than 150 articles in journals, conference proceedings, and book chapters. He is a member of American Society of Mechanical Engineers (ASME), Society of Manufacturing Engineers (SME), and Society of Automotive Engineers (SAE). He has served as an associate editor for the *IEEE Transactions on Automation Science and Engineering* and the ASME *Journal of Computing and Information Science in Engineering.* He has also served on the program committees for the Geometric Modeling and Processing Conference, Computer Aided Design Conference, Product Lifecycle Management Conference, CAD and Graphics Conference, and ACM Solid and Physical Modeling Conference.

F. Patrick McCluskey is an associate professor of mechanical engineering at the University of Maryland, College Park, and a member of the CALCE Center. He has published extensively in the area of materials and materials processing for microelectronics, microsystems (MEMS), and their packaging. He is the coauthor of three books and numerous book chapters, including the book *Electronic Packaging Materials and Their Properties.* He has also served as general or technical chairman for numerous conferences in this area and is an associate editor of the *IEEE Transactions on Components and Packaging Technologies.* He is coordinator for the undergraduate Engineering Materials and Manufacturing Processes and Mechanical Design of Electronic Systems courses. He received his Ph.D. in materials science and engineering from Lehigh University.

Peter A. Sandborn is a professor in the Mechanical Engineering Department and the Research Director in the CALCE Electronic Products and Systems Center (EPSC) at the University of Maryland at College Park. His research interests include technology tradeoff analysis, system life cycle economics, technology obsolescence, and virtual qualification of systems. Prior to joining the University of Maryland, he was a founder and chief technical officer of Savantage, Austin,

Texas, and a senior member of the technical staff at the Microelectronics and Computer Technology Corporation, Austin. He is the author of over 100 technical publications and books on multichip module design and electronic part obsolescence forecasting. Dr. Sandborn is an associate editor of the *IEEE Transactions on Electronics Packaging Manufacturing* and a member of the editorial board for the *International Journal of Performability Engineering.*

1 Product Development at the Beginning of the Twenty-First Century

The current state of the product realization process is summarized and several of its important aspects are introduced.

1.1 INTRODUCTION

The process of creating and making artifacts has been around since the beginning of humankind. It was first applied to the creation of implements for survival: weapons, shelters, clothing, and farming. These implements were improved upon with the appearance of such inventions as fire, the wheel, and steel, and as time went on they became more substantial and more sophisticated. As societies evolved, so did their needs and the artifacts that were required to satisfy those needs. In addition, many societies evolved from being local societies to being regional ones, simultaneously transforming their local economies into regional ones. In the beginning, these transformations took hundreds to thousands of years. Since the start of the industrial revolution about 300 years ago, the pace of development and improvement of devices and artifacts has increased dramatically. During this period of time we have seen companies grow from local entities to global entities, and we have seen in the industrialized nations the economies transition for national economies to interdependent global economies. This has been particularly true in the last half of the twentieth century.

This transformation from primarily local societies to ones that must now compete globally has had a very substantial influence on the product realization process. It is an environment in which one must compete on cost, quality, performance, and time-to-market on a worldwide basis. This requires individuals and companies to reexamine how they go about creating products and services and how these products and services can be brought to the marketplace. During the last 30 years it has become clear that the way to do this is through an integrated approach to the product realization process. This approach tends to do the following: "flatten" organizational structures; involve many more constituencies in the process at the very beginning; place greater emphasis on the customer, product quality, cost, and time-to-market; use a large amount of simultaneity in the realization process; and require organizations to be creative and innovative.

These new approaches have been developed to eliminate situations and conditions that resulted in poor corporate performance and poor customer satisfaction. Some examples of these situations were inconsistent product quality; slow response to the marketplace; lack of innovative, competitive products; noncompetitive cost structure; inadequate employee involvement; unresponsive customer service; and inefficient resource allocation. In its place, these new approaches have transformed many companies into entities that are able to

- Respond quickly to customer demands by incorporating new ideas and technologies into products.
- Produce products that satisfy customers' expectations.
- Adapt to different business environments.
- Generate new ideas and combine existing elements to create new sources of value.

At various stages in the evolution of product development in the last four decades, various descriptors have been used to indicate that an improved method was being implemented to design and manufacture products. The descriptor that will be used here is the **integrated product and process design and development** (IP^{2}D^2) team method. This descriptor is used to indicate, in the broadest sense, the overlapping, interacting, and iterative nature of many of the aspects that impact the product realization process. The method is a continuous process that has the goal of producing products whose cost, performance and features, value, and time-to-market lead to a company's increased profitability and market share.

It is the purpose of this book to present the ways in which one can conduct the integrated product realization process at its various stages in such an environment. This process requires that the IP^{2}D^2 team interact with customers, company management, competitors' products, and suppliers. These interactions strongly influence the design process and require the IP^{2}D^2 team to use certain types of tools and methods to manage these interactions in a constructive manner. For example:

- IP^{2}D^2 teams must interact with customers to understand their needs and preferences and to get their feedback about existing products.
- Quality is very important to customers. Consequently, the IP^{2}D^2 teams need to ensure that the quality of a product meets customer expectations.
- IP^{2}D^2 teams must continually monitor the competitors' products by benchmarking them.
- IP^{2}D^2 teams need to interact with company management to understand how the current product fits in the overall company strategy.
- IP^{2}D^2 teams need to interact with the suppliers to understand their cost structure and to obtain advice on manufacturability.

In order to meet these objectives, this book is divided into 11 chapters, each chapter dealing with a particular aspect of the product realization process from the point of view of an engineer. There is, however, a difficulty that occurs in trying to convey the integrated nature of the product realization process. The product realization process is more or less an overlapping and iterative one; however, when one introduces the method, it must be done sequentially. Therefore, when introducing the method, the way it is learned and the way it is applied in practice, after it has been learned, will differ in this regard. That is, the steps that are learned in a sequential manner will be applied in an overlapping and iterative manner, and with differing time scales.

This chapter, Chapter 1, places in context the environment in which product development engineers have to work. It provides a brief overview of the current state of the manufacturing enterprise and describes several methods that are being implemented successfully in many globally competitive companies. In Chapter 2, the integrated product and process design and development method is described, and suggestions for its successful implementation are presented. In Chapter 3, we present methods that are used to determine a product's total cost—that is, its cost from the time it is conceived to the time it is disposed of as trash or recycled.

Chapters 4, 5, and 6 tackle the heart of the engineers' tasks—how to go about creating profitable products that customers want at a cost that they are willing to pay. Chapter 4 discusses ways in which customer requirements can be determined and translated into product specifications. Chapter 5 introduces methods that can be used to convert the customer requirements into a product's functional requirements and specifications. Chapter 6 suggests ways in which product concepts can be generated, evaluated, and turned into physical entities (embodiments) that satisfy the customer requirements.

Chapters 7 through 10 present many of the important aspects of design for "X"—that is, a design process that produces products that maximize the individual desirable product characteristics, denoted by the X. Chapters 7 to 9 cover assembly methods, materials selection, and manufacturing processes, respectively—three very important and interdependent aspects of the product development cycle that affect the product's cost, time-to-market, producibility, plant productivity, and

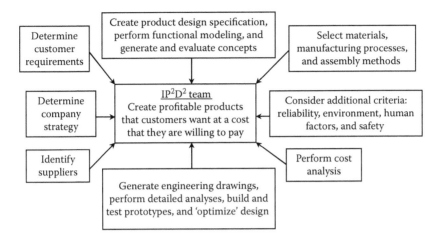

FIGURE 1.1 Major tasks of an IP^{2}D^2 team.

product reliability. Chapter 10 presents specific suggestions on how a product development team can satisfy several manufacturing, marketing, social, life-cycle, and environmental requirements, which sometimes place conflicting constraints on the product. The last chapter, Chapter 11, introduces a powerful statistical technique that can result in products that have fewer defects, reduced variability and closer conformance to target values, reduced development time, and reduced costs.

The topics that have been described above are summarized in Figure 1.1.

1.2   IDEAS AND METHODS CURRENTLY USED IN THE PRODUCT REALIZATION PROCESS

1.2.1   INTRODUCTION

We introduce and briefly discuss the following basic terms: *engineering design*, *manufacturing*, *logistics*, and *producibility*.

1.2.1.1   Engineering Design

Engineering design is a systematic, creative, and iterative process that applies engineering principles to conceive and develop components, systems, and processes that meet a specific set of needs. It is a dynamic and evolutionary process that involves four distinct aspects[*]:

- **Problem definition**—progression from a fuzzy set of facts and myths to a coherent statement of the problem. This is the stage where the idea for the product is formed.
- **Creative process**—a highly subjective means of devising a physical embodiment of the solution that depends greatly on the specific knowledge of the people participating in the process. This is the stage where various concepts for converting the idea into a product are generated.
- **Analytical process**—determines whether the proposed solutions are correct, thereby providing a means of evaluating them. This is stage where prototypes are constructed and evaluated.
- **Ultimate check**—confirmation that the design satisfies the original requirements.

---

[*] N. P. Suh, *Principles of Design*, Oxford University Press, New York, 1990.

These four aspects of engineering design are an integral part of the product realization process, which is described in detail in Section 2.2. It is mentioned that the degree of originality in a design can vary. The design process may be used to create a product that hasn't existed before or it may adapt an existing product to a new application or it may simply improve an existing product. Each of these types of design objectives still requires the integrated product realization process.

Aesthetic design usually refers to the creative act of fashioning an object or device without concern with how, or even if, it can be made. Aesthetic design has now become increasingly important to the product realization process and companies are now seeking professionals who can integrate engineering and aesthetic design. When aesthetic design is specifically integrated into a product's creation for the purpose of improving its usability and marketability, it is usually referred to as *industrial design*. Industrial design emphasizes those aspects of the product or system that relate most directly to human characteristics, needs, and interests such as visual, tactile, safety, and convenience.

1.2.1.2 Manufacturing

Manufacturing is a series of activities and operations that transform raw materials into a product suitable for use.

1.2.1.3 Logistics

Logistics is the time-related positioning of resources. It includes the planning, acquisition, storage, and distribution of goods, energy, information, personnel, and services from the point of origin to the point of consumption in a way that meets the manufacturers' and customers' requirements.

The tracking of resources is an important part of logistics and has been made easier with the use of bar codes. However, a disadvantage of bar codes is that in order for it to be read there must be a direct line of sight between the bar code and the bar code reader. In the last decade or so radio frequency identification (RFID) devices are being employed as a replacement for bar codes. Upon interrogation, these inexpensive microchip-size devices emit a weak radio signal that carries a small amount of information about the item to which it is attached. A radio receiver requests and captures this information. Its advantages over the bar code are that line of sight is not required to read it, it can track moving objects, and the RFID device can contain additional data. These RFID devices are now being used in a wide variety of applications. Some examples[*] are to track and monitor molds in manufacturing plants; to ensure that the components of construction site tower cranes are available in time for their assembly; to track shipping containers from the factory to the storage yards; to manage and track blood in blood banks; to track logs as they move from forest to factory by placing them in plastic nails that are embedded in the logs; and to track cash deliveries to ATMs.

1.2.1.4 Producibility

Producibility denotes the ease with which a product can be made, which is a measure of how easily a design can be manufactured to engineering drawings, on schedule, with the highest level of quality, and at a cost low enough to make a profit. It contains the essence of one of the early contributions during the development of the IP^2D^2 process, which was to use proven design processes that included the following guidelines.

- Simplify by reducing the number and types of parts and part features.
- Standardize by using standard parts, tolerances, part families, and a high degree of interchangeability.
- Select components that employ preferred sizes, weights, materials, near net shapes, etc.
- Ensure testability and reparability by using built-in test features, modularity, test points, and accessibility.

---

[*] Case Studies, *RFID Journal*, http://www.rfidjournal.com/article/archive/4/0.

- Use developmental testing to attain quality improvement, part qualification, and proof of performance during environmental stress screening.
- Minimize the number of different materials.

1.2.2 THE JAPANESE CONTRIBUTION TO THE PRODUCT DEVELOPMENT PROCESS

Many Japanese companies have developed methods that have come to exemplify the successful implementation of several aspects of the product realization process for primarily high-volume manufacturing. These methods were developed as a result of adopting the following principles of Austrian-born academician and management consultant Peter Drucker, first laid out in the late 1940s and early 1950s: corporations must move away from a command-and-control structure and cultivate a true spirit of teamwork at all levels; line workers must adopt a managerial outlook and take responsibility for the quality of what they produce; and the enterprise must be steered by a clear set of objectives while giving each employee the autonomy to decide how to reach those results. Although widely accepted now, many U.S. companies at the time dismissed these notions. Today, of course, these methods have been adopted by many of the U.S. globally competitive companies.

The methods that we shall briefly discuss are just-in-time (JIT) manufacturing, continuous improvement, and lean manufacturing.

1.2.2.1 Just-In-Time (JIT) Manufacturing

Just-in-time manufacturing is an inventory control strategy directed toward minimizing manufacturing in-process inventory and its associated costs. This minimization is frequently accomplished by eliminating such activities as parts inspection, unnecessary movement of materials, shop-floor queues, and rework or repair. The JIT approach places a strong emphasis on the following: the synchronization of the manufacturing process so that assemblies and components are available just when they are needed; the reduction in the number of disruptions to the manufacturing process and their duration; and the physical layout of the factory. It was found, however, that JIT manufacturing will initially expose quality issues with respect to the individual components and subassemblies; that is, one or more attributes of each component/sub-assembly may have unacceptable variation from their expected values and, therefore, cannot be used. Since the JIT approach assumes that each part can be used, this may greatly affect the availability of a sufficient number of parts for a given production run. These unacceptable part variations usually have to be fixed at the source by redesigning the part, by using relaxed tolerances, or by using process control techniques to reduce the variability.

A very effective and simple system for tracking parts movement in a JIT manufacturing environment is the *kanban* system. The kanban system uses a physical token, such as a card, which accompanies a bin of parts. The card is removed when the first item is removed from a bin. The removed card is placed in a collection box. The card contains the item number, the number of parts in the bin, the location to which the bin was delivered, and the number of days after the card is removed that the bin is to be replaced with a full bin. This last piece of information is called the *delay*. The cards are collected once a day and the replenishment of the items denoted on the card are scheduled for delivery to the location as stipulated by the delay cited on the card.

The advantages of this system are the following:

- It is easily understood by all participants.
- It provides very specific information very quickly.
- Its implementation is inexpensive.
- It ensures that there is no overproduction by minimizing inventory.
- It ensures a quick response to any changes.
- It gives responsibility to the shop floor workers.

In a novel implementation of the kanban system, Bosch* has successfully integrated RFID devices with the kanban cards by embedding the RFID device in the kanban card. They interrogate the RFID device four times during the production process and are able to monitor the parts' progress through the manufacturing system.

1.2.2.2 Continuous Improvement

Continuous improvement (*kaizen* in Japanese) is a philosophy that recognizes that industrial competitiveness comes from continually making improvements to the product (or service) realization process to ensure that customer satisfaction levels remain high. The word *continually* means doing the basic things a little better, every day, over a long period of time. Companies that adopt the continuous improvement philosophy have mastered the ability to learn from their mistakes, determine the root cause of the problems, provide effective countermeasures, and empower their employees to implement these countermeasures.

1.2.2.3 Lean Manufacturing†

Lean manufacturing (also known as the Toyota production system) is a manufacturing philosophy that emphasizes the elimination of waste in the production realization process in order to improve customer satisfaction. The elimination of waste includes eliminating the following:

- Production that is ahead of demand (overproduction).
- Unnecessarily moving parts, that is, movement that is not directly related to processing.
- Excess inventory, where too many components/subassemblies will not be used when they are delivered; that is, they are unnecessarily waiting for the next step in the production process.
- More movement of people and machinery than is necessary to perform the processing.
- Defects, which require inspection to locate them and then repair them; can sometimes be due to poor supplier relations.
- Employing too many processes to arrive at the final product; this creates unnecessary activity that is often related to poor product design.

Thus, the aim of lean manufacturing is to get the right things, to the right place, at the right time, in the right quantity, to achieve level work flow. One is to do all this while minimizing waste, being flexible, and being able to change rapidly. These latter two attributes are required in order to attain level work flow. Abnormal production flow increases waste because process capacity must be prepared for peak production.

Waste minimization is attained when

- The production system is pulled by customer demand; that is, JIT techniques are used.
- There are zero defects in the components that comprise the product.
- Continuous improvement is used.
- The manufacturing system is able to produce a mix of products at low production volumes without sacrificing efficiency.
- Good relations exist with suppliers who are willing to share risk, costs, and information.
- Visual monitoring of the actual work in progress takes place.

---

* R. Wessel, "RFID Kanban System Pays Off for Bosch," *RFID Journal*, May 7, 2007, http://www.rfidjournal.com/article/articleview/3293/1/1/.
† For an example of an American manufacturer who successfully applied the lean manufacturing technique, see "Custom Motor? Give us two weeks," *Mechanical Engineering*, September 2008, pp. 52–8.

1.3 INNOVATION

Innovation is the conversion of new or existing knowledge into new or altered products, processes, and services for the purpose of creating new value for customers and for creating financial gains for the innovators. It is almost always a result of what has come before, that is, from the gradual growth of knowledge. However, there is a distinction to be made between innovation and inventiveness, which can lead to a patent. Until recently, one could be issued a patent in some cases if it were found that combining what had been disclosed in prior patents was not obvious to "one skilled in the art."* However, a U.S. Supreme Court ruling in 2007 recognized that engineers routinely use prior devices to solve known problems using obvious solutions. The court held that even if there is no prior teaching, suggestion, or motivation to make a combination, the combination may still be obvious. They noted that a person of ordinary skill in the art is also a person of ordinary creativity. Engineers, in the eyes of the Supreme Court, know that changing one component in a system may require that other components be modified and that familiar items may be used beyond their primary purposes. This may result in improvements and "ordinary innovation," but, now, is most likely not patentable.

There are typically eight challenges that confront innovation.†

1. **Finding an idea**, which can come from anywhere.
2. **Developing a solution**, which is often more difficult that the generation of the idea.
3. **Obtaining sponsorship and funding**, either internally if one is working within an organization or from external sources if one is working independently.
4. **Ensuring that the solution is scalable** so that it can be reproduced in very large quantities.
5. **Reaching the intended customers** by communicating the idea to them and by making it so that the average person can use the innovation.
6. **Beating your competitors** by monitoring them for the purposes of collaboration, inspiration, or tactical awareness.
7. **Timing** the introduction of the innovation so that it matches as closely as possible the peak interests and concerns of the customer.
8. **Keeping the regular business operating**—that is, meet all current obligations while pursuing the innovation.

In the present environment, in order for companies to find opportunities to grow and to innovate they must

- Understand the trends in their industry by recognizing the effects of global competition.
- Know their customers.
- Know the effects of new technologies.
- Understand the implications of the increasing volatility of the availability and cost of natural resources.
- Understand the impact of the increase in environmental concerns.

There are many reasons why companies should innovate. Some companies have used innovation successfully as a means to

- Satisfy customer desires for new products and services.
- Improve the long-range health of a company.
- Become a recognized leader in their industry.

---

* K. Teska, "Ordinary Innovation," *Mechanical Engineering*, September 2007, pp. 39–40.
† S. Berkun, *The Myths of Innovation*, O'Reilly, Sebastopol, CA, 2007.

- Look at solutions and opportunities in new ways, thus staying ahead of the competition.
- Grow, especially with respect to profits, and because of the growth be able to raise investment capital.
- Reveal the interplay of various products lines, which results in expanded product lines.
- Expand to new markets and customers.
- Revitalize its organization, business model, strategy, and processes.
- Create an entrepreneurial environment within the company.
- Learn how to rapidly respond to changes in the marketplace.
- Convert intellectual property into valuable products.
- Retain their most creative employees.

Companies that have adopted innovation as part of their corporate strategy tend to

- Look outside the organization for ideas and opportunities for growth and profit.
- Employ strategic marketing.
- Cultivate and preserve the company's intellectual property.
- Develop the right products, ones that meet customers' needs and get to the market faster; they are market focused rather than product focused.
- Have the right partnerships.
- Make hiring decisions based on their innovation requirements.
- Use the computer to decrease the time it takes to evaluate innovation.
- Employ innovation rapidly.
- Produce products that are easily differentiated from their competitors' products.

Another trend toward innovation—user-centric innovation—has been documented.[*] It is complementing the existing model of manufacturer-centric innovation. User-centric innovation has appeared in software and information products, surgical products, and surf-boarding equipment.

Although a goal of many companies is to be innovative and to rapidly bring these innovations to the marketplace, the reality of which innovations make it to the marketplace and how fast they can get there is not so encouraging. In large companies, few ideas actually make it to the marketplace. In the evaluation process, the initial screening and the business analysis typically eliminate 80% of them. Then, after some development and testing of the remaining 20%, only about 5% of the original concepts survive. After the commercialization process, typically only one idea will make it.[†]

From history, we learn that good ideas take a long time to become successful and that certain marketing shifts and infrastructure changes may have to occur. Consider the introduction of film photography by Kodak. Prior to the introduction of film photography, photographs were taken using glass plates. Celluloid was invented in 1860 and first used for film in cameras in 1889. By 1902, Kodak had 90% of the market when it shifted its focus from the professional photographer to the amateur photographer. Along the way—it took about 20 years—it eliminated the glass plate photography industry. As another example, consider the microwave oven. The first commercial model appeared in 1947 for a cost of around $1,000. In 1955, the first home model went on sale and in 1968 the first countertop model was introduced. In 1971 about 1% of the US households had a microwave oven; in 1986 it was about 25%. Today, it is estimated that 90% of U.S. households have one. As a final example, we note that the prototypes of the Hoover vacuum cleaner first appeared in 1901 and

---

[*] E. Von Hippel, *Democratizing Innovation*, MIT Press, Cambridge, MA, 2005.

[†] In C. Terwiesch and K. T. Ulrich, *Innovation Tournaments: Creating and Selecting Exceptional Opportunities*, Harvard Business Press, Boston, MA, 2009, the authors propose that innovation tournaments be held as a means of identifying exceptional opportunities. By innovation tournaments, they mean that one holds a series of competitive rounds in which ideas are generated and evaluated from increasingly more critical criteria until only a few of them are left for serious consideration.

in 1910 Hoover sold around 2,000 units; in 1920 around 230,000 were sold. It should be noted that only 30 of the U.S. households had electricity at this time.

From history, we also learn that good innovations can also have bad effects. The insecticide DDT controlled malaria but disturbed the ecology and produced DDT-resistant mosquitoes. The automobile personalized transportation and boosted commerce and urban development but creates about one-half the pollution in urban areas and results in around 40,000 traffic fatalities per year in the United States alone. Cell phones provide mobile access, convenience, and a portable safety system, but they have created another source of public annoyance and for some people, dangerous driving habits.

Some innovations that reach the marketplace can be what are called *disruptive* or *discontinuous* innovation. The introduction of Kodak celluloid film mentioned previously is an example of disruptive innovation; it eliminated the glass plate method. An important distinction between conventional product development and disruptive product development is that in conventional product development the markets, customers, and value chain are known. In disruptive innovation this type of information may not be known. For example, when the dry cell battery was introduced in the United States in 1887 by the National Carbon Company, the intended applications were not known and, therefore, what their sizes should be were not yet known. In addition, no one had any idea what price the devices could be because none had been sold before. In 1898, the flashlight was invented by American Ever Ready. In 1914, National Carbon Company created a market by buying American Ever Ready and marketing the dry cell and the flashlight together. Although hard to imagine, at that time there were very few other applications for the dry cell.

1.4 QUALITY

1.4.1 A BRIEF HISTORY OF THE QUEST FOR QUALITY PRODUCTS AND SERVICES

Quality engineering got its start when, in 1924, Dr. Walter A. Shewhart introduced a method that became the basis of statistical quality control. He framed the problem in terms of variations that had assignable causes and variations that were simply due to chance, and introduced the control chart as a tool for distinguishing between the two. Shewhart showed that one could bring a production process into a state of statistical control—that is, where there is only variation due to chance—and keep it in control. In the early 1950s, Dr. W. Edwards Deming started introducing management to methods that improved design, product quality, testing, and sales through various methods, including the application of statistical methods such as analysis of variance and hypothesis testing (see Chapter 11). In addition, Dr. Deming taught that by adopting appropriate principles of management, organizations can increase quality and simultaneously reduce costs by reducing waste, rework, staff attrition, and litigation while simultaneously increasing customer loyalty. He maintained that the key was to practice continual improvement and think of manufacturing as a system, not as bits and pieces. Many companies in Japan embraced his ideas and eventually products made in Japan became synonymous with quality products.

In 1941, Joseph M. Juran discovered the work of Vilfredo Pareto. Juran expanded Pareto's principles by applying them to quality issues and noted that, in general, 80% of the problems are attributable to 20% of the causes. Juran later focused on managing for quality. He went to Japan in 1954 and developed and taught courses in Quality Management, since the idea that top and middle management needed training had found resistance in the United States. For Japan, it would take about 20 years for the training to pay off. In the 1970s, the Japanese began to be seen as world leaders in producing quality products.

Professor Kaoru Ishikawa of the University of Tokyo introduced in 1962 the concept of quality circles, and Nippon Telephone and Telegraph was the first Japanese company to try this new method. Eventually, quality circles would become an important link in the company's total quality management (TQM) system. Dr. Ishikawa also developed what has become known as the Ishikawa diagram (also called a cause-and-effect diagram or a fishbone diagram), which is a graphical tool used to explore the most significant root causes of the failure of a system to perform as expected. (For an example, see Figure 10.1.)

Taiichi Ohno is considered to be the father of the Toyota production system, which, as indicated previously, is also known as lean manufacturing. The Toyota production system was popularized in the West by Shigeo Shingo, who was one of the world's leading experts on manufacturing practices and on the Toyota production system. He wrote several books about the system, and added poka-yoke (mistake-proofing) to the English language (see Section 10.2).

Dr. Genichi Taguchi, who was a Japanese engineer and statistician, developed in the 1950s a methodology for applying statistics to improve the quality of manufactured goods. The methodology was based on the classical design of experiments methods. (See Section 11.4.) Taguchi had realized, just as Shewhart had, that excessive variation lay at the root of poor manufactured quality and that reacting to individual items inside and outside of the specification was counter-productive. To emphasize this, he introduced the concept of a loss function, which is discussed in Section 11.6. Taguchi methods have been controversial among some conventional Western statisticians, but others have accepted many of his concepts as valid extensions to the body of knowledge.

1.4.2 QUALITY QUANTIFIED

Japanese Industrial Standard JIS Z 8101-1981 defines quality as the totality of the characteristics and performance that can be used to determine whether or not a product or service fulfills its intended application. A remark that accompanies the definition states that when determining whether or not a product or service fulfills its application, the effect of that product or a service on society must also be considered. A second remark defines quality characteristics as the elements of which quality is composed. No matter how high the quality result of a product is when based on its quality characteristics, quality objectives have not been met if the product does not fill essential customer requirements. Thus, product quality is evaluated on the basis of whether or not the product carries out its intended functions and the extent to which a product or service meets the requirements of the user—each time the product is used under its intended environment or operating conditions and throughout its intended life.

Imbedded in the definition of quality is the concept of *value*, which can be thought of as the ratio of the aggregation of the attributes of quality to the cost of the product. Thus, the more total quality is perceived per amount spent, the higher its perceived value.

Garvin[*] has proposed the following eight dimensions or categories of quality that can serve as a framework for its estimation.

1. **Performance:** Performance refers to a product's primary operating characteristics. This dimension of quality involves measurable attributes and, therefore, can be ranked objectively on their individual aspects of their primary operating characteristics.
2. **Features:** Features are often a secondary aspect of performance and are those characteristics that supplement a product's basic functions. Features are frequently used to customize or personalize a buyer's purchase. Examples of features are power windows on cars, five different drying cycles on a clothes dryer, etc.
3. **Reliability:** The probability of a product malfunctioning or failing within a specified time period. It is a measure of the freedom of breakdown or malfunction under its specified operating environment. (See Section 10.1.)
4. **Conformance:** Conformance is the degree to which a product's design and operating characteristics meet both customer expectations and established standards, which includes regulatory, environmental, and safety standards, that is, secure and hazard-free operation. Customer expectations include the concept of fitness (or suitability or appropriateness) of

---

[*] D. A. Garvin, "Competing on the Eight Dimensions of Quality," *Harvard Business Review*, pp. 101–109, November–December, 1987.

TABLE 1.1
Performance Criteria and Features of Three Products

Washing Machines	Refrigerators	Self-Propelled Lawn Mowers
Performance	*Performance*	*Performance*
Amount of water used	Efficiency	Motor horsepower
Cleanliness of clothes	Temperature	Handling
Features	Temperature distribution	Maneuverability
Automatic shut-off for unbalanced	*Features*	Sharpness of turns
load	Light	Ease of use
Number of agitator speeds	Size (interior volume)	Starting the engine
Number of spin-dry speeds	Automatic ice-maker	Handling of grass catcher
Number of water fill levels	Location and size of freezer	Shifting gears
Bleach dispenser	Adjustable shelves	Changing cutter height
Automatic control of water	Humidity-controlled crisper	Vacuum action to draw cuttings
temperature	*Conformance*	into catcher
Porcelain lid	Compressor noise level	Cuts evenly
Discharge pump that can lift water	Chlorofluorocarbon (CFC) or	Works well in tall grass
to 2 m above washer	hydrofluorocarbon (HFC)	*Features*
		Grass catcher capacity

use and the ease of use. It is now frequently important to consumers that a product does not pose a high safety risk, and that the product's manufacturing process, its use, and its disposal do not measurably harm the environment. In addition, the product may be expected to produce little or no unpleasant or unwanted by-products, including noise and heat. (See Sections 10.4 and 10.5.)

The dimensions of performance, features, and conformance are interrelated. Thus, when most competing products have nearly the same performance and many of the same features, many customers will tend to expect that all makes of that product will have them. This expectation, then, sets the baseline for that product's conformance. To illustrate this, consider Table 1.1, which gives many of the performance criteria and features that are currently found in three products: washing machines, refrigerators, and self-propelled lawn mowers.* These three examples give a measure of the breadth of characteristics that can influence a customer's buying decision. Another example of conformance is the way *Consumer Reports* makes its "Best Buy" selection for automobiles. In order to be considered for this designation, a vehicle must have an above-average record for reliability, good crash test ratings from both insurance-industry and U.S. government crash tests, and, for SUVs, must not have tipped-up in government rollover tests. In other words, *Consumer Reports* thinks that all cars should have these qualities.

Lastly, an important aspect of conformance is a product's adherence to consensus standards, which define the characteristics of products and services and the way to measure them. The major national and international standards bodies are the International Organization for Standardization (ISO; see Section 1.4.4) [http://www.iso.org/], the American National Standards Institute (ANSI)† [http://www.ansi.org/], and the American Society for Testing

---

* For examples of the numerous attributes that exist for cell phones and vehicular engines, see Chapters 7 and 10, respectively, in F. E. Lewis, W. Chen, and L. C. Schmidt, *Decision Making in Engineering Design*, ASME Press, New York, 2006.

† ANSI administers a search engine for standards through the web site www/nssn.org. It is the largest search engine of its kind, with more than 300,000 records compiled from ANSI, other U.S. private sector standards bodies, government agencies, and international organizations.

Materials (ASTM) [http://www.astm.org]. The U.S. government actively participates in many measurement standards activities through its National Institute of Standards and Technology (NIST) [http://www.nist.gov/]. Standards have the effect of influencing products so that they have higher levels of quality and reliability, safety protection, compatibility between products, and environmental protection.

5. **Durability:** Durability is the measure of product life—that is, the amount of use one gets from a product before it deteriorates. It is also a measure of the amount of use one gets from a product before it breaks down and replacement is preferable to continued repair. Durability and reliability are closely linked.

6. **Serviceability:** Serviceability is the speed with which service is restored, the courtesy and competence of the service personnel, and the ease of repair. It is also related to the degree of maintenance required: simple, infrequent, or none. (See Section 10.3.)

7. **Aesthetics:** Aesthetics—how a product looks, feels, sounds, tastes, or smells—is a matter of personal judgment and individual preference. How a product feels usually includes ergonomics, that is, how well-suited (comfortable) it is for human use. (See Section 10.6.)

Despite its subjectivity, it is becoming more apparent that design aesthetics is very important to customers. Many companies consider aesthetic design as a means of communicating to the customer the differences between their product and those of their competitors.[*] Some argue[†] that good design may be a way to encourage environmental sustainability; that is, if something looks good the purchaser is less likely to throw it away. Other companies have come to realize that some products require a more feminine sensibility to products.[‡] Sony has placed wider spaces between the keys on one of its portable computer notebooks to accommodate longer fingernails that women tend to have. LG Electronics have their cell phone cameras automatic focus calibrated to arm's length after observing that young women are fond of taking pictures of themselves. After noting that customers were shifting away from desktops to laptops, HP realized that design could increase sales.[§] Its designs for the laptop are so successful that HP's average selling price is more than 17% above the industry average price. It appears that in addition to value being important, the market place has also started to value design. However, design by itself is not necessarily sufficient. Using the iPod as an example, one finds that in addition to the product's design being attractive and its performing as expected, it is surrounded by large system that provides content and services, software and interfaces, a good retail experience, and a host of accessories. All these other components are meaningful and relevant to iPod customers.

Bill Gates, former president of Microsoft, is impressed with a person's ability to create aesthetic designs. When asked[¶] what he learned from Steve Jobs, president of Apple, Gates replied, "Well, I'd give a lot to have Steve's taste. You know, we sat in Mac product reviews where there were questions about software choices, how things would be done, that I viewed as an engineering question. That's just how my mind works. And I'd see Steve make the decision based on a sense of people and product that is even hard for me to explain. The way he does things is just different and, you know, I think it's magical."

---

[*] For a summary of the development of the designs of 50 of the world's most successful products see C. D. Cullen and L. Haller, *Design Secrets: Products 2*, Rockport Publishers, Gloucester, MA, 2004.

[†] R. Walker, "Emergency Decor," *The New York Times*, December 9, 2007, http://www.nytimes.com/2007/12/09/magazine/09wwln-consumed-t.html?ref=magazine.

[‡] M. Marriott, "To Appeal to Women, Too, Gadgets Go Beyond 'Cute' and 'Pink,'" *The New York Times*, June 7, 2007, http://www.nytimes.com/2007/06/07/technology/07women.html.

[§] D. Darlin, "Design Helps HP Profit More on PCs," *The New York Times*, May 17, 2007, http://www.nytimes.com/2007/05/17/technology/17hewlett.html.

[¶] L. Grossman, "Bill Gates Goes Back to School," *Time*, June 7, 2007, http://www.time.com/time/magazine/article/0,9171,1630564,00.html.

8. **Perceived Quality:** Reputation is the primary attribute of perceived quality. Its power comes from an unstated analogy: the quality of products today is similar to the quality of products yesterday, or the quality of goods in a new product line is similar to the quality of a company's established products.

There is a strong relationship between a product's quality, its market share, and the company's return on investment. Irrespective of a product's market share, products with the higher quality tend to yield the highest return on investment. A study[*] of 167 automotive companies throughout the world has determined that those companies with poor quality products have an average sales growth of approximately 5.4%, whereas those companies that consistently produce high quality products experience sales growths averaging 16%. Also, a large percentage of those companies that consistently produce quality products report that they use the following techniques: quality function deployment (see Section 4.2), failure modes and effects analysis for products and processes (see Section 10.1.2), design of experiments (see Chapter 11), and poka-yoke (see Section 10.2).

To illustrate these eight dimensions of quality, consider the following excerpts from an advertisement for a laser printer. The corresponding dimensions of quality are given in parentheses.

Reduce wait time: up to 19 pages per minute and produces the first page in 8 seconds [performance].

Get great-looking scanned and copied documents and crisp lines with the 1200 dpi effective output quality [performance].

Reduce intervention, simplify maintenance, and lower costs with the 7,000-page-per-month duty cycle and 2,000-page, single-piece print cartridge [durability, serviceability].

Quickly set up jobs with the control panel, which features a two-line backlit display and a ten-key number pad [features].

Scan, fax, and copy unattended with the 30-sheet automatic document feeder [features].

Keep it out of the way: compact design takes up minimal space [aesthetics].

Stack paper and other media in the 250-sheet input tray and 10-sheet priority tray [features].

Connect your small work group via USB [conformance, feature].

Efficiently handle complex jobs with the 64 MB of RAM [performance, feature].

Expect simple office integration with the built-in support for all popular print languages [conformance].

Stay on top of toner replacement; receive alerts when a cartridge is low, monitor its remaining life, and enjoy easy online ordering or check stock and prices at nearby stores [serviceability].

Get peace of mind with the one-year limited warranty [serviceability, reliability].

Rely on printing excellence: won *PC Magazine*'s Readers' Choice Award for service and reliability for 16 years in a row [quality, reliability].

Get product questions answered toll-free, 24 × 7 or via e-mail [serviceability].

1.4.3 SIX SIGMA

Six Sigma is a methodology or set of practices that systematically improves process performance by decreasing process variation from its stated specification. Six Sigma derives it name from statistics, where sigma (σ) stands for the standard deviation of an attribute of a process. (See also Section 11.7.) Variations in the process that exceed the Six Sigma limits are considered defects. Motorola, which developed and implemented the ideas behind Six Sigma in 1986, had applied it originally as a metric to indicate the number of defects in their manufacturing processes. They have since

---

[*] L. Argote and D. Epple, "Learning Curves in Manufacturing," *Science*, February 1990, pp. 920–924.

extended it to other areas as a methodology that places less emphasis on the literal definition of the metric and has the organization focus on the following: understanding and managing customer requirements; aligning key business processes to achieve those requirements; utilizing rigorous data analysis to minimize variation in those processes designed to meet those requirements; and driving rapid and sustainable improvement to business processes. In addition, it has applied the basic idea of Six Sigma as a top-down management system to do the following: align their business strategy to critical improvement efforts; mobilize teams to attack high impact projects; accelerate improved business results; and govern efforts to ensure that improvements are sustained.

1.4.4   ISO 9000

ISO 9000[*] is a series of international standards on quality management and assurance that provide guidelines for maintaining *quality systems*, which are the organizational structure, responsibilities, procedures, processes, and resources needed to implement quality management.

Some of the benefits of ISO 9000 certification are

- Clear, well-documented procedures.
- Satisfying the needs of the customer is given greater emphasis.
- Evidence of compliance is available with a set of criteria by an independent third party, indicating a quality system is in place.
- The levels of incoming inspection and testing of supplied products are reduced when customers purchase products from registered ISO 9000 organizations. Compliance with the criteria of an international quality standard, coupled with the willingness of the supplier to provide its customers with product certification, indicates an adequate level of product quality and consistency.

1.5   BENCHMARKING

Benchmarking is the search for best practices that will lead to superior performance. It is a process for measuring a company's method, process, procedure, product, and service performance against those companies that consistently distinguish themselves in that same category of performance. Thus, benchmarking is a commitment to continuous process improvement. Benchmarking is different from reverse engineering, which is the systematic dismantling of a product to understand what technology is used and how it is made for the purpose of replication. However, the "tear-down" of a product without the intent of replication is frequently used as part of product benchmarking. (See the discussion of the QFD table in Section 4.2.2.)

Process benchmarking provides an improvement strategy by seeking comparisons that go beyond the boundaries of one industry to find world-class best practices that are independent of the industry in which they are observed, and that can be adapted to provide competitive advantage in one's own industry. By seeking knowledge from outside one's industry, one can develop innovative opportunities that create discontinuities with the accepted industry best practices. If benchmarking is focused only on competitors, it may not lead to superior performance. Benchmarking can also be used to break down in-house myths about certain manufacturing processes, and to stimulate new designs, design approaches, and/or manufacturing processes.

Benchmarking can be started by surveying information that is available in the public domain. A very comprehensive list of vendors for manufacturing-related businesses can be found in the

---

[*] ISO is the International Organization for Standardization, whose objective is to promote the development of standards, testing, and certification in order to encourage the trade of goods and services. The organization consists of representatives from 91 countries. The American National Standards Institute (ANSI) is a standards organization that facilitates the development of consensus standards in the U.S. It neither develops nor writes standards. It provides a structure and mechanism for industry or product groups to come together to establish consensus and develop a standard.

Thomas Register [www.thomasregister.com]. Here, one can determine who one's potential competitors are. Trade magazines often provide comparative studies of products within the area of interest to that magazine. In some cases, these magazines are particularly useful in getting an idea of what critics think about certain performance characteristics and product features. Some examples of these online magazines are as follows. For automotive trends and news, visit http://www.driveusa .net/e-zines.htm for links to numerous online magazines and automotive industry news. For information on a very wide range of consumer appliances, visit http://www.appliancemagazine.com/. This site covers a wide spectrum of appliances from kitchen and laundry appliances to medical appliances to heating and air conditioning equipment.

For consumer electronic products, visit http://www.edn.com/ for information and news geared toward electronics design engineers. For a critical review of a wide range of consumer products, visit *Consumer Reports* at http://www.consumerreports.org/. For access to a host of "free" trade magazines from different industries from agriculture to woodworking, visit http://www.freetrademagazinesource.com/.

1.6   PARTNERING WITH SUPPLIERS—OUTSOURCING

Partnering with suppliers is a business culture characterized by the following: long-term relationships; mutual goals; trust and benefits; candid two-way communications; proactive management support and involvement by both parties; and continuous improvement toward world-class benchmarks, performance, and business growth. Many companies select and develop key suppliers and negotiate long-term agreements based on the results of benchmarking their quality, cost, delivery and lead-time, technology, and commitment to continuous improvement. In addition, they look for suppliers that place customer satisfaction as a top priority, clearly demonstrate that they are the lowest *total cost* producer (see Section 3.1), and maintain organizational and financial stability. One of the important consequences of a successful supplier relationship is that incoming inspection and test of supplier parts and materials are no longer required.

The best way to ensure a successful long-term supplier relationship is to involve the suppliers early in the development of new products and, where practical, make them a member of the design team. Their role in the design process is the following: to help minimize cost and time-to-market; ensure manufacturability; and to test materials and parts to improve the product's overall quality.

In this partnership, a customer for a supplier has the responsibility to

- Communicate to the supplier its responsibilities, strategies, and expectations.
- Be responsive to a supplier's request.
- Provide the supplier a single-point business interface.
- Be willing to listen and change.
- Understand the supplier's culture and problems.
- Reward the supplier's performance with business growth.
- Provide clear and manufacturable specifications.
- Develop common metrics and engage in regular feedback on performance.
- Communicate to the supplier the expectations, needs, and problems clearly, and ensure that corrective actions are taken.

An example of the shift in a company's attitude toward suppliers is Boeing during the development and production of its Boeing 787. Boeing[*] says that about 70% of this plane has been outsourced, so much so that Boeing appears to now act as less of a manufacturer and more like a project manager supervising first- and second-tier contractors. Each of these first- and second-tier contractors may rely on scores of more specialized subcontractors. In addition, each of these contractors

---

[*] P. Hise, "The Remarkable Story of Boeing's 787," July 7, 2007, *Fortune Small Business*, http://money.cnn.com/ magazines/fsb/fsb_archive/2007/07/01/100123032/index.htm.

and subcontractors had been given much more responsibility than in any other plane's development by Boeing. For many of them, their role now includes creating completely new systems, rather than just filling orders to Boeing's specifications.

To some customers, outsourcing has come to mean that all or a substantial portion of a product is made offshore. This has lead to a small movement[*] in the United States of customers who prefer to buy items that carry a "Made in the U.S.A." label. For some of these customers, this label represents a concern for workplace and environmental issues, consumer safety, and premium quality and/or luxury products. Several companies are responding to this market segment by producing a small part of their product line in the United States while continuing to outsource the rest of their product line offshore. Examples of products from these offshore and homeland manufacturers include New Balance 992 running shoes, which are made in Maine, and Fender Custom Shop Stratocaster guitars, which are made in California.

Companies, however, are finding that outsourcing to overseas manufacturers is not necessarily the best way to do things. Keeping the supply chain close to home provides better control over it, which can help a company be more responsive to market demands and to minimize inventories. In addition, sudden increases in transportation costs, for example, can greatly affect any cost savings that were initially realized by going offshore.[†] In actuality, the decision of whether or not to outsource is a function of several factors: exchange rates, consumer confidence, labor costs, government regulation, and the availability of skilled managers.

One of the joint goals of a customer/supplier partnership is for the supplier to become the lowest *total cost* producer. In order to achieve this, the customer must

- Establish realistic cost targets.
- Provide cost benchmarking assistance.
- Share the cost of change.
- Listen and act on cost reduction ideas.
- Provide prompt supplier payment.
- Treat supplier cost and technology ideas confidentially.

At the same time, the customer must strive to

- Minimize schedule changes.
- Develop realistic and achievable schedules and priorities.
- Support just-in-time manufacturing flow.
- Provide forecasts of future requirements.
- Establish electronic data exchange interfaces with supplier for forecasts, releases, invoices, and payments.
- Share benchmarking information and just-in-time know-how.

Regarding supplier delivery and lead times, suppliers must have

- Just-in-time manufacturing processes with short-cycle set-ups and the ability to deal with small production lots.
- The right expertise.
- A high degree of flexibility to accommodate customer schedule changes.
- The ability to deliver on the agreed-upon day.
- Customer's forecasting and scheduling data.

---

[*] A. Williams, "Love It? Check the Label," *The New York Times*, September 6, 2007, http://www.nytimes.com/2007/09/06/fashion/06made.html.

[†] L. Rohter, "Shipping Costs Start to Crimp Globalization," *New York Times*, August 3, 2008.

1.7   MASS CUSTOMIZATION

Mass customization* attempts to provide customized products for individual customers without losing the benefits of mass production—that is, high productivity, low costs, consistent quality, and fast response. The goals of mass production are to develop, produce, market, and deliver goods and services at prices low enough that nearly everyone can afford, whereas the goals of mass customization are to produce products with enough variety and customization so that nearly everyone finds exactly what they want. Mass customization, then, is a synthesis of the two long-competing systems: mass production and individually customized goods and services. In mass production, low costs are achieved primarily through economies of scale—lower unit costs of a single product or service through greater output and faster throughput of the production process. In mass customization, low costs are achieved primarily through economies of scope—the application of a single process to produce a greater variety of products or services more cheaply and more quickly.

There are several ways that mass customization can be implemented.

- **Self-customization**, where the customer alters or combines the product to suit his/her needs. Examples: Microsoft Office, where the programs can be used with varying degrees of sophistication; Lutron Electronics [http://www.lutron.com], which has a family of lighting controls that are customized from a wide variety of shapes, colors, and sizes to meet the needs of interior designers.
- **Customization using a mix of standardized procedures**, where either the first or last activities within the factory are customized while the others are kept standardized. Example: IC3D [http://www.IC3D.com] offers customers the ability to specify many aspects of a pair of jeans, with a wide choice of styles of denim colors and washes, leg shapes, ankle styles, and waistband types.
- **Modular product architecture**, where modular components are combined to produce a customized product. Example: Dell [http://www.Dell.com] computers are customized by selecting within each line of computers the screen size, the amount of memory, the speed of the processor, the capabilities of the CD/DVD device, the capacity and rotational speed of the hard drive, etc. (Also see Section 6.3.)
- **Flexible customization**, where a flexible manufacturing system produces customized products without higher costs. Example: Sovital [http://www.sovital.de] produces customized pills containing a collection of nutrients specified by the customer so that there is no need for the customer to take several pills to acquire the same nutrients.

There are several advantages of mass customization. Mass customization, in some respects, mimics JIT. Products are made after an order is placed, which greatly reduces inventory and the need to do forecasting. Since mass customization tends to integrate the customer early in the production process, the company is essentially ensuring that it is responding to changing market trends. In this respect, fashion cycles are not directly considered since one is producing a steady stream of stylish and modern products. It can also provide a way in which companies can detect trends that can be used to produce product lines using standard mass production techniques. The act of involving the customer in an aspect of the product creation process has been coined "open innovation" or "co-design."

---

* An Internet site that contains numerous discussions and evaluations of up-to-date mass customization artifacts can be found at http://mass-customization.blogs.com/mass_customization_open_i/casesconsumer/index.html.

2 The Integrated Product and Process Design and Development Team Method

The IP²D² team method is described, its agenda outlined, and suggestions for the team's composition and requirements are given.

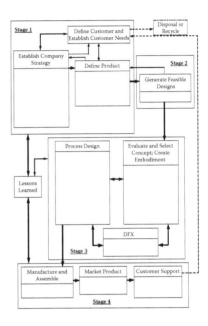

2.1 INTRODUCTION

The way in which products are produced has been evolving rapidly during the last three decades. The start of the evolution in the United States was probably brought about by the realization that within the first 10% of the total time it will take to design, manufacture, and deliver a product, numerous decisions will have been made that will effectively commit 85% of the funds to be expended for the project. However, during this short period of time less than 15% of these funds will actually be spent. In other words, the most influential decisions regarding the eventual expenditures for the product's introduction into the marketplace will occur during the very early stages of its development cycle. This is illustrated in Figure 2.1. Another way of looking at this is to consider the estimates of the cost to perform a change during the following three stages of a product's development: design, process planning and engineering, and production. If the cost of making a change in the design stage is one unit of cost, then the cost of making a change in each of the subsequent stages will be many times that incurred in the design stage.

The overall goal of the IP²D² team is to convert a product concept into a product in such a way that the design of the product and corresponding processes results in

- High customer satisfaction.
- Minimum product cost with high profitability.
- Equaled or surpassed competitively established benchmarks.
- Short time-to-market.
- Lower product development cost.
- High quality.
- High factory throughput with minimum work-in-progress.
- Minimum space, handling, and inventory of raw materials and finished goods.
- Increased utilization of automation and fuller utilization of existing equipment.
- Elimination of redesigns and engineering changes.
- Broadened product line, with considerable variety.
- Early supplier involvement.

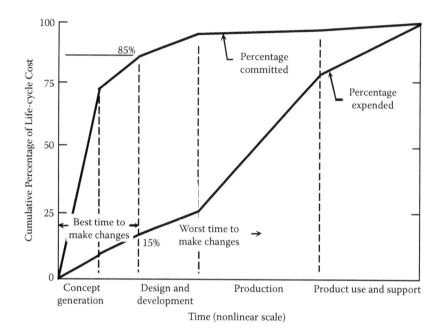

FIGURE 2.1 Cumulative product life-cycle costs at the various stages of the product realization process.

Implicit in the overall goal of the IP²D² team is the idea of producibility. Some approaches used to attain producibility are given in Sections 8.1.2.2 and 9.1.

Lastly, in addition to converting a product concept into a product using the goals listed above, the IP²D² team must place the product under development in its proper context in the marketplace. A good example of putting a product in its larger context is Edison's light bulb. Edison didn't think of just making a working light bulb, he thought of it as making an electrical system that cities could adopt to use his light bulb. He, therefore, avoided making light bulbs that no one at that time could buy. Another example is the iPod, which was mentioned in Section 1.4.2. The iPod itself is not really a "complete" product. It becomes one only when it is realized that it is surrounded by a large system that provides content and services, software and interfaces, a good retail experience, and a host of accessories. All these aspects have proved necessary for the successful adoption of the iPod.

2.2 THE IP²D² TEAM AND ITS AGENDA

An integrated product and process design and development team is multidisciplinary, collaborative, flexible, and responsive. An IP²D² team should include, in addition to the appropriate engineering specialists such as electrical, mechanical, materials, manufacturing, and production engineers, persons from the following areas: industrial design; finance to calculate the cost and to determine whether the project budget can cover it; sales and marketing to represent the customers' needs and desires; and service and parts departments to ensure the interchangeability and reparability of components and systems. The IP²D² team also requires the inclusion of major suppliers in a new product project at its inception. In some organizations, the IP²D² team approach is driven by the availability of standard, high quality, high volume parts, which are available from several worldwide sources that are in their respective businesses "for the long haul."

A product development team tends to develop a product faster if the team meets the following criteria:

- Members have the ability to be the source of most of the ideas and creativity required.
- There is an appropriate maximum number of members on the team.

- Members serve on the team full time from the time of the development of the product concept until the product is shipped to the customer.
- Members report solely to the team leader.
- The marketing, design, engineering, manufacturing, finance, and supplier constituencies have representation on the team.
- Members are located within conversational distance of each other.
- Members accept responsibility and enjoy working in a team environment.

In addition, an IP^2D^2 team needs to have a common vocabulary, agree on a common purpose, and agree on priorities for the team and the individuals. Its members must also be able to communicate, exchange information, and collaborate to create a shared understanding of all issues.

An IP^2D^2 has been compared[*] to that of an American football team in the following way: An IP^2D^2 team functions as a single unit; the team has a common goal; each member has a specific assignment that must be completed; each member's contribution is crucial to the team's success; the combined efforts of the team result in better solutions; the team has a leader; and the organization (management) is on the sidelines supplying guidance and resources.

The environment in which the IP^2D^2 team must function is now described. First, we state the obvious: design involves numerous subjective judgments. In this context, it is noted[†] that when dealing with design solutions there are always a number of different solutions, an optimal solution may not exist, and if it exists, finding it may not be practical. The first point is due to the realization that design problems cannot be comprehensively described, are always subjectively interpreted, and are frequently hierarchically organized. The second point relates to the fact that in order to measure optimality one has to define a measure of performance against which the design solution can be measured. Design, however, involves tradeoffs, choices, and compromises. Good performance in one area is sometimes achieved at the cost of performance in another, and statements of objectives might be contradictory. Thus, there may be no optimal solutions, only a range of acceptable solutions. The appraisal and evaluation of solutions are largely a matter of judgment.

In addition, there are frequently resource limitations on time, money, and manpower. Because of these limitations, it is not possible to "complete" a design before the process must stop. The objectives are met, but perhaps not to one's satisfaction. Also, design problems are all different, and their solutions are a function of the problem definition and many other factors, such as legal considerations, politics, fashion, and so on. Therefore, there is no sequence of operations that will always guarantee a result. Furthermore, design involves problem finding as well as problem solving, because the problem statement often evolves along with the problem solution, and as the process continues, the problem and solution become clearer and more precise.

The role of each member of an IP^2D^2 team is to participate in a decision-making process[‡] where each decision helps to bridge the gap between an idea and its physical realization. In this type of environment, one finds that

- A principal role of an IP^2D^2 team member is to make decisions, some of which may be made sequentially and others that may be made concurrently. In addition, these decisions are often hierarchical, and the interaction between the various levels of the hierarchy must be taken into account.

[*] D. R. Hoffman, "Concurrent Engineering," in W. G. Ireson, C. F. Coombs, Jr., and R. Y. Moss, Eds., *Handbook of Reliability Engineering and Management,* 2nd ed., McGraw-Hill, New York, 1996.

[†] B. Lawson, *How Designers Think,* Architectural Press, London, 1980.

[‡] F. Mistree, W. F. Smith, B. A. Bras, J. K. Allen, and D. Muster, "Decision-Based Design: A Contemporary Paradigm for Ship Design," *SNAME Transactions,* Vol. 98, pp. 565–597, 1990.

- Decisions are often made from information that comes from different sources and disciplines, and may be governed by multiple measures of merit and performance; however, all the information required to arrive at a decision may not be available.
- Some of the information used in arriving at a decision may be based on scientific principles, and some information may be based on a team member's judgment and experience.

The development of a new product is often one of balancing three factors: development speed, product cost, and product performance and quality. Development speed, or time to market, is measured as the time between the first instant someone could have started working on the product development program and the instant the final product is available to the first customer. It has been found that, when possible, the use of past designs, standardization, and computerization have helped to speed up the product realization process.

Product cost is the total cost of the product delivered to the customer. It is important not to use merely the term *manufacturing cost*, because total cost will determine profits. Cost includes one-time costs associated with manufacturing start-up, one-time development costs, and recurring costs. These aspects are discussed in detail in Chapter 3.

Product performance is how well a product meets its market-based performance specification. A design that meets the needs of marketplace has achieved good product performance, which is rated by the customer based on how well it meets his/her needs. An excellent design is one that meets performance targets using cost-effective technologies and design approaches. For some products, a majority of the time that goes into a design is spent on achieving a cost-effective solution.

The product realization process can be represented several ways. The way that we choose to represent it is by considering it to be made up of four stages[*]: product identification, concept development, design and manufacturing, and product launch. Each of these stages must satisfy certain criteria before proceeding to the next stage. When an evaluation of the criteria for any stage indicates that it is infeasible to proceed to the next stage, the project should be abandoned.

Within each stage there are specific tasks that are performed. A description of the four stages and their evaluation criteria will now be discussed. The specific tasks that the IP^{2}D^2 team must address are shown in context with the four stages in Figure 2.2. A list of the tasks needed to be considered in the development of a product is given in Table 2.1. Not all of the items listed in Table 2.1 have equal importance; the importance varies as a function of the product itself, the company's strategy, and the intended market. However, for many products it will be found that function, performance and features, safety and the environment, reliability and durability, cost, manufacturability (producibility), appearance, and timeliness to the marketplace are among the most important.

2.2.1 STAGE 1: PRODUCT IDENTIFICATION

The purpose of the product identification stage is to generate a product idea that is a good business investment. The output of this stage is

- Demonstration of a strong customer need
- Determination of its market potential
- A business model that shows that the company can make a profit from the sale of the product
- Identification and evaluation of the risks of undertaking the project
- A determination of whether a *sustainable* competitive advantage can be gained
- An estimate of the resources that it will take to develop the product

---

[*] R. L. Kerber and T. M. Laseter, *Strategic Product Creation*, McGraw-Hill, New York, pp. 78–100, 2007.

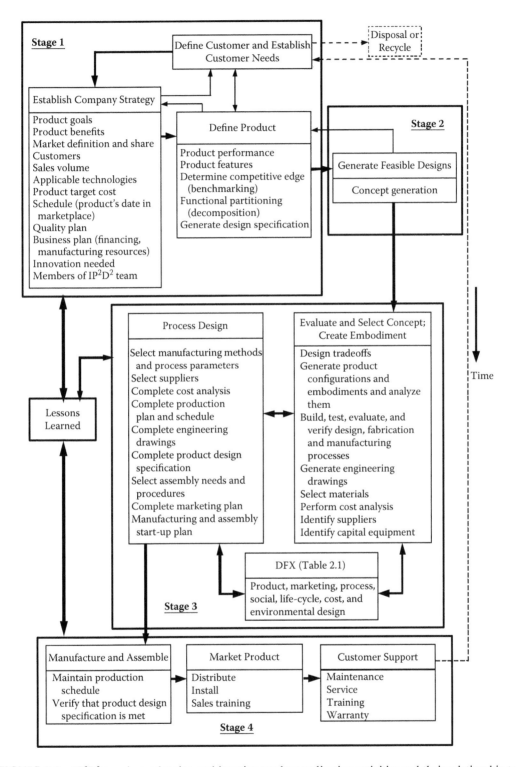

FIGURE 2.2 IP²D² team's overlapping and iterative product realization activities and their relationship to the four stages of product development.

TABLE 2.1
Elements of the Product Realization Process That Are Considered in a More or Less Overlapping Fashion in the IP²D² Method

Product Design

Function
　How will it be used?
　Usability
Performance characteristics
Environment—factory floor, packaged, stored, during
　transportation and use
Features
Technology/innovation required
Analysis and simulation
Prototypes
Producibility
　Material selection
　Manufacturing methods
　Handling and presentation
　Fabrication/assembly
Number of parts and part variation
Documentation/engineering drawings
Human factors/ergonomics
Appearance and style (aesthetics)
Configuration
Modularity—room to incorporate future technological
　enhancements
Degree of standardization
Reliability
Patents

Quality Plan

Fit, feel, and finish
Supplier quality and certification
Cost and elimination of internal failures (scrap, rework,
　etc.)
Service call rates
ISO 9000
Process capability

Social Design

Safety
Legal issues: disclaimers, liability, user warnings
Standards (U.S./global)

Marketing Design

Company strategy
　Customers' needs and who they are
　Cost
　Market share
　Competitors (benchmarking)

Process Design

Responsiveness
　Delivery date to customers
　Packaging and shipping
　Adaptability to variability in materials and process
　　conditions
　Integration of new process technology into existing
　　system with minimum disruption and cost
　Maximum responsiveness to surges in demand
　Minimum changeover time and cost
　Maximum production flexibility
　Quick turnaround capability
　Maximum product family
Design-dependent
　Assembly methods
　Waste
　Manufacturing methods
　Materials
Factory characteristics
　Material handling and flow
　Work station design (ergonomics) and operator
　　training
　Manufacturing equipment capabilities and
　　reliability
　Floor layout
　Plant location
　Safety
　Training of factory personnel
　Waste management
　　　Process pollution/toxicity
Production
　Quality control
　Capacity/production rate
　Production planning, scheduling, and purchasing
　Transition into production of existing products
　Outsourcing of parts and subassemblies
　Suppliers
　Documentation

Life-Cycle Design

Testability/accessibility
Inspectability
Reliability
Durability (product life)
Spare parts availability

TABLE 2.1 (CONTINUED)
Elements of the Product Realization Process That Are Considered in a More or Less Overlapping Fashion in the IP²D² Method

Marketing Design (Continued)

  Breadth of product line (number of generations/
    versions of product)
  Time to market
Product price/volume/mix
  Quantity
  Financial plan for total cost
  Purchasing
Product features
  Expansion and upgrading (planned product
    improvements and future designs)
Eliminate/simplify adjustments
  Product options
  Product's useful life
Product support
  Warranties
  Spare parts
  Product servicing and servicing frequency
End user
  Customer assembly/installation required
  User training required
  Company installation required
Documentation
  User instruction manuals and documentation
  Warnings and legal issues
Product identity: logo, trademark, brand name
Packaging and labels
Advertising literature (catalogues and brochures)

Life-Cycle Design (Continued)

  Maintainability/serviceability/supportability
  Logistics
  Upgradability
  Shelf life and storage
  Installability
  Environmental design
  Disassembly
  Federal and state regulatory requirements
    and compliance
  Product pollution/toxicity
  Recycling and disposal
  Reuse/remanufacture

Cost Analysis

  Make/buy(outsource)
  Target pricing
  Cost model
  Start-up costs
  Investments required

In many cases, these estimations and judgments are made on preliminary data. When the totality of these preliminary studies is favorable, one can proceed to the second stage. However, the preliminary studies should be explored more completely while the next stage is in progress.

It is during this stage that the answers to the following questions are sought.[*]

- What is the business plan for the new product?
- How does this product relate to all of the company's other products?
- To what degree will existing designs be used?
- When is the product needed in the marketplace?
- How long is it planned to be in production?
- How many units will be made?
- Where will the new product be manufactured?
- What regulations, standards, and safety requirements will be met?
- Are there sufficient financial resources?
- How will the product be differentiated from its competitors?
- Does the company have extensive knowledge and experience with this type of product?

[*] D. Clausing, *Total Quality Development*, ASME Press, New York, 1994.

- Will there be any sourcing problems?
- Will there be any need for capital investment?
- Will there be any potential patent infringements?
- Will this be a long-term or short-term market opportunity?

2.2.2   STAGE 2: CONCEPT DEVELOPMENT

The purpose of the concept development stage is to generate and develop candidate concepts that will satisfy the product's performance goals. The candidate concepts are then subjected to a set of evaluation criteria that must be satisfied. These criteria include the following:

- The requirements to produce the product lie within the company's core competencies.
- There is a relatively small technical risk in its development since the project began.
- There is neither substantial change in the market conditions nor in the competition.
- The manufacturing resource requirements are close to those predicted.
- The product prototypes indicate that the creation of the product will be economically viable and that its manufacture is feasible.

In addition, any studies initiated for the first stage should be completed by the end of this stage and it should be found from them that the conditions on which the preliminary conclusions were based have not changed.

2.2.3   STAGE 3: DESIGN AND MANUFACTURING

The purpose of the design and manufacturing stage is to produce the engineering drawings for the product, complete a test production run that demonstrates that the product meets all customer requirements and all quality criteria, and, if the test run is successful, enter into full production.

2.2.4   STAGE 4: LAUNCH

The purpose of the launch stage is to deliver the product to the marketplace. Success of the launch is determined by meeting the quality goals, the customer satisfaction goals, and the business plan goals.

From each of these four stages, the participants in the process have gained experience and learned from them. What they have learned and how this knowledge will influence the process the next time it is used to create a product that is often termed "lessons learned." This collection of experiences is a very important part of a company's ability to evolve into a company that can continually create and rapidly introduce into the marketplace products that are highly valued by the customer.

We now return to Figure 2.2 and Table 2.1. The totality of the factors in Table 2.1 defines the product realization process and comprises the elements of the IP^{2}D^2 team's agenda. Each of the items in Table 2.1 can be considered either as a goal, in the DFX context, or as constraints on the product. Hence, the organization of the factors listed in Table 2.1 does not imply that each factor, or group of factors, is considered sequentially. In fact, they should be treated in a more or less concurrent and overlapping manner, except for those factors that set the goals for the product, such as company strategy, compliance to regulation, and product performance and features. It is noted that it is important to overlap the product design portion of the product development cycle with the process design portion, even though when the product design is in its early stages it is subject to changes. The advantage of this approach, which is the satisfaction of the market requirements with a timely product, often outweighs the risk and cost of introducing product changes that may affect the process design.

The way in which the IP^{2}D^2 team frequently carries out its mission is summarized in Figure 2.2. The customer needs are first established. The company then identifies a strategy and an IP^{2}D^2 team is established to help define a product, or a family of products, that will meet these needs.

The feasibility of the product as defined is determined, and then compared to the company strategy to ensure that it conforms to it. The competition is identified and their products benchmarked. Adjustments are made to the product definition, if required, and a preliminary product specification is drawn up. The IP^2D^2 team then generates candidate concepts that could satisfy these needs, and the concepts are evaluated to determine how well they satisfy the customer requirements and how well they meet the most important goals and constraints among those listed in Table 2.1. Adjustments are made, if required. The most promising concepts are then explored further. For each of the most promising concepts, different configurations and embodiments are evaluated based again on the relevant factors selected from those listed in Table 2.1. Additionally, high-risk areas and alternative approaches are identified and predictions of system performance, reliability, producibility, etc. are made. It is during this period that the IP^2D^2 team attempts to resolve the most important issues that appear to be the most difficult to achieve.

The best configuration and embodiment are then determined. Preliminary engineering drawings are generated and prototypes built, tested, and evaluated, along with any special manufacturing and assembly processes that are required. In addition, suppliers are identified, preliminary production plans and schedules are created, marketing plans initiated, and a manufacturing start-up plan is created. The evaluations of the prototypes are compared to the preliminary product specification, the satisfaction of the major performance factors and customer requirements are verified, and a refined cost analysis is performed. Any changes that are required to the design, manufacturing processes, and assembly procedures are made at this point. Once the final decisions are made, the details of the product's design, materials, manufacturing and assembly processes, process plans and scheduling are completed, the final costs are determined, and the product enters production and, shortly thereafter, the marketplace.

2.3　TECHNOLOGY'S ROLE IN IP^2D^2

Product designs are often rejuvenated when there has been a technological breakthrough or major improvement in the performance or manufacturing process of one or more of a product's components. A good example is the laptop computer, which has undergone a series of changes and improvements since its inception that include its screen, which went from black and white to color; improved battery life; a large increase in storage density of the hard disk; and a decrease in its overall weight.

However, any attempt to design a product before the necessary technologies are ready almost always leads to low-quality products that are behind in schedule and over in cost. To determine if the technology is mature enough to be used in a product's design the following issues need to be addressed.[*]

- Can the technology be manufactured with known processes?
- Are the critical parameters that control the new technology's function(s) identified?
- Are the safe operating ranges of the technology's parameters known?
- Have the failure modes been identified and evaluated?
- Has the technology's life-cycle been evaluated and are its environmental affects known?

The role of technology is also complicated by the fact that small random events at critical moments in a product's development and introduction can determine the success of that technology in such a way that once adopted it becomes a standard, thereby making it very expensive to change. A classical example is the layout of a typewriter's keys, which was chosen initially for reasons that have little relevance today, but which cannot be changed. As an example of how random events can influence the future course of technology, consider the automotive engine in the very early part of the 20th century. The customers had a choice of gasoline and steam engines. In 1909, the Stanley Steamer reached speeds of 200 km/hr, but was marketed as a luxury car, never trying to reach mass

---

[*] D. G. Ullman, *The Mechanical Design Process*, 4th ed., McGraw-Hill, New York, 2009.

production quantities. What hastened its demise was a brief outbreak of hoof and mouth disease in 1914, which forced the closing of the horse troughs thereby denying steam cars their convenient supply of water. With better technology or simply more steam-driven cars on the road a quick solution would have been found. However, with other alternatives readily available, the buying public was not motivated to wait until a solution presented itself.

Sometimes an error in a company's strategy can cause it to lose its competitive advantage, even if it is the first in the marketplace. This has happened with the public's acceptance of Matsushita's VHS video cassette format over Sony's Betamax, and with Microsoft's DOS operating system over Apple's. Both Apple and Sony decided not to license its technology to its competitors, whereas Matsushita and Microsoft did. This encouraged more videos and more application software to appear in the marketplace, giving customers a good reason for buying these products.

2.4 IP^{2}D^2 TEAM REQUIREMENTS

2.4.1 TEAM REQUIREMENTS

For an IP^{2}D^2 team to be successful, there are two interdependent sets of requirements.[*][†][‡] The first requirement pertains to the individual members of the team who, in addition to having specific knowledge in their particular area of expertise, should have a positive attitude and should be creative. Maintaining a positive attitude permits everyone to feel free to contribute and permits ideas to flow freely. This type of atmosphere can be maintained if the team members refrain from using "initiative-killing" remarks, a representative sampling of which are given in Table 2.2.

Creative people typically have the following traits. They have the ability to

- Recognize a problem when it exists.
- Develop a large number of alternative solutions to a problem.
- Develop a wide variety of approaches to solve a problem.
- Develop unique or original solutions to a problem.

In addition, team members must be active listeners. An active listener does the following:

- Listens before speaking.
- Evaluates what is being said.
- Tests his/her understanding of what has been said by paraphrasing.
- Does not talk down to anyone.
- Listens for the contribution that the person is trying to make.

The second requirement for a successful IP^{2}D^2 team pertains to its composition, the way it functions and the members interact, and the characteristics of the team leader. To maximize the success of the team's effort, the following guidelines of IP^{2}D^2 team work have been identified:[§]

- Team members must have respect for each other's expertise, which will represent many areas.
- The members must share a common vision of the team's goals.

---

[*] N. Cross, *Engineering Design Methods*, John Wiley & Sons, Chichester, 1984.

[†] L. B. Archer, "Systematic Methods for Designers," in *Developments in Design Methodology*, N. Cross, Editor, John Wiley & Sons, Chichester, 1984.

[‡] G. Pahl, W. Beitz, J. Feldhusen, and K.-H. Grote, *Engineering Design: A Systematic Approach*, Springer-Verlag, Berlin, 2007.

[§] E. Morley, "Building Cross-Functional Design Teams," in *Proceedings of the First International Conference on Integrated Design Management,* London, pp. 100–110, June 13–14, 1990.

TABLE 2.2
Examples of Initiative-Killing Remarks

"But our customers want it this way."	"You can't fight success."
"Process mastery has not been proven."	"No."
"We don't do it that way here."	"We're different."
"I'm sure that's been considered."	"We're too busy."
"We've been through all this before."	"I don't like it."
"Let's think about it for awhile."	"It's impractical."
"That was ruled out a long time ago."	"That's old technology."
"Last time we tried it, it didn't work."	"I didn't budget for that."
"There are no personnel to put on it."	"That's been tried before."
"It must be interchangeable."	"Be practical."
"Underwriters Laboratory would never approve it."	"We can't risk that."
"Our customers won't buy it."	"That's not my job."
"This isn't the right time for it."	"It's been tried already."
"That's not to our standard XYZ."	"We'll get to that later."
"That's manufacturing's problem."	"We tried that many years ago."
"We have the best system already."	"Our business is different."
"Designing our products is different."	"It's not a proven design."
"They are not on our qualified suppliers list."	"We can't change that, Industrial Design won't allow it."
"We're too small for that."	"We can't afford it."
"We are already using design for assembly!"	"Sure it costs more, but we'll have a better product."
"It's a good idea, but I don't like it."	"The schedule won't allow it."
"It has never been tried before."	"It will not work in our industry."
"Let's sleep on it."	"It's against company policy."

- There should be a controlled convergence to solutions that everyone understands and accepts.
- Open-minded thinking should be encouraged and premature consensus avoided.
- A proper balance between individual and group work should be maintained.
- Both formal and informal communication should be employed.
- Leadership should be given to those who are willing to accept responsibility and who are willing to empower the team members.

Another way to look at the composition of the IP²D² team is as a "wise crowd." It has been conjectured[*] that under the right circumstances groups are often smarter than the smartest people in them. There are four key qualities that make the collective intelligence of the group produce a better outcome than a small group of experts. The group needs to be diverse, so that people are bringing different pieces of information to the table. It needs to be decentralized, so that no one at the top is dictating the crowd's answer. It needs a way of summarizing member's opinions into one collective verdict. And, lastly, the people in the group need to be independent, so that they pay attention mostly to their own information, and they do not worry about what everyone around them thinks.

An effective IP²D² team leader has been found to have many of the following characteristics:

- Accepts the overall responsibility for the project for its duration, and has the requisite knowledge and experience in a variety of disciplines to be able to communicate effectively with all team members.

[*] J. Surowiecki, *The Wisdom of Crowds*, Random House, New York, 2004.

- Accepts responsibility for generating the various outputs for the team, such as specifications, translating product concept into technical detail, costs, and schedule.
- Communicates and interacts frequently and directly with all IP^{2}D^2 members.
- Maintains direct contact with customers.
- Possesses market imagination, and has the ability to discern the true voice of the customer.

In addition, it is highly desirable for each team member to strive for the following intellectual standards regarding reasoning and oral and written communications:[*]

- **Clarity**—Understandable; the meaning can be grasped.
  This is the most important standard, for we cannot tell anything about something if we don't understand what it means.

- **Accuracy**—Free from errors or distortions; true.
  A statement can be clear but inaccurate. (All primates weigh less then 5 kg.)

- **Precision**—Exact to the necessary level of detail.
  A statement can be both clear and accurate, but not precise. (The solution in the beaker is hot. [How hot?])

- **Relevance**—Relating to the matter at hand.
  A statement can be clear, accurate, and precise, but not relevant to the issue. (There was a full moon when we made the measurements. [So what?])

- **Depth**—Containing complexities and interrelationships.
  A statement can be clear, accurate, precise, and relevant, but be superficial. (Radioactive waste from nuclear reactors threatens the environment. [Superficial.])

- **Breadth**—Encompassing multiple viewpoints.
  When two theories exist and both are consistent with available evidence, if you only choose one of them you are lacking breadth in your discussion.

- **Logic**—The parts make sense together, no contradictions.
  The use of logical thinking produces conclusions that are supported by one's propositions or supporting data.

- **Fairness**—Justifiable; not self-serving or one-sided.
  Give all relevant perspectives a voice, while realizing that not all perspectives may be equally valuable or important.

2.4.2 TEAM CREATIVITY

Creativity is the ability to produce through imaginative skill ideas that can work. It is an important component of the IP^{2}D^2 process and, as discussed in Section 1.2, it is becoming a more necessary component of engineering design. Before proceeding, we note the distinction between creativity and design methods introduced and discussed in the subsequent chapters. Design methods are called rational methods, which are methods that encourage a systematic approach to design. There is a wide range of rational design methods covering all aspects of the design process

---

[*] R. Paul, R. Niewoehner, and L. Elder, *Engineering Reasoning*, Foundation for Critical Thinking, Dillon, CA, 2006.

TABLE 2.3
Six Design Stages

Stage in the Design Process	Aim of Method Used[a]
Clarifying objectives	To clarify design objectives and sub-objectives and the relationships among them. [Chapters 4 and 5]
Establishing functions	To establish the functions required and the system boundary of a new design. [Chapters 4 and 5]
Setting requirements	To make an accurate specification of the performance required of a design solution. [Chapter 5]
Generating alternatives	To generate the complete range of alternative design solutions for a product, and hence to widen the search for potential new solutions. [Chapter 6]
Evaluating alternatives	To compare the utility values of alternative design proposals on the basis of performance against differentially weighted objectives. [Chapter 6]
Improving details	To increase or maintain the value of a product to its purchaser while reducing its cost to its producer. [Chapters 6 to 9]

[a] Numbers refer to chapters in which aim is illustrated.
Source: Adopted, in part, from N. Cross, ibid.

from problem clarification to detail design. A selection of six of the most relevant and widely used methods is summarized in Table 2.3. Also given in this table are the chapters where these methods are illustrated.

Creative, original ideas can occur spontaneously, either to an individual or to a group. This spontaneous appearance of an idea has been identified by psychologists as a five-stage process that involves recognition, preparation, incubation, illumination, and verification. Recognition is the realization or acknowledgment that a problem exists. Preparation is the application of a deliberate effort to understand the problem. Incubation is a period of leaving it to mull over in the mind and allowing a person's subconscious to work on it. Illumination is the (often quite sudden) perception or formulation of the key idea. Verification is the work of developing and testing the idea.

When ideas do not spontaneously occur, one usually resorts to group techniques where the members of the group try to use their collective imagination to address the task. When a group is assembled to engage in creative thinking an appropriate atmosphere must be created. In many cases, it has been found that the following attributes will help to establish a creative atmosphere.

- Welcome and encourage constructive nonconformity, initiative, individuality, and diversity.
- Allow as many team members as possible to have a say in decision-making and in planning.
- Provide personal recognition for accomplishments and reward them appropriately.
- Permit as much freedom as possible by allowing team members to guide themselves.
- Keep the organization flexible to cope with changing situations.
- Encourage the constant interchange of ideas and information.
- Lead by persuasion.
- Provide stimulation by exposure to outside experiences.

2.4.2.1 Brainstorming

The most widely known creative method is brainstorming. This is a method for generating a large number of ideas from which a few ideas will be identified as worth following up. The essential guidelines for a successful brainstorming session are as follows:

- **Focus:** Start with a clear statement of the problem with the right level of specificity.
- **No criticism:** All ideas are accepted and are considered as equals without any initial judgment, debate, or critique.
- **No constraints:** Independent and wild ideas and startling opinions should be encouraged.
- **Build on other people's ideas:** Taking off from, extending, varying, or expanding on someone else's idea is a goal of the session.
- **Encourage participation:** People should be encouraged to speak out often; the more ideas suggested, the more successful the session.
- **Record ideas:** Number each idea and write them so that the entire group can see them; it also provides a means of returning to ideas without losing track.
- **Be visual:** Use as many visual techniques as practical: sketching, diagrams, physical artifacts, and model-building material such as foam, tubing, and tape.
- **Warm up:** Use warm-up techniques for groups that have not worked together before, or do so infrequently, or seem distracted by unrelated work issues.

Some suggestions[*] for avoiding brainstorming session pitfalls are

- Don't let the boss speak first—this may inadvertently limit the scope.
- Don't require that everyone take a turn—it is free-flowing ideas that are wanted.
- Don't only include experts—outside views often stimulate the flow of ideas.
- Don't do it off-site—creativity and inspiration can happen anywhere, and should be able to happen at work.
- Don't restrict "silly stuff"—let the participants say anything that they want; the session should also be fun.
- Don't let participants write anything down, except sketches of their ideas—taking notes severely limits that individual's participation.

2.4.2.2 Enlarging the Search Space

A common form of mental block to creative thinking is to assume rather narrow boundaries within which a solution is sought. Several creativity techniques have been used to enlarge the search space. These are: transformation; random input; why? why? why?; and counter-planning.

Transformation is a technique that attempts to transform the search for a solution from one area to another. This often involves applying verbs that will transform the problem in some way, such as[†]

magnify	minify	modify	unify	subdue	subtract
add	divide	multiply	repeat	replace	relax
dissolve	thicken	soften	harden	roughen	flatten
rotate	rearrange	reverse	combine	separate	substitute
eliminate	invert				

---

[*] T. Kelly, *The Art of Innovation*, Doubleday, New York, pp. 56–62, 2001.

[†] In some respects, this transformation technique has some commonality with the TRIZ method discussed in Section 6.2.3. See also Table 6.6.

Creativity also can be triggered by random inputs from many sources. This can be applied as a deliberate technique, e.g., opening a dictionary or other book and choosing a word at random and using that to stimulate thought on the problem at hand.

Another way of extending the search space is to ask a string of why-questions about the problem, such as "Why is this device necessary?" "Why can't it be eliminated?" etc. Each answer is followed with another why-question until a dead end is reached or an unexpected answer prompts an idea for a solution.

The last method is counter planning, which is based on the concept of pitting an idea (the thesis) against its opposite (the antithesis) in order to generate a new idea (the synthesis). It can be used to challenge a conventional solution to a problem by proposing its deliberate opposite and seeking a compromise. Alternatively, two completely different solutions can be deliberately generated with the intention of combining the best features of each into a new synthesis.

3 Product Cost Analysis

Factors that determine manufacturing and life cycle costs of products and the methods that are used to estimate these costs are presented. Several examples are provided to illustrate the various ways in which design decisions and tradeoffs influence cost.

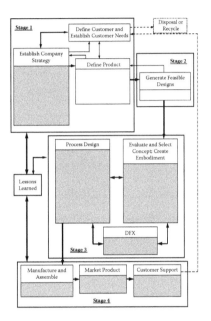

3.1 INTRODUCTION

The value of a product is the ratio of perceived quality to its cost. Therefore, the goals of design activities are to create a product that satisfies the customers' requirements and to maximize profit. Profit is the difference between the price at which the product is sold and the cost of providing the product to the customer. Profit is required for a business enterprise to remain viable. Cost is also important because it is the most accessible and universally understood measure of resource consumption, and can be readily used as an important metric on which to base engineering design decisions. The effects of engineering decisions on costs were shown in Figure 2.1, where decisions made during the earliest portions of the design process set the majority of the product's ultimate cost.

3.1.1 ENGINEERING ECONOMICS AND COST ANALYSIS

Engineering economics* is the application of accumulated knowledge in engineering and economics to identify alternative uses of resources and to select the best course of action from an economics point of view. Resources refer to time, labor, expertise, materials, money, equipment, etc. Generally, engineering economics deals with capital allocation; that is, which of many available investment alternatives should be selected in order to maximize the long-term wealth of the enterprise. Common key attributes of engineering economics are the time value of money, evaluation of assets, depreciation, taxes, and inflation. The problems addressed by engineering economics are very important to the enterprise; however, the cost analysis necessary for determining the manufacturing and life cycle costs of specific products is usually outside the scope of traditional engineering economics.

3.1.2 SCOPE OF THE CHAPTER

The factors that influence cost analysis are shown in Figure 3.1. For low-cost, high-volume products, the manufacturer of the product seeks to maximize the profit by minimizing its cost. For

---

* D. G. Newman, T. G. Eschenback, and J. P. Lavelle, *Engineering Economic Analysis*, 9th ed., Oxford University Press, New York, 2004.

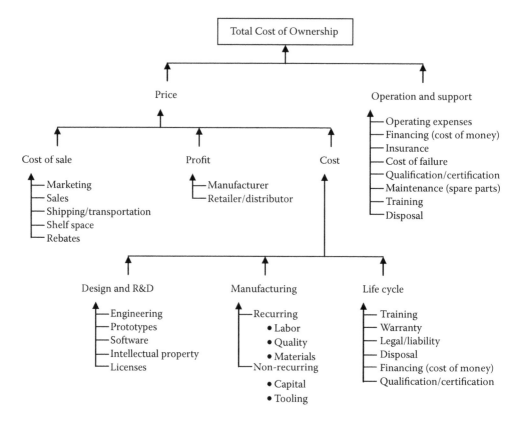

FIGURE 3.1 The scope of cost analysis. Generally, the items contributing to the price of the product are the responsibility of the manufacturer, while operation and support are the responsibility of the customer. However, depending on the product, there may be life-cycle costs that are shared by the customer and manufacturer. In some cases (e.g., large military systems), the design and R&D costs may be paid for by the customer.

higher-cost products, a more important customer requirement for the product may be the minimization of the *total cost of ownership*[*] of the product. The total cost of ownership includes not only the cost of purchasing the product, but the cost of maintaining and using the product, which for some products can be significant. Consider an inkjet printer that sells for $100. A replacement ink cartridge may cost $40 or more. Although the cost of the printer is a factor in deciding what printer to purchase, the cost and number of pages printed by each ink cartridge contributes much more to the total cost of ownership of the printer. In Section 3.6.5, we provide a detailed cost-of-ownership analysis for printers. For products such as aircraft, the operation and support costs can represent as much as 80% of the total cost of ownership of the product.

---

[*] In product development, the sum of all costs that are incurred in the production and use of a product by the manufacturer and the customer is called *life-cycle cost* and is usually broken down into manufacturing cost and user cost. This definition is typically applied to high-volume products where manufacturing cost also includes all nonmanufacturing costs borne by the manufacturer. This definition is not well-suited for "sustainment-dominated" products such as aircraft, telecommunications infrastructure, and military systems where the nonmanufacturing costs borne by the original equipment manufacturer and the operation and support costs borne by the customer can dwarf the manufacturing costs. Therefore, in this chapter, we refer to *cost of ownership* or *total ownership cost* as the sum of everything: manufacturing cost borne by the original equipment manufacturer, life-cycle cost borne by the original equipment manufacturer, and operation and support costs borne by the product's owner.

TABLE 3.1
Summary of Cost Modeling Factors Introduced in Chapter 3

Cost Contribution	Description	Section	Relevancy High-Volume Consumer Products[a]	Low-Volume Complex System[b]
Overhead and indirect manufacturing costs	Accumulated costs of running a business	3.2.2	Medium	Low
Hidden	Costs that are difficult to quantify	3.2.3	High	Low
Design and development	Non-recurring costs of product development	3.3.1	Low	Medium
Manufacturing	Recurring costs: labor, materials, tooling, and capital	3.3.2	High	Low
Quality	Costs that account for manufactured items that do not meet specifications	3.3.3	Low (depends on product)	Medium
Test, diagnosis, and rework	Recurring costs of detecting and fixing defects during manufacture	3.3.4	Low	High
Spare parts, availability, and reliability	Costs of maintaining a product after it is delivered to the customer	3.4.1	Low	High
Warranty and repair	Cost of providing warranty coverage	3.4.2	High	Not applicable[c]
Qualification and certification	Cost of obtaining and maintaining required certifications	3.4.3	Low (depends on product)	High
Cost of money	Cost of the money needed to finance manufacture of products	3.5.2	Low	High

[a] Examples: toasters, golf balls, pencils, and laptop computers.
[b] Examples: airplanes, ships, and 911 systems for cities.
[c] For low-volume complex systems, these costs are included in the spare parts, availability, and reliability contributions.

In this chapter, we shall introduce the general aspects that influence a product's total cost of ownership, and several tools that can be used to keep an IP^{2}D^2 team focused on the important aspects of cost. In Table 3.1, we have summarized the significant contributions to cost modeling addressed in this chapter and their relevancy to several types of products. The evaluation of concepts, processes, and embodiments with respect to costs should be performed throughout the product realization process; however, cost analysis has the largest potential impact during the embodiment stage, which is discussed in Section 6.5.

3.2 DETERMINING THE COST OF PRODUCTS

3.2.1 THE COST OF OWNERSHIP

In this section, we introduce the basic contributions to the cost of a product from two perspectives, that of the manufacturer of the product and that of the owner of the product. The selling price P to the customer is obtained from the following relation:

$$P = \frac{1}{N_{pm}}(C_{pm} + C_{sa} + P_r)$$

(3.1)

where

N_{pm} = Total number of units produced during the lifetime of the product

C_{pm} = Manufacturer's total cost to produce N_{pm} units (Equation 3.2)

C_{sa} = Cost of making the sale to the customer; it includes the costs of marketing (advertising), transportation, shelf space, sales personnel salaries, and rebates

P_r = Accumulation of all the profits for all units charged by the individual entities involved in the distribution chain: manufacturer, distributor, and retailer

From the view of the manufacturer, a product's *total cost* is given by

$$C_{pm} = N_{pm}(C_M + C_L + C_c + C_W) + C_T + C_{OH} + C_D + C_{WR} + C_Q \tag{3.2}$$

where

C_M = Material costs on a per unit basis (Equation 3.6 summed over all activities associated with manufacturing the product)

C_L = Labor cost for manufacturing and assembly on a per unit basis (Equation 3.5 summed over all activities associated with manufacturing the product)

C_c = Capital costs on a per unit basis not included in overhead (e.g., equipment and facilities) (Equation 3.7 summed over all activities associated with manufacturing the product)

C_W = Waste disposition cost on a per unit basis, which includes the management of hazardous and nonhazardous waste generated during the manufacturing process

C_T = One-time costs not included in overhead costs (e.g., tooling costs)

C_{OH} = Overhead (indirect) costs; traditional cost accounting may include this in C_L or C_M (Equation 3.4a)

C_D = Design and development cost; Section 3.3.1

C_{WR} = Life cycle support costs; Equation 3.21 and Section 3.4

C_Q = Qualification and certification costs (e.g., FCC certification, UL approval); Section 3.4.3

The number of products manufactured N_{pm} can be a factor that influences the cost of materials C_M, the cost of making the sale C_{sa}, and overhead costs C_{OH}. In addition, the overhead costs can depend on N_{pm} if these include the cost of inventory.

The product owner has a different view of the product's cost C_{pc} that includes the price paid for the product P and other contributions to the product's total cost of ownership given by

$$C_{pc} = N_{pc}(P + C_x + C_O + C_S) + C_{sp} + C_t + C_Q \tag{3.3}$$

where

N_{pc} = Total number of units purchased

P = Price on a per unit basis

C_x = Applicable taxes (sales tax, tariffs, import taxes, etc.)

C_O = Cost of operation on a per unit basis

C_S = Cost of support on a per unit basis (regular oil changes, maintenance contracts, etc.)

C_{sp} = Cost of spare parts to maintain N_{pc} units

C_t = Training costs

The number of products purchased by the customer N_{pc} can influence the amount of profit P_r that various entities in the supply chain choose to charge and thereby impact the price P paid by the customer.

Equations 3.1 through 3.3 are generally applicable across a broad range of products ranging from pencils to airplanes. However, Equations 3.1 and 3.3 assume that the customer has purchased the product. These formulas are not applicable if the product is leased from the manufacturer.

3.2.2 Overhead or Indirect Costs

Overhead costs are the portion of the costs that cannot be clearly associated with particular operations, products, or projects and must be prorated among all the product units on some basis.[*] Overhead costs include labor costs for persons who are not directly involved with a specific manufacturing process such as managers and secretaries; various facilities costs such as utilities and mortgage payments on the buildings; noncash benefits provided to employees such as health insurance, retirement contributions, and unemployment insurance; and other costs of running the business such as accounting, taxes, furnishings, insurance, sick leave, and paid vacations.

In traditional cost accounting, overhead costs are allocated to a designated base. The base is often determined by direct labor hours, but it can also be determined by machine time, floor space, employee count, material consumption, etc. When overhead is allocated based on direct labor hours, it is often called a *burden rate* and is used to determine either the overhead cost C_{OH} or a burdened labor rate L_{RB} as, respectively,

$$C_{OH} = N_{pm}bC_L \tag{3.4a}$$

or

$$L_{RB} = L_R(1+b) \tag{3.4b}$$

where
   b = Labor burden rate (typical range: $0.3 \leq b \leq 2$)
   L_R = Labor rate (often expressed in dollars per hour), which, when converted to an annual basis, is an employee's gross annual wage

One particular method of cost analysis called *activity-based costing* specifically allocates overhead costs to the various activities necessary to manufacture a product or run an enterprise. This method is discussed in Section 3.3.2.

3.2.3 Hidden Costs

Hidden costs are those costs that are difficult to quantify. They are not explicitly identified in C_{pm} and C_{pc}, and may even be impossible to connect with any one product. Examples of hidden costs include:

- Product's gain or loss of market share.
- Company's stock price changes.
- Position in the market for future products.
- Impacts on competitors and their response.
- Future value of engineering, manufacturing, and support experience associated with using new technologies or materials in the current product.
- Long-term health, safety, and environmental impacts that may have to be resolved in the future.

The impacts listed above are difficult to quantify in terms of cost because they require a view of the enterprise (i.e., entire organization or company) that is broader than one product and has an analysis horizon that is longer than the manufacturing and support life of one product. However, these costs are real and may make significant contributions to product cost.

---

[*] P. F. Ostwald and T. S. McLaren, *Cost Analysis and Estimating for Engineering and Management*, Pearson Prentice Hall, Upper Saddle River, NJ, 2004.

3.3 DESIGN AND MANUFACTURING COSTS

In this section, we shall discuss nonrecurring design and development costs, and the recurring costs associated with manufacturing and assembly. The relationship between cost and quality is discussed in the context of manufacturing yield.

3.3.1 DESIGN AND DEVELOPMENT COSTS

Design and development costs are referred to as nonrecurring costs because they are a one-time charge no matter how many units of the product are manufactured. For small-volume manufacturing, design costs usually have a larger impact on the product's cost than for high-volume products. The specific elements of design and development costs that must be included are

- Development of the product specification.
- Conceptual design.
- Intellectual property acquisition and protection such as licensing costs and patent filing costs.
- Design of the product including the creation of engineering drawings.
- Software development.
- Creation of prototypes.
- Functional testing.
- Environmental testing to determine or verify reliability.
- Product qualification and certification such as meeting Underwriter Laboratory requirements, U.S. Federal Aviation Administration requirements, and U.S. Food and Drug Administration requirements (see Section 3.4.3).

The design and development costs can vary widely, depending on whether a new product is being developed or the product is based on an existing product. Table 3.2 provides examples of product development costs and their development times for a range of products.

3.3.2 MANUFACTURING COSTS

Manufacturing costs form the basis for determining the actual recurring cost of making a product. Manufacturing costs are generally the sum of the costs from four primary sources: recurring labor costs, recurring material costs, the allocation of nonrecurring tooling, and capital costs. In order to gain a more complete picture of manufacturing costs, one must also consider the cost of quality and the cost of testing, which are discussed in Sections 3.3.3 and 3.3.4, respectively.

We shall discuss each of these four contributions in detail.

Labor Costs (C_L): Labor costs refer to the cost of the people required to perform specific activities. The labor cost per unit associated with an activity performed during manufacturing is determined from either

$$C_L = \frac{N_L T L_R}{N_p}$$

(3.5a)

or

$$C_L = \frac{N_L T L_{RB}}{N_p}$$

(3.5b)

where
   N_L = Number of people associated with the activity: it can have a value < 1
   T = Length of time taken by the activity
   N_p = Number of units that can be treated simultaneously by the activity

TABLE 3.2
Examples of Product Development Costs and Development Times

Product	Design and Development Costs ($)	Design and Development Time (years)
Ford's first automobile[a]	4,000 (1902 dollars)	—
Stanley Tools *Jobmaster* Screwdriver[b]	150,000	1
Polaroid color photo printer (CI-700)[b]	5,000,000	1
Video console games[c,d]	15–20,000,000	2–3
Mainframe central processing unit[e]	100,000,000	5
Pharmaceuticals[f,g]	897,000,000	6–12
Boeing 787 airliner[h,i]	10–12,000,000,000	5

[a] D. Hochfelder and S. Helper, "Suppliers and Product Development in the Early American Automobile Industry," *Business and Economic History*, Vol. 25, No. 3, Winter 1996. http://h-net.org/~business/bhcweb/publications/ BEHprint/v025n2/p0039-p0052.pdf.

[b] K. T. Ulrich and S. D. Eppinger, *Product Design and Development*, 4th ed., McGraw-Hill, Boston, 2007.

[c] http://en.wikipedia.org/wiki/Video_game_publisher/

[d] http://en.wikipedia.org/wiki/Game_development

[e] http://en.wikipedia.org/wiki/CPU_design.

[f] M. Moran, "Cost of Bringing New Drugs to Market Rising Rapidly," *Psychiatric News* August 1, 2003, Vol. 38, No. 15. http://pn.psychiatryonline.org/cgi/content/full/38/15/25.

[g] S. A. Bernhardt and G. A. McCulley, "Knowledge Management and Pharmaceutical Development Teams: Using Writing to Guide Science," *IEEE Transactions on Professional Communication*, February/March 2000. http://ieeexplore.ieee.org/iel5/47/17890/00826414.pdf?arnumber=826414.

[h] D. G. Greising and J. Johnsson, "Behind Boeing's 787 Delays," *Chicago Tribune*, December 8, 2007. http://www.chicagotribune.com/services/newspaper/printedition/saturday/chi-sat_boeing_1208dec08,0,528945.story

[i] http://en.wikipedia.org/wiki/Boeing_787.

Equation 3.5a is used when the overhead costs are computed separately from labor costs, and Equation 3.5b is used when the overhead costs are incorporated into the labor costs.

The product $N_L T$ is sometimes referred to as the *touch time*. For example, if a process step takes 5 minutes to perform, and one person is sharing his/her time equally between this step and another step that takes 5 minutes to perform, then $N_L = 0.5$ and $T = 5$ minutes for a touch time $N_L T = 2.5$ minutes.

Material Costs (C_M): The cost of the materials associated with an activity is given by

$$C_M = U_M C_m \qquad (3.6)$$

where

U_M = Quantity of the material consumed as indicated by its count, volume, area, or length

C_m = Unit cost of the material per count, volume, area, or length

Material costs may include the purchase of more material than what ends up in the final product due to waste generated during the manufacturing process, and it may include the purchase of consumable materials that are used and completely wasted during manufacturing, such as water.[*]

Tooling Costs (C_T): Tooling costs are nonrecurring costs associated with activities that occur only once or only a few times. Examples of tooling costs are programming and calibration costs for

---

[*] P. A. Sandborn and C. F. Murphy, "Material-Centric Modeling of PWB Fabrication: An Economic and Environmental Comparison of Conventional and Photovia Board Fabrication Processes," *IEEE Transactions on Components, Packaging, and Manufacturing Technology*—Part C, Vol. 21, April 1998, pp. 97–110.

manufacturing equipment, training of people, and the purchase or manufacture of product-specific tools, jigs, stencils, fixtures, masks, etc.

Capital Costs (C_c): Capital costs are the costs of purchasing and maintaining the manufacturing equipment and facilities. In general, capital costs associated with an activity are determined from

$$C_c = \frac{TC_e}{N_p T_{op} T_d} \qquad (3.7)$$

where T and N_p are as defined for Equation 3.5, and

C_e = Purchase price of the capital equipment or facility

T_{op} = Operational time of the equipment or facilities expressed as the number of hours per year

T_d = Depreciation life in years[*]

In some cases, the capital costs associated with a standard manufacturing process are incorporated into the overhead rate. Even if the capital costs are included in the overhead, Equation 3.7 may still be used to include the cost of unique equipment or facilities that must be created or purchased for a specific product.

There are several different ways to model manufacturing costs. Any of the modeling methods introduced subsequently can be used for any product; however, the primary contribution to the cost—that is, the largest of the labor, material, tooling, or capital costs—often determines the type of model used.

Process Flow Model: Manufacturing processes can be modeled as a sequence of process steps that take place in a specific order. The steps and their order are referred to as a process flow. In process flow models, a product unit accrues cost as it moves through the sequence of process steps. Each process step starts with the state of the unit after the preceding process step. The current step then modifies the unit and its output is a new unit state, which forms the input to the next process step. For the purposes of cost modeling, the input and output of each step are the cost per unit as described by labor, material, tooling, and capital equipment and facilities costs, and the costs of quality described as the number of defects per unit.[†] Usually, process flow models are constructed so that the form of the process step input matches the form of their output. The advantage that process flow models have over the other manufacturing modeling approaches discussed in this section is that they capture the sequence of the manufacturing activities. The sequence is important whenever units have to be removed at some step in the process; that is, they are scrapped. When a unit is removed from the process, the amount of money spent up to the point of removal must be allocated to the units remaining in the process. The procedure for doing this is discussed in Section 3.3.4.

Cost of Ownership (COO) Model: In the cost of ownership modeling approach, a sequence of process steps is of secondary interest; the primary interest is the answer to the question, What proportion of the lifetime cost of a piece of equipment or facility does the manufacture of a product unit consume? Accumulating the fractional lifetime costs of all the equipment and facilities for one unit of a product gives an estimate of the cost of a unit of the product. Cost of ownership is fundamentally different from process flow cost modeling. In process flow models, the actual path of a unit through a manufacturing process is emulated with the product unit accruing cost as it moves through a sequence of process steps. COO calculates an effective total

---

[*] Depreciation is the lessening of the value of a physical asset over time. In Equation 3.7, the quantity T_d is the depreciation life, which is the length of time that the purchase price of the asset is spread over for accounting purposes. Equation 3.7 assumes that a "straight line" method is used to model depreciation; that is, depreciation is linearly proportional to the length of time of service.

[†] Many other properties of the unit can be accumulated through process steps as well, including manufacturing time, various material and energy inventories for life cycle assessment, environmental impact analysis, mass, and scrap, which is sometimes referred to as fallout. For detailed process step definitions that include various material and waste inventories, see P. A. Sandborn and C. F. Murphy, 1998, ibid.

cost of ownership for each piece of equipment in the manufacturing process and then charges each unit a fraction of that cost based on the portion of the lifetime of the equipment used up by the unit.

Cost of ownership was originally developed to model integrated circuit (IC) fabrication costs. IC costs are dominated by equipment and facilities since labor, tooling, and material contributions to the cost are very small compared to the cost required to construct and maintain the IC fabrication facility. The nature of the COO model makes it best suited for those products where the equipment and facilities cost dominate.

Activity Based Cost (ABC) Model: Activity-based costing is a method of assigning an organization's resource costs through activities to the products and services provided to its customers. In traditional cost accounting, overhead costs are most often allocated to products in proportion to labor hours. In ABC, distinct activities associated with the manufacture of a product are identified and the primary cost drivers behind each of the activities are found. Activities are the collection of actions performed by an organization to design, manufacture, and support a product. The drivers behind activity costs are often transactional—that is, the number of holes to drill, the number of layers to laminate, the number of boxes to pack, the number of setups, and so on. Once activities and their associated cost drivers are identified, an *activity rate* A_R (with units of cost per activity) is determined from the relation

$$A_R = \frac{\text{Activity cost pool}}{\text{Activity base}} \tag{3.8}$$

where the *activity cost pool* is the total amount of overhead required by the activity during some period of time and the *activity base* is the number of times the activity was performed on all products during that period of time. The total cost of the ith activity for a single product is determined from

$$C_{A_i} = A_{R_i} N_{A_i} + C_{L_i} + C_{M_i} \tag{3.9}$$

where N_{A_i} is the number of times the activity must be performed to manufacture one unit of a product. The product $A_{R_i} N_{A_i}$ in Equation 3.9 is the overhead allocated to one unit of the product by the activity. The sum of C_{A_i} over all activities associated with the manufacture of a product gives the manufacturing cost of one unit of the product.

The advantage of ABC models over other approaches is that they more accurately allocate overhead costs to products. The disadvantage is that a history of accounting data that has tracked total costs associated with various activities over time is required to calculate the activity rates.

Parametric Cost Model: Parametric cost models form the basis for many top-down cost models that seek to establish an estimate of the cost of a product from high-level design parameters that define the product's performance, functionality, and physical attributes. The relationship between design parameters and costs is called a *cost estimating relationship* (CER). Design parameters such as weight, tolerance, speed, and force are commonly used. The simplest cost estimating relationships are mathematical relationships obtained from fitting curves of particular design parameters as a function of the product's cost. As an example of parametric cost modeling, consider the following CER developed for the average annual operating cost of military aircraft[*]

$$\text{Cost} = T_f^{\beta_2} P_p^{\beta_3} e^{(\alpha + \beta_1 A_g)} \tag{3.10}$$

---

[*] G. G. Hildebrandt and M.-B. Sze, "An Estimation of USAF Aircraft Operating and Support Cost Relations," The RAND Corporation, May 1990.

where

    T_f = Average annual flying hours
    P_p = Average procurement cost
    A_g = Average aircraft age divided by 100
    α, β_1, β_2, and β_3 = Parameters determined from the curve-fitting procedure

Parametric models can be very accurate for well-known and well-defined products. For example, the most accurate cost models for fabricating printed circuit boards are parametric models. However, parametric models are only valid when used to determine the cost of products that fall within the bounds of the original data used to create the model. In the case of Equation 3.10, only U.S. Air Force cargo and fighter aircraft that logged flying hours from 1981 to 1986 were used to generate the model. Therefore, attempting to use Equation 3.10 to determine the annual operating cost of a commercial airliner, a helicopter, or the F-22 fighter aircraft that entered service in December 2005 may be inappropriate.

Parametric cost models that have been applied to the determination of the cost of mechanical and solid objects are referred to as *feature-based cost modeling*. Feature-based cost modeling involves the identification of a product's cost-driving features such as the number of holes, edges, folds, corners, etc., and the determination of the costs associated with each of these features.[*] Feature-based cost models can be incorporated into CAD systems to automatically estimate manufacturing costs of objects based on their features.

Neural network based cost estimation is an extension of parametric modeling, and it can potentially represent more complex relationships between process and product design parameters than simple CERs used in most parametric approaches.[†]

Technical Cost Modeling (TCM): The process of predicting the primary cost contributions from the physical parameters associated with a manufacturing process and product-specific details is called *technical cost modeling* (TCM).[‡] Technical cost modeling can be used in conjunction with any of the modeling approaches that have been discussed so far. For example, TCM would use algorithms relating product characteristics to the physical parameters associated with a process (e.g., temperature, pressure, and flow rate) to predict values such as process cycle time or materials consumption, which are in turn directly related to the labor or material costs of the process step or activity. TCM marries physical process requirements derived directly from the product details to cost models.

3.3.3 COST OF MANUFACTURING QUALITY

Minimizing the manufacturing cost of a product is not sufficient to ensure that a product can be produced in a cost-effective way. The likelihood that a manufacturing process introduces defects into the product being manufactured, and the costs associated with minimizing and correcting those defects, must be considered as well. For example, suppose process A manufactures a product for $20 per unit and produces no defective products and suppose that process B manufactures the same product for $11 per unit but half of the products produced by this process are defective and cannot be reworked or salvaged. For process A, the effective cost per good unit is $20 per unit, while for process B the effective cost per good unit is $11/0.5 = $22 per unit.

---

[*] Cost Analysis Improvement Group (CAIG), "Operating and Support Cost-Estimating Guide," http://www.dtic.mil/pae/, May 1992. Examples of manufacturing process-step-specific, feature-based cost estimators applicable to injection molding, machining, and casting operations can be found at http://www.custompartnet.com/.

[†] A. S. Ayed, "Parametric Cost Estimating of Highway Projects Using Neural Networks," Memorial University of Newfoundland, Engineering and Applied Science, M.S. degree engineering thesis, 1997.

[‡] J. Busch, "Cost Modeling as a Technical Management Tool," *Research Technology Management*, Vol. 37, No. 6, November/December 1994, pp. 50–56.

The cost of quality[*] is defined as the cost incurred because less than 100% of the products produced can be sold. Generally, quality costs are composed of the following four elements:

1. **Prevention costs**, which is the cost of preventing defects and includes such items as education, training, process adjustment, screening of incoming materials and components, etc.
2. **Appraisal costs**, which are the costs of tests and inspections to assess if defects exist in manufactured or partially manufactured products.
3. **Internal failure costs**, which are the costs of defects detected prior to delivery of the product to the customer.
4. **External failure costs**, which are the costs of delivering defective products to the customer.

In this section, we shall discuss internal failure costs through the introduction of the concepts of yield and yielded cost. In Section 3.3.4, we address appraisal costs by discussing test, diagnosis, and rework, and in Section 3.4.2 we discuss warranties, which is one aspect of external failure costs.

Yield is a measure of quality that represents the fraction of good units resulting from an operation or process. Manufacturing step yield Y_{step} is given by

$$Y_{step} = \frac{N_u}{N_I} \tag{3.11}$$

where

N_u = Number of usable or non-defective units after a manufacturing process step
N_I = Number of units that start or complete the manufacturing process step

If N_I is the number of units that start the manufacturing process step, then Y_{step} is the process step yield. If N_I is the number of units that complete the manufacturing process, then Y_{step} is the yield of the product after the process step. The quantity Y_{step} is the probability of obtaining a product with zero defects out of a manufacturing process step. If the yields of the individual steps are independent from each other, then the yield of an entire process is given by

$$Y_{Process} = \prod_{i=1}^{m} Y_{step_i} \tag{3.12}$$

where Y_{step_i} is the yield of the ith step in the process, and m is the total number of steps in the process. For the special case where the yield in each step is equal, $Y_{step_i} = Y_a$ for all i and Equation 3.12 simplifies to $Y_{Process} = Y_a^m$. Thus, if $Y_a = 0.9$ (90% yield at each step), and $m = 20$ process steps, then the product's yield without rework is 0.12 (= 0.9^{20}); that is, only 12 out of 100 units manufactured will be free of defects.

3.3.4   Test, Diagnosis, and Rework

For products whose constituent parts, materials, or processing contain a high incidence of defects, recurring functional testing[†] is required, and this testing may significantly affect the total cost of manufacture. As an example, consider functional testing used in the fabrication of integrated circuits on a semiconductor wafer. Depending on the minimum feature size and unpackaged circuit die size,

---

[*] M. Sakurai, *Integrated Cost Management*, Productivity Press, Portland, OR, 1996.
[†] Recurring functional testing is testing that occurs as part of the manufacturing process and takes place for every unit manufactured. Inspection is a form of recurring functional testing. Recurring functional testing differs from nonrecurring verification tests that may be conducted as part of the development of the product and nonrecurring environmental testing that is done to determine the reliability or to qualify the product.

production yield may be as low as 10%, and extensive testing is required to determine which fabricated devices are good and which are bad. In some cases, greater than 60% of a product's recurring cost can be attributed to testing costs[*]; for integrated circuits, testing costs are approaching 50% of the total product cost.[†] When the products that result from a manufacturing process are imperfect, four costs are potentially involved:

1. **Testing cost**, which is the cost of determining whether a given unit is good or bad.
2. **Diagnostic cost**, which is the cost of determining what defect caused the unit to be bad and where the defect is located.
3. **Rework cost**, which is the cost to fix a defect during the manufacturing process.
4. **Continuous improvement cost**, which is the cost incurred by all efforts to eliminate the causes of the defects.

Depending on the maturity of the product, its placement in the market, its cost to manufacture, and the profit associated with selling it, all, some, or none of the four activities may be undertaken. Understanding the test–diagnosis–rework costs may determine the extent to which the product designer can control and optimize the manufacturing cost, and the extent to which it makes sense to do so.

The ultimate goal of any functional test strategy is the determination of the following:

- When in the manufacturing process the product should be tested.
- How thorough the testing should be.
- The steps that should be taken to make the product easier to test.

These three goals would be easy to realize if one had unlimited time, resources, and money. Since this is not the case, one usually must determine how to obtain the best test coverage possible for the least cost.

The specific goal of test economics is to minimize the cost of discarding good product and the cost of shipping bad product. This goal is enabled through the development of models that allow the yield and cost of units that pass through test operations to be predicted as a function of the properties of the unit entering the test and the characteristics of the testing operations, which are its cost, yield, and ability to detect defects in the unit being tested.

Shown in Figure 3.2a is a simple functional test process step that could be used in a process flow model, discussed in Section 3.3.2, that has no diagnosis or rework associated with it; that is, units that do not pass the test are scrapped. The test step is characterized by the following relations.[‡] Assuming that all units that fail the tests are scrapped, the total effective cost per unit that passes the test is given by

$$C_{out} = \frac{C_{in} + C_{test}}{Y_{in}^{f_c}} \tag{3.13}$$

where
C_{in} = Total cost spent on the unit up to the start of the test
Y_{in} = Yield of units as they enter the test and which can be determined from Equation 3.12 if the test is the $m + 1$ process step

---

[*] J. Turino, *Design to Test—A Definitive Guide for Electronic Design, Manufacture, and Service*, Van Nostrand Reinhold, New York, 1990.
[†] W. Rhines, Keynote address at the Semico Summit, Phoenix, AZ, March 2002.
[‡] P. Sandborn, *Course Notes on Manufacturing and Life Cycle Cost Analysis of Electronic Systems*, CALCE EPSC Press, College Park, MD, 2005.

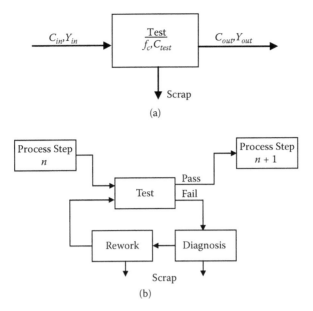

FIGURE 3.2 (a) A simple test process step and (b) test, diagnosis, and rework between two steps in a manufacturing process flow.

C_{test} = Cost of performing the test on one unit
f_c = Fraction of faults successfully detected by the test; called *fault coverage*.[*]

The quantity C_{out} incorporates the money spent on scrapped units into the cost of units that continue in the process—that is, those that have passed the test. It is noted that C_{out} represents whatever mixture of costs are accumulated through the process prior to the test; for example, it could be all or a portion of Equation 3.2. The yield of units that pass the test is given by

$$Y_{out} = Y_{in}^{1-f_c} \tag{3.14}$$

Lastly, the fraction of units S_s entering the test that did not pass the test is given by

$$S_s = 1 - Y_{in}^{f_c} \tag{3.15}$$

To illustrate the test–diagnosis–rework aspect of cost analysis, consider the testing of a product between two manufacturing steps as shown in Figure 3.2b. In this example, all of the units coming from step n are tested. For those units that have failed the test, a diagnostic test is performed to identify what the defects are and where they are located. All units that are identified as fixable are reworked and retested. The diagnosis and rework activities themselves are not perfect and can

---

[*] A defect is a flaw that causes a product not to work correctly under certain conditions, and a fault is the effect of a defect on the product. Tests and testers measure or detect faults, not defects. For example, a defect in an electrical system might be a broken connection. The fault detected by the tester due to this defect would be an electrical open circuit, instead of the expected short circuit. A diagnosis activity isolates the fault and relates it to an actual defect; i.e., a diagnosis would determine where the open circuit is. Fault coverage f_c is a measure of the ability of a set of tests to detect a given class of faults that may occur in a product. A defective product may have more than one defect in it, but the test only needs to successfully detect one fault in order to remove the product from the process.

introduce defects, make misdiagnoses, and fail to correct the defect. Therefore, a product may have to go through the testing, diagnosis, and rework process multiple times. Detailed cost models for complex test–diagnosis–rework operations have been developed.[*]

3.4   SUSTAINMENT COSTS: LIFE CYCLE, OPERATION, AND SUPPORT

While traditional cost analysis focuses on manufacturing costs, which are extremely important for inexpensive high-volume products, operation costs can be substantial for some products and sustainment[†] costs can dominate for low-volume complex systems. For products where the operation and sustainment costs are significant, it is important for the customers to understand these costs when making purchase decisions. As an example, consider the representative *EnergyGuide* label shown in Figure 3.3, which provides the estimated annual operating cost for electricity for operating the product to which it is attached. In Figure 3.4a, we have shown the breakdown of costs for an F-16 aircraft. It is seen in this figure that its operation and support costs represent 78% of the overall cost of the aircraft. However, one need not consider such a complex system to see the dominance of operation and support costs. Consider a network of 25 personal computers managed by one full-time system administrator. As shown in Figure 3.4b, we see that the percentage operation and support costs for this network are comparable to that shown for the F-16 aircraft.

Next, we shall discuss the following major sustainment costs: reliability, availability, spare parts, warranty, and qualification and certification.

3.4.1   SPARE PARTS AND AVAILABILITY: IMPACT OF RELIABILITY ON COST

Reliability can be the most important attribute for many products. High reliability may be necessary in order for one to realize value from the product's performance, functionality, or low cost. The ramifications of reliability on the product's life cycle are linked to sustainment cost through spare parts requirements and warranty return rates. The combination of how often a system fails and the efficiency of performing maintenance when the system does fail determine the system's *availability*.

Reliability is the probability that a unit will not fail. Maintainability is the probability that the unit can be successfully restored to operation after failure. Availability provides information about how effectively the unit is managed and it is a function of reliability and maintainability.

When a unit encounters a failure while in use, one of the following takes place:

- Nothing happens because a workaround is implemented, the unit is disposed of and its functionality is accomplished another way, or the system operates without the functionality.
- The unit is repaired.
- The unit is replaced.

The replacement of failed parts of a product requires that one have spare parts. The issues that arise concerning spare parts are the following:

- Estimating the total number of spare parts that will be needed.
- Determining when the spare parts will be used.

T. Trichy, P. Sandborn, R. Raghavan, and S. Sahasrabudhe, "A New Test/Diagnosis/Rework Model for Use in Technical Cost Modeling of Electronic Systems Assembly," *Proc. International Test Conference*, pp. 1108–117, November 2001.

[†] *Sustainment* refers to all activities necessary to (a) keep an existing system operational so that it can successfully complete its intended purpose, (b) continue to manufacture and install versions of the system that satisfy the original requirements, and (c) manufacture and install revised versions of the system that satisfy evolving requirements. See P. Sandborn and J. Myers, "Designing Engineering Systems for Sustainability," in *Handbook of Performability Engineering*, K.B. Misra, Editor, Springer, London, 2008, pp. 81–103.

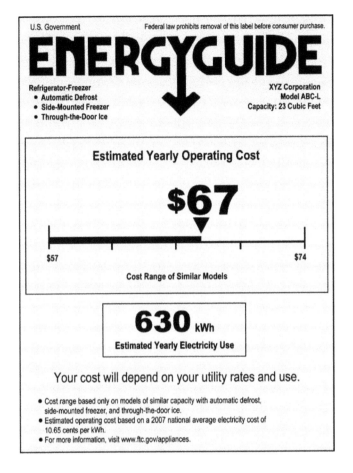

FIGURE 3.3 A representative *EnergyGuide* label, which provides an estimate of the annual cost of electricity for this particular product model; that is, it provides an estimate of its yearly operational costs. In addition, it indicates how this annual amount compares to similar models of this product. (From: http://www.ftc.gov/opa/2007/08/energy.shtm.)

- Deciding where the spare parts should be kept, which implies two critical issues[*]: (1) who will install the spare parts and what facilities are required to install them, and (2) for what portion of the product should spare parts be purchased.

To determine the number of spare parts required for a certain level of confidence that a system will survive to a time t, we need to consider a product's reliability. Many mechanical systems are well-characterized by a constant failure rate λ. The mean time between failure (MTBF) is related to the constant failure rate by MTBF $= 1/\lambda$. The number of spare parts k that are required for a nonrepairable item to last until a time t is given by

$$k = n_s \lambda t \tag{3.16}$$

where n_s is the number of units in service.

---

[*] It makes sense to carry a spare tire in the trunk of a car, but it does not make sense to carry a spare transmission in the trunk for several reasons: transmissions do not fail as frequently as tires, transmissions are large and heavy to carry, and one doesn't have the tools or expertise to install a new transmission on the side of the road.

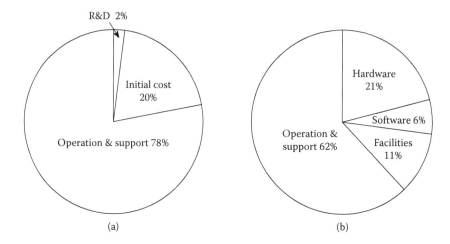

FIGURE 3.4 Total cost breakdowns for (a) an F-16 aircraft and (b) personal computer network with one full-time system administrator. [Data for Figure (a) from: Cost Analysis Improvement Group (CAIG), "Operating and Support Cost-Estimating Guide," Office of the Secretary of Defense, http://www.dtic.mil/pae/, May 1992. Data for (b) from: Gateway Inc., www.gateway.com, December 2001 and Shields, P., "Total Cost of Ownership: Why the price of the computer means so little," http://www.thebusinessmac.com/features/tco_hardware.shtml, December 2001.]

The assumption of a constant failure rate implies that the times to failure are exponentially distributed, and the reliability $R(t)$ of one unit at time t is given by

$$R(t) = e^{-\lambda t} \tag{3.17}$$

The reliability $R(t)$ is the probability of encountering exactly 0 failures in time t.[*] Equation 3.17 can be generalized to give the probability of exactly x failures in time t as

$$P(x) = \frac{(\lambda t)^x e^{-\lambda t}}{x!} \tag{3.18}$$

From Equation 3.18, we see that $P(0) = R(t)$. Therefore, for a product with no spare parts, the probability of surviving to time t is $P(0)$. For a system with one spare part, the probability of surviving to time t is $P(0) + P(1) = e^{-\lambda t} + \lambda t e^{-\lambda t}$. Consequently, the probability of surviving to time t with k spare parts is

$$P(x \leq k) = \sum_{x=0}^{k} \frac{(n_s \lambda t)^x e^{-n_s \lambda t}}{x!} \tag{3.19}$$

Equation 3.19[†] gives the probability $P(x \leq k)$ that the system will survive to time t when one has k spare parts. For a repairable product, where spare parts are only needed to keep the product operating while repairs are performed, t represents the time that it takes to perform repairs and $P(x \leq k)$ determined from Equation 3.19 gives the probability that k is a sufficient number of spares.

---

[*] For exponentially distributed times to failure, 63.2% of the units have failed by $t = 1/\lambda$, which is the MTBF.
[†] Equation 3.19 is called the Poisson cumulative distribution function.

The costs associated with carrying spare parts are:

* Manufacturing or purchasing the spare parts.
* Money, which is tied up in spare parts that are for future use and is known as "the cost of money" (see Section 3.5.2).
* Transportation of spare parts to where they are needed or transportation of the product to the location where the spare parts are kept.
* Warehousing the spare parts until they are needed (inventory costs).
* Replenishment of the spare parts when they are depleted.
* Unavailability of the system when spare parts are not in the right place at the right time.

Availability is the probability that the product will be able to function when called upon to do so. As indicated previously, availability is a function of a product's reliability and its maintainability—that is, how quickly the product can be replaced or repaired.

For many types of products, availability is a very important requirement. Examples of products requiring virtually 100% availability are bank ATM machines, communications systems such as 911 systems, and military systems. The cost of unavailability can be very high. For a large customer using a point-of-sale verification system, it has been estimated that system downtime, its unavailability, costs $5,000,000 per minute.[*] In this case, the availability of the point-of-sale verification system is probably a more important characteristic than the system's price. Examples of other systems where there can be significant costs directly associated with availability are loss of capacity for manufacturing operations, loss of customer confidence in airline operations, and loss of mission in military operations. For a repairable system, the inherent availability $A_{inherent}$ of a product is given by

$$A_{inherent} = \frac{\text{MTBF}}{\text{MTBF} + \text{MTTR}}$$ (3.20)

where MTBF is the mean time between failures, and MTTR is the mean time to repair. For products where availability is critical, purchase agreements and contracts may include clauses that stipulate the portion of the contract price that will be paid to the provider as a function of the availability of the product that the customer actually experiences.

3.4.2 WARRANTY AND REPAIR

A warranty is a manufacturer's assurance to a buyer that a product or service is as represented. A warranty is considered a contractual agreement between the buyer and the manufacturer that is entered into upon sale of the product or service. In broad terms, the purpose of a warranty is to establish liability between the manufacturer and the buyer in the event that the product or service fails. This contract specifies both the performance that is to be expected and the redress available to the buyer if a failure occurs.[†]

Warranty cost analysis is performed to estimate the cost of servicing a warranty so that it can be properly accounted for in the sales price or maintenance contract for the product. Similar to the analysis of spare parts, warranty analysis is focused on determining the number of expected product failures during the warranty period that will result in a warranty action.

---

[*] McDougall, R., "Availability—What It Means, Why It's Important, and How to Improve It," Sun BluePrints OnLine, October 1999, http://www.sun.com/blueprints/1099/availability.pdf.

[†] D. N. P. Murthy and I. Djamaludin, "New Product Warranty: A Literature Review," *International Journal of Production Economics*, Vol. 79, No. 3, 2002, pp. 231–260.

Warranty analysis differs from spare parts analysis in several ways. First, warranty analysis determines a warranty reserve cost, which is the total amount of money that has to be set aside to cover the warranty of a product. Warranty analysis does not base its spare parts needs on maintaining specific system availability, but only servicing warranty claims. Another way in which warranties differ from spare parts analysis is that the period of performance, the period in which expected failures need to be counted, can be defined in a more complex way. For example, the automotive industry warranties typically stipulate two conditions such as "4 years or 48,000 miles, whichever comes first."

The cost of servicing an individual warranty claim varies depending on the type of warranty provided. The simplest case is a free replacement warranty in which every failure prior to the end of the warranty period is replaced or repaired to its original condition at no charge to the customer. In this case, the warranty reserve fund amount C_{WR}, if we ignore the cost of money, is given by

$$C_{WR} = C_{fr} + nM(t_W)C_{rc} \qquad (3.21)$$

where

$\quad C_{fr}$ = Fixed cost of providing warranty coverage, such as the cost of maintaining a toll free
$\qquad$ number for customers to call
$\quad n$ = Number of products sold; equal to N_{pm} if all the units manufactured are sold
$\quad C_{rc}$ = Average recurring replacement and/or repair cost on a per unit basis
$\quad t_W$ = Warranty length of time
$M(t_W)$ = Expected number of replacements in the interval from 0 to t_W

The quantity $M(t_W)$ is equal to λt_W for a product with a constant failure rate λ assuming "good-as-new" replacement or repair upon failure.[*]

Many other types of warranties exist, including ordinary free replacement and pro rata warranties.[†] Another type of warrant response is a product recall, which offers some level of repair or replacement to all product owners. Product recalls are generally very expensive responses to product problems and are usually employed only when critical safety issues are involved.

3.4.3 QUALIFICATION AND CERTIFICATION

Many types of products require extensive qualification and/or certification in order to be sold or used. An example of one such device is a power supply for a laptop computer, which, as the picture in Figure 3.5 shows, must pass numerous certifications. Qualification is the process of determining a product's conformance with specified requirements. The specified requirements may be based on performance, quality, safety, and/or reliability criteria. Certification is the procedure by which a third party provides assurance that a product or service conforms to specific requirements. The terms *qualification* and *certification* are sometimes used interchangeably. Examples of certifications required for some products in the United States include the following.

- **Food and Drug Administration** (FDA) requires that certain standards be met for food, cosmetics, medicines, medical devices, and radiation-emitting consumer products such as microwave ovens and lasers. Products that do not conform to these standards are banned from being sold in the United States and from being imported to the United States.

---

[*] In general, $M(t) \neq 1 - R(t)$ because $M(t)$ accounts for the possibility that units could fail more than once during the time interval; $M(t)$ is called a renewal function.
[†] E. A. Elsayed, *Reliability Engineering*, Addison Wesley, Reading, MA, 1996.

FIGURE 3.5 Certifications indicated on a Dell laptop computer AC power supply.

- **Federal Communications Commission** (FCC) requires certification of all products that emit electromagnetic radiation such as cell phones and personal computers. Devices that intentionally emit radio waves cannot be sold in the United States without FCC certification.
- **Environmental Protection Agency** (EPA) certification is required for every product that exhausts into the air or water including all vehicles (cars, trucks, boats, ATVs), heating, ventilating and air conditioning systems (air conditioners, heat pumps, refrigerators, refrigerant handling and recovery systems), landscaping and home maintenance equipment (chain saws and snow blowers), stoves and fireplaces, and even flea and tick collars for pets.
- **Federal Aviation Administration** (FAA) certification certifies the airworthiness of all aircraft operating in the United States. The FAA also certifies parts and subsystems that are used on the aircraft.

It is difficult to assign a specific cost to certifications since in addition to the cost of performing the qualification testing, substantial cost is incurred in the process of designing the product so that it will meet the requirements. The direct cost of certification includes application fees, time to manage the appropriate paperwork, and the cost of legal and other expertise necessary to navigate the certification requirements processes. The indirect costs of certification, which are usually the larger portion of the certification costs, are the cost of performing required qualification testing prior to seeking certification, product modifications and redesign if certification is not granted, and the time required to gain the certification, which can be years in some cases. On one end of the spectrum, the cost for an FCC certification of a new personal computer by an approved third party can range from $1,500 to $10,000 and can be obtained in a few days. On the other end of the spectrum, the average time for approval of a new drug from the start of clinical testing to obtaining FDA approval was approximately 90 months in 2003, with estimated costs that can exceed $500,000,000.

Other certifications may not be required by law, but may be required by the customers or the retailers of the product. One of the more common certifying organizations is the Underwriter Laboratories (UL), which provides certification regarding the safety of products. The cost of obtaining a UL certification can range from $10,000 to $100,000 for one model of one product. In addition, there are annual fees that are required to maintain the certification. Another example of an optional approval is the EPA's Energy Star program for products that meet specific energy efficiency guidelines.

General certifications (UL, FDA, FCC, etc.) are usually nonrecurring costs borne by the manufacturer. However, qualification of products for specific uses may be borne by either the manufacturer or the customer. For example, the manufacturer of a new electronic part will run a set of qualification tests that correspond to a common standard and then market the part as compliant with that standard. When a customer decides to use the part they may perform additional qualification tests to ensure that the part functions appropriately within their usage environment. Manufacturer and customer qualification testing can range from a few thousand dollars to hundreds of thousands of dollars for simple parts. For complex systems, such as aircraft, qualification testing costs millions to tens of millions of dollars. Generally, these are one-time nonrecurring expenses; however, they may have to be partially or completely repeated if changes to the part are made or changes to the system using the part are made.

3.5 MAKING A BUSINESS CASE

A business case is a structured proposal for new business or business improvement that is part of a decision-making process within an organization. The purpose of a business case is to provide a comprehensive evaluation of the reasons that a proposed action should be considered. One very important attribute of most business cases is the development of an economic justification. Two important attributes of an economic justification introduced in this section are the return on investment and the cost of money.

3.5.1 RETURN ON INVESTMENT

Return on investment (ROI) is a useful quantitative means of gauging the economic merits of a decision. ROI measures the cost savings, profit, or cost avoidance that result from a given use of money.* At the enterprise level, ROI may reflect how well an organization is managed with regard to specific organizational objectives such as gaining market share, retaining more customers, or improving availability. The ROI may be measured in terms of how a change in practice or strategy results in meeting these goals. The ROI allows for enhanced decision-making regarding the use of investment money and research and development efforts by enabling comparisons of alternatives. However, the quantities used to calculate the ROI must be accurate and inclusive in order for the calculation to be meaningful. In the case of a new product, the investment includes the costs necessary to develop, manufacture, and support the product; the return is a quantification of the benefit realized by offering the product. In the simplest case, the return is the revenue from selling the product; in more complex cases, the return could be the revenue plus a quantification of increased market share and the experience gained by offering the product, and other hidden costs and returns discussed in Section 3.2.3.

---

* G. T. Friedlob and F. J. Plewa Jr., *Understanding Return on Investment*, John Wiley & Sons, New York, 1996.

In general, ROI is the ratio of gain to investment. One way of defining the ROI over a product's life cycle is given by either

$$\text{ROI} = \frac{\text{Return} - \text{Investment}}{\text{Investment}} \tag{3.22a}$$

or

$$\text{ROI} = \frac{\text{Avoided cost} - \text{Investment}}{\text{Investment}} \tag{3.22b}$$

Equation 3.22a is the classical ROI definition and Equation 3.22b is the form of the ROI that is applicable to investments made to enhance the maintainability of a product. An ROI > 0 indicates that there is a cost benefit.

Constructing a business case for a product does not necessarily require that the ROI be greater than zero; in some cases, the value of a product is not fully quantifiable in monetary terms, or the product is necessary in order to meet a system requirement that could not otherwise be attained such as an availability requirement discussed in Section 3.4.1.

3.5.2 THE COST OF MONEY

Financial costs are part of the economics of both producing the product and buying the product. Whether manufacturers have to borrow money to fund the development and manufacture of a product or have to use resources they already have in lieu of investing those resources elsewhere, there is a cost associated with the use of money involved. For a customer, there are similar costs of money. If a buyer has to obtain financing to purchase the product or has to use cash on hand that could otherwise be earning money in another investment, these are costs that are associated with acquiring the product.

One way to determine the cost of money is to obtain its present value, and to compare this value to its value in the future. The premise for computing the present value is that money available today can be invested and grow whereas money spent today cannot. If we ignore the effects of inflation or deflation and we denote the present value of an investment as V_n, then the present value of V_n that is n_t time units from the present is given by

$$\text{Present value} = \frac{V_n}{(1+r)^{n_t}} \tag{3.23}$$

where r is the discount rate per time unit. For example, if n_t is in months, then r is the monthly rate.[*]

Equation 3.23 should be used on the various terms in Equations 3.2 and 3.3 to convert them to their present value at a predetermined time. For example, if you wish to calculate all costs in 2010 dollars, then a $100,000 qualification and certification cost C_Q that will be incurred in 2012 will have a present value in 2010 dollars of $85,734 = 100,000/(1 + 0.08)^2$ when an annual discount rate of 8% is used.

There are other forms of the present value calculation that have been used for various assumptions about the growth of money over time.[†] The effective after-tax[‡] discount rates r vary, depending on the business sector: public, private, nonprofit, or government. In 2007, the rate

---

[*] The discount rate is the interest rate paid to borrow money or the interest rate that could be earned by investing the money.
[†] D. G. Newman et al., 2004, ibid.
[‡] After all applicable taxes have been deducted.

for the U.S. government varied between 3% and 4% and for moderate growth public companies, the rate varied from 10% to 12%. The effective discount rate assumed by an organization is not just the interest rate that could be earned if the money was placed in a bank or invested in the stock market; it is the return on the money that could be obtained if the money was used for another purpose. For a high-growth rate company with limited resources, if $1 million can be invested in opportunity A and the investment returned $1.5 million at the end of one year, then the discount rate that must be used for the cost of using the $1 million for a different opportunity is 50%.

3.6   EXAMPLES

In this section, we give examples of the different aspects of cost modeling discussed in this chapter by considering different types of products.

3.6.1   PROCESS FLOW MODEL: THE MANUFACTURE OF A BICYCLE[*]

This example is the first of a two-part example. In this part, we only model the manufacturer's cost per bicycle built. In the second part, given in Section 3.6.2, we determine the selling price of the bicycle and the cost of ownership for a resort that purchases and maintains 50 of these bicycles.

To determine the cost to manufacture a bicycle, we model the manufacturing process by using a simple process flow model that includes the effects of defects that occur at specific steps in the process. We assume that the process flow for the manufacture of the bicycle is that shown in Figure 3.6. The process starts with the cutting, welding, and painting of aluminum tube stock to form a frame. It then proceeds to assembling the prefabricated parts to the frame to form the bicycle. It ends with the shipping of the bicycle to a dealer.

In Table 3.3, the data characterizing the process steps shown in Figure 3.6 are given. To determine the quantities that appear in the columns of the table and to be able to interpret them, we shall examine the cutting process step in detail. We first note that the values appearing in columns 1 though 4 and column 6 are assumed to be based on the company's experience in building bicycles and from building prototypes of the current bicycle. We also assume that the capital equipment is to be depreciated over a five year period; thus, $T_d = 5$ years in Equation 3.7. In addition, we shall assume that the equipment is fully occupied; that is, if this product is not being built on this piece of equipment, then the equipment is being used to build another product. We will also assume that the operational time per year for the bicycle manufacturing process is given by $T_{op} = $ (40 hr/wk) $\times$ (50 wk/yr) = 2000 hr/yr.

With respect to labor rates L_R, it is assumed that welders receive $23/hour and that all other tasks are performed by personnel who receive $14/hour. A burden rate $b = 0.3$ (30%) on labor hours is assumed to account for all overhead costs except the capital costs associated with specific process steps as indicated in column 6 of Table 3.3. We shall assume that the contribution of tooling costs is negligible and, therefore, it will not be considered.

The process flow model emulates the actual process used to manufacture the bicycle. In modeling the process, we have to make a distinction between the actual time that it takes to complete a process step and the time that people spend on that same process step. If we examine, for example, the cleaning process step that is shown in the third row in Table 3.3, we see that this step takes 1 hour to complete; 1/3 of an hour of labor (or 1/3 of a person for one hour); and that during this 1-hour period six objects are cleaned. In the second part of this example presented in Section 3.6.2, we will need to determine the throughput of the manufacturing process. In order to do this, we will need to identify the process step that limits the number of bicycles that can be produced in a unit time. The

---

[*] The idea for this example came from G. J. van Ryzin, "XTM Bike Corporation: An Exercise in Process Analysis," Columbia Business School, December 11, 2000: http://www.columbia.edu/~gjvl/XTM%20Bike.pdf.

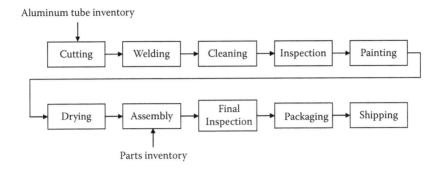

FIGURE 3.6 Process flow for the manufacture of a bicycle.

only way to determine this is to know how long a process step takes to complete, which may not equal the amount of labor time required by the step.

We now proceed to examine the cutting process step of Table 3.3. The labor cost of the cutting step is calculated by combining Equations 3.4a and 3.5b to obtain

$$C_L = \frac{N_L T L_R (1+b)}{N_p} = \frac{3 \times (1 \text{ hour}) \times (\$14/\text{hour})(1+0.3)}{6 \text{ frames}} = \$9.10/\text{frame} \qquad (3.24)$$

The capital cost for the cutting step is calculated by using Equation 3.7 as

$$C_c = \frac{TC_e}{N_p T_{op} T_d} = \frac{(1 \text{ hr}) \times (\$2,000)}{(6 \text{ frames}) \times (2000 \text{ hr/yr}) \times (5 \text{ yr})} = \$0.03/\text{frame} \qquad (3.25)$$

TABLE 3.3
Process Flow Cost Analysis of Bicycle Manufacture

	1	2	3	4	5	6	7	8
	N_L	N_p	T	C_M	C_L	C_e	C_c	C_{step}
Process Step	(Number of People)	(capacity; Number of Objects)	(Time, hr)	(Material Cost, $/unit)	(Labor Cost, $/unit)	(Capital Equipment Cost, $)	(Capital Cost, $/unit)	(Total Cost,[a] $/unit)
Cutting	3	6	1	200.00	9.10	2,000	0.03	209.13
Welding	6	4	1	5.00	44.85	5,000	0.13	49.98
Cleaning	0.33	6	1	0.50	1.00			1.50
Inspection	0.67	6	1	0.00	2.03			2.03
Painting	3	6	1	20.00	9.10	7,500	0.13	29.23
Drying	0	30	8	0.00	0.00	6,000	0.16	0.16
Assembly	7	4.67	1	300.00	27.28			327.28
Final inspection	1	6	1	0.00	3.03			3.03
Packaging	1	6	1	6.00	3.03			9.03
Shipping	0	1		25.00	0.00			25.00
Total				556.50	99.43	20,500	0.45	656.37

[a] $C_{step} = C_M + C_L + C_c$.

TABLE 3.4
Process Flow Cost Analysis of Bicycle Manufacture Continued

	9	10	11	12	13	14
Process Step	f_c (Fault Coverage)	Y_{step} (Step Yield)	Y_{in} (Incoming Yield)	Y_{out} (Outgoing Yield)	C_{in} (Incoming Cost, $)	C_{out} (Outgoing Cost, $)
Cutting		1.0000	1.0000	1.0000	0.00	209.13
Welding		0.9500	1.0000	0.9500	209.13	259.11
Cleaning		1.0000	0.9500	0.9500	259.11	260.61
Inspection	0.98	1.0000	0.9500	0.9990	260.61	276.18
Painting		0.9900	0.9990	0.9890	276.18	305.41
Drying		1.0000	0.9890	0.9890	305.41	305.57
Assembly		0.9800	0.9890	0.9692	305.57	632.85
Final inspection	0.985	1.0000	0.9692	0.9995	632.85	655.78
Packaging		1.0000	0.9995	0.9995	655.78	664.81
Shipping		0.9900	0.9995	0.9895	664.81	689.81

The sum of the material, labor, and capital costs gives the total cost for each process step. Adding the total costs of all of the process steps gives a total cost of $656.37/bicycle as shown in the last row of column 8 in Table 3.3.

Regarding the two inspection steps in Table 3.3, the model described so far does not address what is actually occurring during these inspections. If these two steps are detecting and correcting (reworking) all the defects in the process up to the point of the process where they appear, then $656.37/bicycle is correct. The model of Table 3.3 also assumes that all the defects are correctable; that is, no product has to be scrapped because it can't be reworked.

Next, a more detailed model that includes the yield of each individual process step and the effects of imperfect inspection is presented.

3.6.1.1   Consideration of Manufacturing Yield

From a cost analysis point of view, the total cost computed in Table 3.3 is independent of the order of the process steps because none of the steps removed units from the process. In order to take into account the yield of the process steps, we consider the yield data presented in Table 3.4. In this table, two additional inputs for each process step have been defined: fault coverage (column 9) and step yield (column 10). The step yield represents the fraction of units passing through the process step for which the process step does not add defects. For example, for the welding step, the step yield is 0.95; that is, 95% of the units will not incur additional defects in this step. The two inspection steps have been assigned fault coverage[*] since they are test steps. Fault coverage of less than 1.0 means that the inspection is not perfect; that is, some defects escape detection. In this case, we are accumulating defects in addition to costs as a unit moves through the process flow. The incoming yield Y_{in} to a process step is the outgoing yield Y_{out} of the previous step. For nontest/noninspection steps, the outgoing yield is given by Equation 3.12 as

$$Y_{out} = Y_{in}Y_{step} \qquad (3.26)$$

---

[*] P. Sandborn, ibid, 2005.

Thus, for example, the outgoing yield of the assembly step is $Y_{out} = (0.989)(0.98) = 0.9692$. For the inspection steps, the outgoing yield is given by Equation 3.14. Thus, for example, the outgoing yield of the second (final) inspection step is $Y_{out} = 0.9692^{(1-0.985)} = 0.9995$. In our process flow model, the step yield of the inspection steps has been assumed to be 100%; in reality, test steps can introduce defects.

The incoming cost C_{in} to a process step is the outgoing cost C_{out} of the previous step, and for nontest/noninspection steps, the outgoing cost is given by

$$C_{out} = C_{in} + C_{step} \tag{3.27}$$

where C_{step} is given in column 8 of Table 3.3. For the inspection steps, the outgoing cost is given by Equation 3.13. For example, the outgoing yield of the second inspection step is

$$C_{out} = \frac{C_{in} + C_{step}}{Y_{in}^{fc}} = \frac{\$632.85 + \$3.03}{0.9692^{0.985}} = \$655.78, \tag{3.28}$$

which is the amount that appears in column 14. The cost after the completion of all the process steps is given in the last row of column 14 of Table 3.4 and the final product yield after the completion of all the process steps is given in the last row of column 12; they are, respectively, $C_{out} = \$689.81$ and $Y_{out} = 0.9895$. The effective cost per nondefective bicycle that completes the process (called the "yielded cost") for the model of Table 3.4 is given by[†]

$$\text{Yielded Cost} = \frac{C_{out}}{Y_{out}} = \frac{\$689.81}{0.9895}$$

$$\tag{3.29}$$

$$= \$697.13/\text{non-defective bicycle that completes the process}$$

The difference in the cost determined by the model represented in Table 3.4, and that represented in Table 3.3 is $\$689.81 - \$656.37 = \$33.44$; that is, the model that includes yield gives higher unit cost that the model that doesn't. This difference is due to the fact that the model in Table 3.3 did not include the effects of defects,[‡] whereas the model used in Table 3.4 assumes that the inspection steps scrap the units in which defects are found and that some defects escape detection; that is, just over 1% of the bicycles shipped to the dealer are defective in some way ($1 - 0.9895 = 0.0105$). Scrapping defective bicycles is very expensive. By the time the first inspection step is encountered, we see from the fourth row of column 13 of Table 3.4 that over $260 has been spent on each bicycle, and for the assumptions of this example, this money is lost on 5% of the bicycles, as indicated in column 11.

3.6.2 THE TOTAL COST, SELLING PRICE, AND COST OF OWNERSHIP OF A BICYCLE

In the example in Section 3.6.1, we computed only a portion of the bicycle manufacturer's total cost C_{pm}. In this example, we shall compute the manufacturer's total cost of the bicycle, its selling price, and the cost of ownership for the purchaser of the bicycle. The assumptions that have been made for this example are listed in Table 3.5.

---

[*] D. Becker and P. Sandborn, "On the Use of Yielded Cost in Modeling Electronic Assembly Processes," *IEEE Trans. on Electronics Packaging Manufacturing*, Vol. 24, No. 3, pp. 195–202, July 2001.

[†] It is noted that C_{out} incorporates the money spent on scrapped bicycles but does not reflect the fact that the yield of the remaining bicycles is not 100%; therefore, C_{out} does not incorporate yield.

[‡] Alternatively, the inspection steps in Table 3.3 detected and corrected all the defects at no additional charge.

TABLE 3.5
Bicycle Manufacturing and Ownership Data

Symbol	Parameter	Value
C_T	Tooling	0
C_D	Design and development	$50,000
C_W	Waste disposition: costs to manage the waste created during bicycle manufacture	$200/month
C_Q	Qualification/certification	0
	Duration of bicycle manufacturing	3 years
t_W	Warranty length with unlimited free replacement	1 year
MTBF	Mean time between failure ($=1/\lambda$)	500 months
	Cost of maintaining a toll-free number for warranty claims	$35/month
C_{ship}	Cost of round trip shipping of a bicycle from the customer to the manufacturer	$50
	Manufacturer's profit	20%
	Average repair cost per bicycle repaired under warranty	$30
	Fraction of bicycles returned under warranty that can be repaired	0.8
	Fraction of bicycles returned under warranty that must be replaced	0.2
	Dealer's profit	25%
C_{sa}	Cost of sale	$15,000
	Sales tax	8%
C_L	Resort employee labor cost	$15/hour
b	Resort employee labor burden	0.15
	County bicycle license	$10/year/bicycle
	Out of warranty repair charge per bicycle[a]	$150[a]

[a] Does not include the cost of shipping the bicycle to the manufacturer.

In order to use Equation 3.2 to evaluate the manufacturer's total cost, we need the values of C_M, C_L, and C_c from the last row in Table 3.3; they are $C_M = \$556.50$/bicycle, $C_L = \$99.43$/bicycle, and $C_c = \$0.45$/bicycle. As given in Table 3.5, we will assume that the tooling costs C_T and the qualification/certification costs C_Q are zero, and that the design and development costs $C_D = \$50,000$. The overhead cost C_{OH} was included as a labor burden in C_L (see Equation 3.24); therefore, $C_{OH} = 0$. With these assumptions, Equation 3.2 becomes,

$$C_{pm} = N_{pm}(\$556.50 + \$99.43 + \$0.45 + C_W) + 0 + 0 + \$50,000 + C_{WR} + 0$$
$$= N_{pm}(\$656.38 + C_W) + \$50,000 + C_{WR} \tag{3.30}$$

We see from Equation 3.30 that we also need to determine N_{pm}, C_W, and C_{WR}. We start by first obtaining the total number of bicycles that can be manufactured in three years, N_{pm}. From the data in columns 2 and 3 of Table 3.3, the number of units per hour that can be manufactured is determined by the process step with the smallest ratio of the values in column 2 divided by the corresponding values in column 3. This number is 3.75 units/hr ($= 30/8$); it corresponds to the drying step and is referred to as the process "bottleneck." The number of hours that is available to manufacture these bicycles over a 3-year period based on a 40-hour work week for a 50-week work year is 6,000 hours. Then, the quantity of bicycles that can be made in this period is

$$N_{pm} = (6,000 \text{ hr}) \times (3.75 \text{ bicycles/hr}) = 22,500 \text{ bicycles} \tag{3.31}$$

The warranty reserve fund cost C_{WR} is given by Equation 3.21. As indicated in Table 3.5, 80% of the bicycles that are returned for warranty repair can be repaired for $30/bicycle and the remaining 20% must be replaced. The replacement cost of one bicycle is C_{pm}/N_{pm}. Thus, the average cost of a warranty claim per bicycle is

$$C_{rc} = C_{ship} + C_{repair} + C_{replace}$$

$$= \$50 + 0.8 \times \$30 + 0.2 \times \frac{C_{pm}}{N_{pm}} \tag{3.32}$$

$$= \$74 + 0.2 \times \frac{C_{pm}}{22,500}$$

The failure rate for the product is $\lambda = 1/500 = 0.002$ failures/month; therefore, from Equation 3.21, the warranty reserve fund is

$$C_{WR} = C_{fr} + N_{pm}M(t)C_{rc} \tag{3.33}$$

From Table 3.5, we see that $C_{fr} = (\$35/month) \times (12\ months/year) \times (4\ years) = \$1680$ and $M(t) = \lambda t = (0.002)(12\ months) = 0.024$ replacements (failures); therefore, Equation 3.33 becomes

$$C_{WR} = \$1680 + 22,500 \times 0.024 \times \left(\$74 + 0.2 \times \frac{C_{pm}}{22,500} \right) \tag{3.34}$$

$$= \$41,640 + 0.0048 C_{pm}$$

From Table 3.5, the waste disposition cost per bicycle C_W for 3 years of manufacturing at $200/month is calculated to be

$$C_W = \frac{\$200 \times 12\ months/yr \times 3\ years}{22,500\ bicycle} = \$0.32/bicycle \tag{3.35}$$

The quantity C_W is not the cost of disposing of a bicycle; it is the cost of managing the waste created during the manufacturing of a bicycle and includes transportation, disposal, record keeping, and labeling. Upon inserting Equations 3.31, 3.32, 3.34, and 3.35 into Equation 3.30, solving for C_{pm} gives

$$C_{pm} = N_{pm}(\$656.38 + C_W) + \$50,000 + C_{WR}$$

$$C_{pm} = 22,500(\$656.38 + \$0.32) + \$50,000 + \$41,640 + 0.0048 C_{pm} \tag{3.36}$$

$$C_{pm} = \frac{\$14,867,390}{0.9952} = \$14,939,098$$

Thus, the cost per bicycle is $663.96 (= $14,939,098/22,500). Consequently, the nonmanufacturing portion of the total cost per bicycle is $663.96 − $656.37 = $7.59/bicycle.

The cost of the bicycle to the customer is computed as follows. From Table 3.5, the manufacturer's profit is assumed to be 20% and the dealer's profit is assumed to be 25%. Then, the cost per bicycle to the dealer is $C_d = 1.2(C_{pm} + C_{sa})/N_{pm}$ and the cost of the bicycle to the customer is $P = 1.25C_d$.* From Table 3.5, $C_{sa} = \$15,000$; hence,

$$P = 1.25 \times 1.2 \times (\$14,939,098 + \$15,000)/22,500$$

$$= \$996.94 \tag{3.37}$$

This analysis assumes that all the warranty claims made by customers are legitimate. This may not always be the case. Customers may attempt to file claims for bicycles whose damage is not the fault of the manufacturer (i.e., not included in λ) and not covered under the terms of the warranty. It is likely that the bicycle manufacturer will have to cover some of these fraudulent claims and even in cases where the claims are denied there is still a cost incurred by the manufacturer in denying the claim.

3.6.2.1 Cost of Ownership

Let us assume that a resort wants to provide bicycles to its patrons and wishes to purchase 50 bicycles and maintain them for 5 years. We make the following assumptions. The resort plans on dedicating 10% of one employee's time to managing and maintaining the bicycles. Since the employee will work 40 hours per week, 52 weeks per year for the next 5 years, the total labor required to maintain the bicycles is $0.1 \times (40 \text{ hours/week}) \times (52 \text{ weeks/year}) \times (5 \text{ years}) = 1,040$ hours. From Table 3.5, we determine the burdened labor rate as $1.15 \times \$15 = \$17.25$ per hour. This labor burden is assumed to include all applicable resort overhead costs. The operating cost per bicycle over 5 years is

$$C_O = \frac{\$17.25 \times 1,040 \text{ hours}}{50 \text{ bicycles}} = \$358.80/\text{bicycle} \tag{3.38}$$

The cost of support per bicycle is calculated as follows. During the first year after purchase, the bicycles are covered by the manufacturer's warranty. After that, bicycle repairs are the responsibility of the resort. Assuming that the MTBF for the bicycles remains the same for years 2 through 5, then $M(t) = \lambda t = 0.002 \times 12 \text{ months/year} \times 4 \text{ years} = 0.096$ failures/bicycle. From Table 3.5, the cost per repair is \$150 for an out-of-warranty repair plus \$50 for shipping. Then $C_{rc} = \$50 + \$150 = \$200$ per out-of-warranty repair; hence, the average cost of replacement and repair per bicycle is

$$C_S = M(t)C_{rc} = 0.096 \times \$200 = \$19.20 \tag{3.39}$$

We further assume that the resort purchases \$500 worth of spare parts for the bicycles a year, which includes such items as inner tubes for flat tires. Thus, $C_{sp} = \$500/\text{year} \times 5 \text{ years} = \$2,500$. In addition, the resort has to pay \$10 per year to license the bicycle; therefore, $C_Q = (\$10/\text{bicycle}) \times (50 \text{ bicycles}) \times (5 \text{ years}) = \$2,500$. From Table 3.5, the sales tax on each bicycle purchase is $C_x = 0.08 \times \$996.94 = \$79.76$. Then, the cost of ownership is computed from Equation 3.3 as

$$C_{pc} = N_{pc}(P + C_x + C_O + C_S) + C_{sp} + C_t + C_Q$$

$$= 50 \times (\$996.94 + \$79.76 + \$358.80 + \$19.20) + \$2500 + 0 + \$2500 \tag{3.40}$$

$$= \$77,735$$

---

* The accumulation of all the profits P_r appearing in Equation 3.1 can be determined by setting P from Equation 3.37 equal to P from Equation 3.1; that is, $\$996.94 = (\$14,939,098 + \$15,000 + P_r)/22,500$. Upon solving for P_r we find that $P_r = \$7,477,049$ or an amount that is 50% of $C_{pm} + C_{sa}$.

FIGURE 3.7 Distribution of ownership costs for 50 bicycles.

Thus, the cost of purchasing and maintaining each of the 50 bicycles is $77,735/50 = $1,554.70/ bicycle, which is about 56% more than the purchase price of the bicycle. A breakdown of the costs is shown in Figure 3.7. The percentages in the figure are found by dividing each cost element by $1,554.70 and multiplying by 100; for example, Price = 100 × $996.94/$1,554.70 = 64.1%.

The analysis in this section assumed that the bicycle dealer did not provide any volume discount for buying 50 bicycles. We also did not explicitly include the cost of insurance, although it could be part of the labor burden. It is possible that the insurance premiums may increase because the resort could be held liable for a guest's injuries sustained while riding one of their bicycles.

3.6.3 PARAMETRIC COST MODEL: FABRICATION OF APPLICATION-SPECIFIC INTEGRATED CIRCUITS

In this example, we use a parametric cost model for estimating the cost of products based on a manufacturer's history with similar products. We consider an application specific integrated circuit (ASIC), which is an electronic chip that is custom designed for a specific application. ASICs are often designed by companies in order to obtain improved performance or cost, or for proprietary reasons.

We shall assume that the primary factor that influences the cost of ASICs is the number of logic gates that they contain. The company has previously designed the eight ASICs shown in Table 3.6,

TABLE 3.6
ASICs Developed by One Company in the Last 2 Years

Number of Logic Gates (N_{gate})	Die Area (mm²) (A_{die})
200,000	48.4
250,000	59.4
100,000	27.7
147,000	38.7
500,000	129
457,000	129
321,000	83.9
256,000	58.1

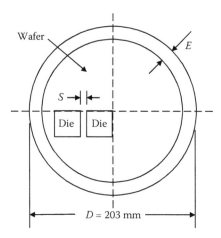

FIGURE 3.8 Wafer and die geometry.

which were used in various products. The ASIC fabricator with whom the company has subcontracted uses 203 mm diameter wafers that cost $900/wafer to process.

To obtain a parametric cost model, we first need to develop a relationship between the number of logic gates and the cost of the die as a function of die layout on the wafer. Die are unpackaged integrated circuits fabricated on silicon wafers, which are thin round disks of silicon. Many rectangular die are fabricated concurrently on the wafer by arranging them in rows with a minimum spacing S between adjacent die as shown in Figure 3.8. There is a border of width E around the wafer's perimeter that is needed for handling and for the placement of test structures. The usable area of the 203 mm diameter die is

$$A_{usable} = \pi(101.5 - E)^2 \text{ mm}^2 \tag{3.41}$$

In order to simplify the problem, we assume that the die is square with an area of A_{die}. Then the effective wafer area required by each die must include the minimum spacing S. Hence,

$$A_{eff} = \left(\sqrt{A_{die}} + S\right)^2 \tag{3.42}$$

The number of die that can be fabricated on a wafer, which is called the gross die per wafer (GDW), is approximately

$$N_{GDW} = \frac{A_{usable}}{A_{eff}} \tag{3.43}$$

Equation 3.43 overestimates the number of die that can be fabricated because the wafer is round and the die are square.[*] The cost per die fabricated on the wafer is given by

$$C_{die} = \frac{\$900}{N_{GDW}} \tag{3.44}$$

---

[*] Equation 3.43 is most accurate when $A_{eff} \ll A_{usable}$. More accurate calculations of the number of rectangular die that can be fabricated on a round wafer can be found in D. K., DeVries, "Investigation of Gross Die Per Wafer Formulas," *IEEE Trans. on Semiconductor Manufacturing*, Vol. 18, No. 1, February 2005, pp. 136–139.

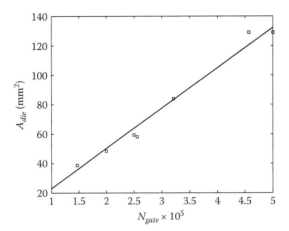

FIGURE 3.9 Fit of ASIC data in Table 3.6 for die area as a function of the number of logic gates.

The last remaining element required for the model is the relationship between the die area A_{die} and the logic gate count N_{gate}, which can be found by curve fitting the data given in Table 3.6 as shown in Figure 3.9. A least-squares fit to the data provides the following relationship between the number of logic gates and the die area

$$A_{die} = 2.74 \times 10^{-4} N_{gate} - 4.5 \quad \text{mm}^2 \tag{3.45}$$

It is noted that the relationship between N_{gate} and A_{die} is not derived from any physical principle or geometric argument, it is a curve fit to observed data. Upon combining Equations 3.41 through 3.45, we obtain the following cost estimating relationship (CER) that relates the logic gate count N_{gate} to die cost C_{die} as a function of the edge scrap allowance E and minimum distance between die S

$$C_{die} = \frac{900\left(\sqrt{2.74 \times 10^{-4} N_{gate} - 4.5} + S\right)^2}{\pi(101.5 - E)^2} \quad \text{\$/die} \tag{3.46}$$

Equation 3.46 provides a lower bound on the cost per die because of the overestimation of the number of die per wafer. Equation 3.46 may be an accurate and useful model for the company whose data are presented in Table 3.6, but it also highlights the potential problems with cost estimating relationships. Equation 3.46 is only valid for the following conditions.

- ASICs made on 203 mm diameter wafers.
- ASICs fabricated using whatever technology and minimum feature size corresponded to the data in Table 3.6 (unspecified in this example).
- ASICs with logic gate counts between 100,000 and 500,000.
- The timeframe associated with the data in Table 3.6 (unspecified); that is, it does not account for any increase or decrease over time in ASIC fabrication costs. In this particular example, \$900/wafer appears as a constant multiplier in Equation 3.46 and could be adjusted up or down. However, in general, the quantities appearing in Equation 3.46 need not have any direct connection to the assumptions buried in the development of the CER.

As long as these assumptions are satisfied, Equation 3.46 is an appropriate model; however, none of these assumptions is obvious when looking at Equation 3.46 and could only be known if they are documented and accompanied Equation 3.46.

TABLE 3.7
Web Banner Advertising Data

Cost for 50,400 impressions	$1,500
Click-through rate[a] (click-through/impression)	1%
Sales rate (sales/click-through)	1/15
Profit per sale	$30
Average lifetime sales per customer	2
Number of impressions per month	4,200
Annual cost of money (r)	7 %

[a] This value of 1% indicates that the banner is clicked once (one visitor) for each 100 impressions.

3.6.4 THE RETURN ON INVESTMENT ASSOCIATED WITH WEB BANNER ADVERTISING

In this example, we calculate the return on investment that is associated with a business decision. It demonstrates the various types of life cycle details that need to be considered to accurately assess ROI, including an assessment of the cost of money. The business decision in this example is to decide whether to spend money on a manufacturing process improvement that would increase the profit per sale or on Web banner advertising that would increase the number of sales.

Banner advertisements on Web sites have become commonplace. However, before one enters into such an advertising venture one must first estimate what the potential ROI is and then decide if the ROI justifies the investment. To assess the ROI, one must compare the cost of purchasing the banner to the total value that you expect to obtain from the banner. If a customer clicks on the banner that you have purchased, they are taken to your Web site where they have the opportunity to purchase your product if they desire.[*] The analysis assumes that the persons who purchase your product via the banner would not have found and purchased your product without having seen the banner. The data that have been collected regarding banner advertising are given in Table 3.7. In order to determine the ROI, we first need to determine the cost per sale. Referring to Table 3.7, the cost per visitor to the Web site is

$$\text{Cost/visitor} = (\text{cost/impression})/(\text{visit/impression})$$

$$= \frac{\$1,500}{50,400 \text{ impressions}} \frac{100 \text{ impressions}}{\text{visit}} \tag{3.47}$$

$$= \$2.976/\text{visit}$$

According to the data in Table 3.7, the sales rate is 1 sale for every 15 click-throughs. However, each new customer tends to purchase an average of two times from a Web site. Therefore, the effective sale rate is 1 sale for every 7.5 click-throughs. Thus, the effective cost per sale is

$$\text{Cost for each banner-initiated sale} = (\text{cost/visit}) \times (\text{visit/sale})$$

$$= \frac{\$2.976}{\text{visit}} \times \frac{7.5 \text{ visit}}{\text{sale}} \tag{3.48}$$

$$= \$22.32/\text{sale}$$

---

[*] A Web banner is displayed when a Web page that references the banner is loaded into a Web browser. This event is known as an "impression." When the Web page viewer clicks on the banner, the viewer is directed to the Web site advertised in the banner. This event is known as a "click through."

The ROI for the banner advertisement is obtained from Equation 3.22a as

$$\text{ROI} = \frac{\text{Profit/sale} - \text{Banner cost/sale}}{\text{Banner cost/sale}}$$

$$= \frac{\$30/\text{sale} - \$22.32/\text{sale}}{\$22.32/\text{sale}} \tag{3.49}$$

$$= 0.344$$

Thus, the ROI is 34.4%. However, the $1,500 to pay for the banner advertisement most likely had to be borrowed from a bank or an investor, both of whom expect a rate of return on their money. Let us assume that a bank charges 7% annual interest rate for loans. If it is assumed that there are 4,200 impressions per month, the $1,500 banner lasts for 12 months ($12 \times 4{,}200 = 50{,}400$). Solving Equation 3.23 for V_n, the cost of $1,500 to advertise with the banner for 12 months is

$$V_1 = (1+r)^1(\text{Present value}) = (1+0.07)^1(\$1{,}500) = \$1{,}605.00 \tag{3.50}$$

In other words, V_1 is the amount of money required to pay off the loan in one year ($n_t = 1$) at an interest rate of 7% ($r = 0.07$).

Using $1,605.00 for the cost of the banner, the cost per visitor given by Equation 3.47 yields

$$\text{Cost/visitor} = \frac{\$1{,}605}{50{,}400 \text{ impressions}} \frac{100 \text{ impressions}}{1 \text{ visit}} \tag{3.51}$$

$$= \$3.19/\text{visit}$$

Therefore, upon substituting the value from Equation 3.51 into Equation 3.48, we find that the cost per sale increases to $23.89. Consequently, the ROI computed in Equation 3.49 decreases to 0.256—that is, to 25.6%.

Having computed the ROI for this scenario, the company is now in a position to determine if the $1,500 can be spent another way, one that provides a higher ROI. Let us suppose that the company has the $1,500 available (it doesn't have to borrow it) and that it could be spent to improve the manufacturing process for the product so that the profit per sale would increase from $30 to $34. To determine the conditions under which the company should pursue this option, we need to determine the ROI for this case. If there are N products that will be sold without the banner advertisements, then the ROI is determined from the following relation:

$$\text{ROI} = \frac{\text{Total Additional Profit} - \text{Banner cost}}{\text{Banner cost}}$$

$$= \frac{(\$34/\text{sale} - \$30/\text{sale})N - \$1{,}500}{\$1{,}500} \tag{3.52}$$

$$= 0.002667N - 1$$

From Equation 3.52, we find that the ROI is greater than 0.344 when $N > 503$.[*] Therefore, if one expects to sell more than 503 units without using banner advertising, then the $1,500 is better spent on improving the manufacturing process. If the number of sales without banner advertising is expected to be less than 503 units, then the money is better spent on the banner advertising.

---

[*] The banner ad will result in (50,400 impressions)/[(100 impressions/visit) × (7.5 visits/sale)] ≈ 67 additional sales.

3.6.5   COMPARING THE TOTAL COST OF OWNERSHIP OF COLOR PRINTERS

In this example, we shall determine the total cost of ownership of three printers that are used to print black-and-white and color pages and show that the purchase price of printers is not the best way to assess their real cost.

In Table 3.8, we have listed the assumptions associated with an inexpensive color inkjet printer, a home color laser printer, and a business color laser printer. All the printers are manufactured and marketed by the same company. We shall determine which of these printers has a lower cost of ownership.

To determine the cost of ownership, we shall use a simplified version of Equation 3.3 in which only the price $P_{printer}$ and the operating costs are considered. Thus,

$$C_{pc} = C_{printer} + C_{paper} + C_{ink/toner} \tag{3.53}$$

The cost of the printers is determined from

$$C_{printer} = N_{printers} P_{printer} \tag{3.54}$$

where
  $N_{printer}$ = Number of printers needed: equals $\lceil N_{pages}/L_{printer} \rceil$
  N_{pages} = Total number of pages printed
  $L_{printer}$ = Lifetime of the printer in terms of the number of printed pages
  $P_{printer}$ = Purchase price of the printer
    $\lceil\ \rceil$ = Ceiling function, which rounds up to the nearest integer

For the purposes of this example, we assume that each printer is scrapped after $L_{printer}$ pages have been printed and that each printer does not malfunction during the printing of these pages.

The cost of the paper per printed page is $3/500 = \$0.006$/page; therefore, $C_{paper} = \$0.006 N_{pages}$. The cost of the ink/toner is given by

$$C_{ink/toner} = N_{refill} I_{ink/toner} \tag{3.55}$$

where N_{refill} is the number of ink refills needed; that is,

$$N_{refill} = \lceil (N_{pages} - N_{printers} N_{withprinter})/Z \rceil$$

where $N_{refill} \geq 0$, and
  $I_{ink/toner}$ = Cost of an inkjet cartridge set or toner cartridge set
      Z = Number of pages that can be printed with one cartridge set
$N_{withprinter}$ = Number of pages that can be printed with ink/toner cartridge set that comes with the printer.

The quantity N_{refill} gives the number of ink cartridges sets or toner cartridge sets that need to be purchased, and it accounts for the amount of ink or toner that is included with each printer when purchased.

Using the data in Table 3.8, we have summarized in Table 3.9 the cost calculations corresponding to printing 15,000 pages on each of the printers. From Table 3.9, we see that while the business laser color printer is the most expensive to purchase, it is the least expensive printer to own if 15,000 pages are to be printed. Equations 3.53 to 3.55 can be used to compare the total cost of these three printers as a function of the total number of pages printed. An alternative measure of cost of

TABLE 3.8
Comparison Data for Three Color Printers

Symbol	Description	Inkjet Printer	Home Laser Color Printer	Business Laser Color Printer
$P_{printer}$	Printer purchase price[a]	$67.18	$210.94	$952.94
$L_{printer}$	Printer lifetime (pages/warranty period)[b]	12,000	12,000	90,000
$I_{ink/toner}$	Ink/toner cartridge cost per set[c]	$76.32	$297.82	$934.88
Z	Cartridge set life (pages printed)[d]	500	2,200	7,500
$N_{withprinter}$	Number of pages printed with cartridges included with printer when purchased	125	550	7,500
	Paper cost[a]	$3/500 sheets	$3/500 sheets	$3/500 sheets

[a] Includes 6% sales tax.
[b] Printer lifetime is the manufacturer's maximum suggested pages/month multiplied by the warranty length in months.
[c] Set includes black, cyan, yellow, and magenta; price included 6% sales tax.
[d] Cartridge life is based on standard pages as defined in ISO/IEC 19798.

ownership that might be useful to organizations that provide printing services is the cumulative average cost per page. This value is obtained from C_{pc}/N_{pages}.

Several important effects have not been considered in this analysis. We have assumed that the quality of the printed page and speed of printing are not issues. We have not considered what is being printed; for example, it takes more ink/toner to print photos than text. In this example, we have used the pages counts cited by the printer manufacturer on their ink/toner cartridges. We have also not considered the option of refilling the ink cartridges rather than purchasing new cartridges. Refilling is an option that may reduce the ink costs at the risk of decreasing the lifetime of the printer. Lastly, Equations 3.54 and 3.55 do not assume that there is any credit provided for unused printer life or unused ink/toner after the specified number of pages is printed.

TABLE 3.9
Cost Calculations for Three Color Printers for $N_{pages} = 15,000$

Quantity	Color Inkjet Printer	Home Laser Color Printer	Business Laser Color Printer
$N_{printer}$	$\lceil 15,000/12,000 \rceil = 2$	$\lceil 15,000/12,000 \rceil = 2$	$\lceil 15,000/90,000 \rceil = 1$
$P_{printer}$ (Table 3.8)	$67.18	$210.94	$952.94
$C_{printer}$ (Equation 3.54)	$2 \times \$67.18 = \$134.36$	$2 \times \$210.94 = \$421.88$	$1 \times \$952.94 = \$952.94$
N_{refill}	$\left\lceil \dfrac{15,000 - 2 \times 125}{500} \right\rceil = 30$	$\left\lceil \dfrac{15,000 - 2 \times 550}{2,200} \right\rceil = 7$	$\left\lceil \dfrac{15,000 - 1 \times 7,500}{7,500} \right\rceil = 1$
$C_{ink/toner}$ (Equation 3.55)	$30 \times \$76.32 = \$2,289.60$	$7 \times \$297.82 = \$2,084.74$	$1 \times \$934.88 = \$934.88$
C_{paper}	$\$0.006 \times 15,000 = \$90$	$\$0.006 \times 15,000 = \$90$	$\$0.006 \times 15,000 = \$90$
C_{pc} (Equation 3.53)	$\$134.36 + \$90 + \$2,289.90 = \mathbf{\$2,514.26}$	$\$421.88 + \$90 + \$2,084.74 = \mathbf{\$2,596.62}$	$\$952.94 + \$90 + \$934.88 = \mathbf{\$1,977.82}$

TABLE 3.10
Factors in Voting Machine Warranty Cost Calculation

Symbol	Factor	Value
	Purchase price per voting machine	$3,500
n_s	Number of voting machines supported simultaneously	7,300
	Operating time per election day	15 hours
	Election days per year	4
	Length of warranty	10 years
$MTTR_{actual}$	Actual time to repair a failed machine	2 hours
L	Labor rate	$50/hour
	Labor burden rate	15%
MTBF	Mean time between failure	163 hours

3.6.6 Reliability, Availability, and Spare Parts of New York City Voting Machines

This example assesses a product whose reliability and availability are critical to its success. The example determines how the product's spare parts must be considered and maintained in order to meet the customer's requirements. The parameters that will be used in this example are listed in Table 3.10.

The EAC VVSG[*] specifies that the mean time between failure (MTBF) for voting system devices has to be at least 163 hours.[†] Assuming that failures are exponentially distributed, this MTBF corresponds to a constant failure rate of $\lambda = 1/163 = 0.00614$ failures/hour. If one assumes that an election day is 15 hours long, then the probability of a voting machine not failing during this period is given by Equation 3.17 with $t = 15$ hours. Thus,

$$R(15) = e^{-15/163} = 0.912 \tag{3.56}$$

In other words, 8.8% of the voting machines will fail during a 15-hour day.

If the voting machine customer requires 99.8% availability of the voting machines on Election Day, then one must determine how long a failed machine can be unavailable when it fails. To make this determination, we first use Equation 3.20 to determine the required mean time to repair. Thus,

$$MTTR_{required} = \frac{MTBF}{A_{inherent}} - MTBF = \frac{163}{0.998} - 163 = 0.3267 \quad \text{hours} \tag{3.57}$$

We now apply these results to a specific example. In the 2004 general election, the City of New York had 7,300 voting machines.[‡] Assume that New York City requires a minimum of 99.8% availability of the voting machines during a 10-year warranty period. The manufacturer knows that it takes an average of 2 hours to repair a failed voting machine. But, from Equation 3.57, a 99.8% availability requirement means that failed machines can only be out of service for a maximum of 0.3267 hours. Therefore, in order to meet the required availability, the voting machine manufacturer must provide spare voting machines that can be used while repairs to failed machines are being made. Let us

---

[*] EAC stands for Election Assistance Commission and VVSG stands for Voluntary Voting System Guidelines. See "Voluntary Voting System Guidelines Recommendations to the Election Assistance Commission," August 31, 2007, http://www.eac.gov/files/vvsg/Final-TGDC-VVSG-08312007.pdf.

[†] In 2008, the reliability specification was the same for both mechanical and direct recording electronic (DRE) voting machines.

[‡] H. Stanislevic, "DRE Reliability: Failure by Design?" http://www.votetrustusa.org/pdfs/DRE_Reliability.pdf.

assume that every time a machine fails a spare machine needs to be provided for $2 - 0.3267 = 1.673$ hours. We further assume that the voting machine supplier desires a 90% probability that they will have enough spare machines on hand to support the customer on election days. To determine how many spare machines are needed, we use Equation 3.19 to determine the value of k, which is the number of spare machines necessary. Then, with $n_s = 7,300$, $\lambda = 1/163$ per hour, and $t = 1.673$ hours, we have that

$$0.9 \le \sum_{x=0}^{k} \frac{1}{x!} \left(\frac{7300 \times 1.673}{163} \right)^x e^{-(7300 \times 1.673/163)} = e^{-74.93} \sum_{x=0}^{k} \frac{1}{x!} (74.93)^x \tag{3.58}$$

The solution* to Equation 3.58 is $k \ge 86$; that is, at least 86 spare machines are needed. Note that this analysis assumes that each spare machine is used for 1.673 hours.

To determine how much it will cost to provide this support, we have to take into account the labor and materials required to repair the voting machines and the cost of providing 86 spare machines. The purchase price of a voting machine is $3,500. To determine the fixed cost of providing the warranty coverage, we use Equation 3.21. The fixed cost of providing the 86 spare voting machines is

$$C_{fr} = (86 \text{ spare machines}) \times (\$3,500/\text{machine}) = \$301,000$$

In addition, it is assumed that each machine is used four times a year and during each of these days it is used for 15 hours. Then the total number of hours that each machine is used per year is 60 hours. Therefore, the total number of hours that each spare machine can be expected to be used in the warranty period of 10 years is 600 hours. Hence, the expected number of replacements in the 10-year period is

$$M(t) = \lambda t = 600/163 = 3.68 \text{ repair incidents per machine per 10-year period}$$

We assume that the burdened labor cost to fix a machine is $50/hr and that the profit on labor is 15%. Then, from Equation 3.4b, the burdened labor cost including profit is $50(1 + 0.15) = \$57.50/$ hr. Since it takes 2 hours to repair a unit, the labor repair cost is $C_{rc} = (\$57.50/\text{hr}) \times (2 \text{ hr/machine}) = \$115.00/$repaired machine. From Equation 3.21, the cost to support all the units is

$$C_{WR} = C_{fr} + nM(t)C_{rc}$$

$$= \$301,000 + (7300 \text{ units}) \times (3.68 \text{ repairs/uint}) \times (\$115.00/\text{repair}) \tag{3.59}$$

$$= \$3,390,360$$

From Equation 3.59, we see that the additional cost to the manufacturer per unit sold is $C_{WR}/7,300 = \$464.43$. This cost must be added the selling price of each machine to cover the cost of supporting New York City.

The analysis in this example assumes that none of the spare machines fail and that the repairs to the failed machines result in repaired machines that are "as good as new." We have also not included any repair costs beyond labor. We have also assumed that the spare voting machines are located where they are needed. Voting machines are also a good example of a product whose cost of ownership may be significant and includes the cost of programming the machines for every election; the cost of storing the machines between elections; the cost of transporting and setting up the

---

* The value of k can be obtained using `poissinv` from the Matlab Statistics toolbox.

machines for elections; the cost of training poll workers and employees to use, test, and maintain the voting machines; educating the public on how to use the machines; and the cost of modifying polling places to accommodate the machines.

BIBLIOGRAPHY

W. R. Blischke and D. N. P. Murthy, *Warranty Cost Analysis*, Marcel Dekker, New York, 1994.

B. Boehm, *Software Engineering Economics*, Prentice Hall PTR, Upper Saddle River, NJ, 1981.

G. Boothroyd, W. Knight, and P. Dewhurst, *Product Design for Manufacture and Assembly*, 2nd ed., Marcel Dekker, New York, 2002.

W. J. Fabrycky and B. S. Blanchard, *Life-Cycle Cost and Economic Analysis*, Prentice Hall, Upper Saddle River, NJ, 1991.

P. F. Ostwald and T. S. McLaren, *Cost Analysis and Estimating for Engineering and Management*, Person Prentice Hall, Upper Saddle River, NJ, 2004.

P. Sandborn, *Course Notes on Manufacturing and Life Cycle Cost Analysis of Electronic Systems*, CALCE EPSC Press, College Park, MD, 2005.

M. U. Thomas, *Reliability and Warranties: Methods for Product Development and Quality Improvement*, CRC Press, Boca Raton, FL, 2006.

4 Translating Customer Requirements into a Product Design Specification

Several methods of obtaining the voice of the customer are given, the quality function deployment method is introduced, and the elements comprising a product design specification are presented

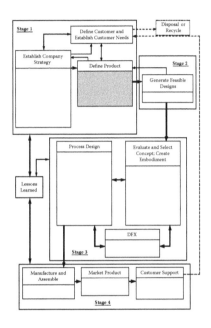

4.1 VOICE OF THE CUSTOMER

The central belief of the product development process is that products should reflect the customers' desires and tastes. One of the most important aspects of the product development cycle, then, is to understand and learn from the customer. Some of the reasons that people buy products are because they perform well, have features that they seek, and are reliable, durable, safe, easy to use, comfortable to use, aesthetically pleasing, and familiar to them. Not all of these reasons are applicable at one time to one product, nor is this list exhaustive. However, it does give some insight as to what customers may consider when forming their decision to buy a product.

In observing customers' buying behavior, it has been found that their decisions about products are based on attributes that can be arranged into eight categories as follows[*]:

1. **Cost**—Can I afford it?
   The cost of the product can be used to gain a competitive advantage.

2. **Availability**—Can I find it?
   A product should be available when and where the potential customer wants it.

3. **Packaging**—Does it look attractive?
   Packaging is what the customer sees, and frequently has a powerful influence on the buyer's selection.

4. **Performance**—Can it do what I want it to do?
   Performance must directly satisfy the customers' most important requirements.

---

[*] P. Marks, "Defining Great Products," 5th International Conference on Design for Manufacturability and Concurrent Engineering: Building in Quality and Customer Satisfaction, Management Roundtable, Orlando, Florida, *Design Insights*, Los Gatos, CA, November 1991.

5. **Ease of use**—Do I know how to use it?
   The product should be easy to use and operate. An easy-to-use product may be preferable to one that has more features and capabilities.

6. **Assurances**—Will it last?
   A reputation for durability, reliability, and support will influence many customers' choices.

7. **Life-cycle costs**—Will it cost too much to maintain?
   A product whose total cost of maintenance, repairs, energy use, supplies, and downtime is considerably less than similar products may have a competitive edge.

8. **Social standards**—What do others think of it?
   A product that becomes a de facto standard is clearly the best.

There are four levels of customer requirements that must be satisfied at each level before addressing those of the next level. The four levels are (1) *expecters*, (2) *spokens*, (3) *unspokens*, and (4) *exciters*.

1. **Expecters** are the basic qualities one must offer to be competitive and to remain in business. They are characteristics that customers assume are part of the product or service; that is, they expect them as standard attributes. Expecters are attributes that are frequently easy to measure and, therefore, are used in benchmarking.
2. **Spokens** are specific features customers say they want in a product. They are items a company must be willing to provide as they represent the aspects of the product that define it for a customer.
3. **Unspokens** are product characteristics that customers do not talk about, but they are important and cannot be ignored. It is the IP^{2}D^2 team's job to discover what they are. To do this, the IP^{2}D^2 team should make use of market surveys, customer interviews, and brainstorming. Unspokens are omitted responses that usually fall into one of three groups:

   Didn't remember to tell you. Customers forget to say everything they want.
   Didn't want to tell you. Customers simply do not want to give details about their requirements.
   Didn't know what it was. Customers most likely mean "I'll know it when I see it."

4. **Exciters** are unexpected features of a product that make the product unique and distinguish it from the competition. An exciter is also referred to as a *delighter*, because it defines product attributes or features that are pleasant surprises to customers when they first encounter them. However, if they are not present, then the customers will not be dissatisfied, since they will be unaware of what they are missing.

Customer satisfaction increases as a product fulfills successive customer requirement levels. Expecters must be satisfied first, because they are the basic qualities a product must have. Spokens increase customer satisfaction to a higher level because they go beyond the product's basics and satisfy specific customer requirements. Unspokens are the elusive items customers are not yet aware they want. Exciters are product features that no other vendor offers, and can make a product unique.

Customer satisfaction attributes are not equal. Not only are some more important to the customer than others, but some are important to the customer in different ways than others. It is also noted that customer complaints are mostly linked to expecters and, therefore, a strategy based on solely removing them may not result in satisfied customers. This is because responding to customer complaints is a passive strategy. To obtain customer satisfaction, one must actively determine what the customer wants.

4.1.1 RECORDING THE VOICE OF THE CUSTOMER

There are several methods that can be used to obtain customer preferences about products. The customer, in this case, is the end user. Which method is used depends on the amount and type of information required, its availability, and the time and cost allotted to collect the data. However, irrespective of the method used, it is very important that the data correctly reflect the customers' wants and needs. One of the purposes of customer surveys is to clear up any misconceptions about the customers' desires. The sources fall into two broad categories: existing information and new information. Existing information is most readily available for a redesigned product and can be obtained from (1) company sales records, including repair and replacement parts; (2) complaints, both written and verbal; (3) warranty data; (4) publications from the government, trade journals, and the consumer; (5) the company's designers, engineers, and managers; and (6) benchmarked products.

New information can be obtained from (1) surveys, including mail, telephone, comment cards, and at the point of purchase; (2) interviews, both face-to-face and telephone; (3) focus groups; (4) observation, using clinics and displays; (5) field contacts, using sales meetings, service calls, and trade shows; and (6) direct visits with the users.

Each of these approaches has its advantages and disadvantages. For example, in monitoring complaints one can identify specific problems. However, many people do not complain to the company, and many former customers may not have had complaints, but simply like the competition's product better. On the other hand, observation of the product in use provides the opportunity to both observe how the product is used and to question the user directly about his/her preferences and the product's attributes. This method can be expensive, however, and may introduce the bias of the surveyor.

When asking customers questions there are some simple guidelines to follow; these are illustrated in Table 4.1. First, the questions that are to be asked, no matter how the information is to be obtained, should be identified and prioritized by the IP^2D^2 team. Then the target customers are identified. For example, for a tool, is it for the hobbyist or the professional; or for a toaster, is it for the individual or for a restaurant. Third, the participants are selected. These selections should be confined to those who play important roles in the selection, installation, use, and servicing of the product; that is, they should be representative of the target market. Fourth, the right environment for

TABLE 4.1
Guidelines for Eliciting Useful Information About Customer Preferences

Don't Ask Questions—	Instead, Ask Questions—
That are self-serving or biased	**That will help you reach your goals**
What do you like most about our product?	What do you like about this product?
What do you dislike most about our competitor's product?	What do you dislike about this product?
With one obviously correct answer	**That allow customers to voice their opinions**
Is low cost an attractive feature?	What do you consider when purchasing the product?
	What improvements would you make to the product?
That contain more than one question	**That will differentiate customer preference**
Would you prefer a blue sports car or a red convertible?	Would you prefer a red or blue car?
	Would you prefer a sports car or a sedan?
For which customers can't be expected to have a reasonable response	**That help you understand what customers might want or need**
How often would you travel in space if you had your own rocket?	Do you want a device to travel in space?
That are leading	**That are open-ended**
Are you satisfied with this product?	What have your experiences been with this product?
Do you use a seat belt when you drive?	What are the steps you take in starting your car?

the interview is established. Last, it is important to realize that the questions asked may both reveal and conceal what the interviewee thinks about the product. Thus, in one-on-one and focus group interviews, one should be ready to refocus the direction of the questions.

To illustrate the type of questions that should be asked, consider those listed below, which were some of the questions asked by an IP^2D^2 team in the initial stages of the development of a palm grip orbital sander. (The team members visited over 500 cabinetmaker shops in the U.S. and Germany over a two-month period.)

- What are your applications?
- What do you like about your present product?
- What don't you like about your present product?
- What would you change?

The answers to these questions resulted in the customer requirements shown subsequently in Figure 4.2, along with their corresponding order of importance.

In general, before conducting a survey, the IP^2D^2 team should first have the answers to the following three questions:

1. What do we want to know?
2. From whom do we want to know it?
3. What will we do with the information once we have it?

The answers to these questions will help keep the survey focused.

A survey typically asks sets of questions that cover different areas, depending on whether the product is a new product or an improvement to an existing product or one that is entering into an existing market. The areas usually are a combination of questions about background, product features, product controls, product displays, product usability, product cost, and satisfaction with existing products. Some sample questions in each area are listed below. Many times the survey will provide a choice of answers or a range of values from which the respondent chooses. These ranges are indicated by the $\leftrightarrow$ symbol.

Background
> Age, gender, income
>> Do you own the product?
>>> Where did you purchase it?
>>> How often do you use it?
>>> How satisfied are you with it?
>>> How much did you pay for the product?
>>> Who purchased the product?
>>> How long have you owned the product?
>>> Did the product come with instructions? Did you read them?

Product features
> From the following list of features, which are the three that are most important to you?
>> [Provide list of features.]
> What do you like most about these features?
> List the typical attributes, the relevancy of which greatly depends on the type of the product—
>> Appearance/look/style: interesting/nice/elegant $\leftrightarrow$ boring/ugly/drab
>> Shape: sleek $\leftrightarrow$ bulky
>> Weight: light $\leftrightarrow$ heavy
>> Texture: smooth $\leftrightarrow$ rough

Size: small ↔ large
Quality: poor ↔ excellent
Feel: comfortable ↔ uncomfortable
Usability (ease-of-use): easy ↔ difficult
Layout: poor (illogical) ↔ good (logical)
Function: obvious ↔ ambiguous
Color: appropriate/suitable ↔ inappropriate/unsuitable
Packaging: attractive ↔ unattractive
Labels: poor ↔ good

Product performance and use

Main performance characteristic(s): poor ↔ good
Ergonomic attribute(s): poor ↔ good
Safety: poor ↔ good
Usability (ease-of-use): easy ↔ difficult
Reliability: poor ↔ good

4.1.2 ANALYZING THE VOICE OF THE CUSTOMER

After having recorded the voice of the customer, the IP^2D^2 team takes their responses, sometimes called *verbatims*, and writes them on individual cards. These responses are then "scrubbed," which is the process of editing the verbatims, eliminating those verbatims from other interviewees that mean the same thing, and writing more cards for those verbatims that express more than one idea. These cards are then sorted into groups that seem to be different descriptions of a more broadly expressed attribute. It is possible that a few of the verbatims may actually express this higher level description.

These *verbatims* have to be ranked in their order of importance. There are several ways to do this. A passive way is to count the number of times the equivalent attribute is mentioned by the interviewees. Those that are mentioned the most often are assumed to be the most important. Another way is to take those attributes that have been ranked the most important in the manner just described, and to ask another group of interviewees to rank their importance. Each ranking is assigned a number from either 1 to 10 or from 1 to 5, the higher values being the most important. See the subsequent discussion of Figures 4.2 and 4.3, respectively.

Consider a drywall taping that is discussed in Section 5.2.5. A small survey of drywall specialists led to the list of *verbatims* given in Table 4.2. This list was scrubbed, and the results of the

---

TABLE 4.2
Customer Verbatims from a Survey of Drywall Taping Professionals

Low price	Doesn't break
Low maintenance cost	Doesn't leak
Easy to tape joints	Isn't messy to use
Easy to lift and hold overhead	Operates in a dirty environment
Fast taping of joints	Works well
Doesn't corrode	Long life
Easy to transport	Can tape inside corners
Long reach	Can use fiberglass tape
Small size	Embeds tape fully in compound
Easy to assemble	Little operator training
Easy to repair	Low maintenance
Easy to clean	

---

scrubbed list appear as the customer requirements presented subsequently in the customer requirements column of Figure 4.3.

4.2 QUALITY FUNCTION DEPLOYMENT (QFD)[*,†,‡,§,¶,**]

4.2.1 Introduction

Quality function deployment (QFD) is a formalized method of matching the expressed needs of the customer to the features and functions of the product. It is a very powerful method that helps to define the product in terms of the customers' requirements, which, as shown in Figure 2.2, is the most important element in the product realization process. In addition, QFD takes customer requirements and provides a means of converting them into product specifications as discussed in Section 4.3. Lastly, QFD is an IP^2D^2 team decision-making process that is used to achieve a common understanding of the product and to obtain a consensus and commitment necessary to convert that understanding to an operational level.

As discussed in Section 2.2, an IP^2D^2 team needs to have a common vocabulary, agree on a common purpose and on priorities, and be able to communicate a shared understanding of all issues. The QFD method greatly facilitates meeting these requirements. In order to facilitate the discussion of the QFD table and the material in the subsequent chapters, we introduce the following terms: customer requirements (CRs), functional requirements[††] (FRs), and engineering characteristics (ECs). Customer requirements are phrases customers use to describe products and their characteristics. Functional requirements are those requirements that specify a mandatory action of a product or system. Engineering characteristics are the relevant *measurable* characteristics that describe each of the product's functional requirements.

The quality function deployment method, in general, addresses the following:

1. **What the customers want:** QFD begins with the customer's requirements. Not all customers are end users and the CRs can include the demands of regulators ["safe in a side collision"], the needs of retailers ["easy to display"], the requirements of vendors ["satisfy assembly and service organizations"], and so forth.
2. **The customer preferences that are important:** QFD records the relative importance to the customer of the CRs based on either the IP^2D^2 team members' direct experience with customers or from surveys.
3. **How to convert the perceived needs of the customer to yield a competitive advantage:** QFD employs engineering characteristics to describe each of the product's functional requirements. The engineering characteristics are related by the IP^2D^2 team to the customers' requirements as discussed in Section 4.2.2. The product's ECs are also benchmarked to identify those characteristics that must be matched or exceeded in order to be competitive. It is emphasized that the EC is an attribute of the FR and not the CR.

---

[*] J. R. Hauser and D. Clausing, "The House of Quality," *Harvard Business Review*, Vol. 66, No. 3, pp. 63–77, May–June 1988.

[†] R. Ramaswamy and K. Ulrich, "Augmenting the House of Quality with Engineering Models," in *Design Theory and Methodology* DTM-92, DE-VOL. 42, D. L. Taylor and L. A. Stauffer, Eds., ASME, New York, pp. 309–316, 1992.

[‡] Y. Akao, Ed., *Quality Function Deployment: Integrating Customer Requirements into Product Design*, Productivity Press, Portland, OR, 1990.

[§] L. R. Guinta and N. C. Praizler, *The QFD Book*, AMACOM Books, New York, 1993.

[¶] D. Clausing, *Total Quality Deployment*, ASME Press, New York, 1994.

[**] B. A. Bicknell and K. D. Bicknell, *The Road Map to Repeatable Success, Using QFD to Implement Change*, CRC Press, Boca Raton, FL, 1995.

[††] See Section 5.1 for a detailed discussion of functional requirements.

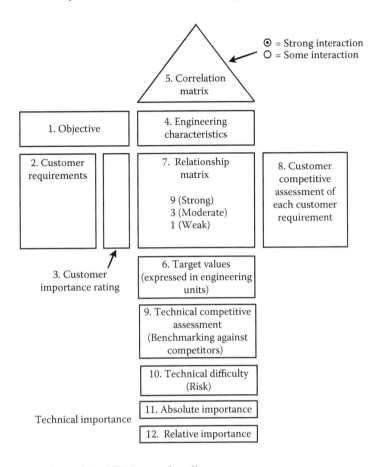

FIGURE 4.1 Twelve regions of the QFD house of quality.

4.2.2 QFD AND THE HOUSE OF QUALITY[*]

The house of quality is a multidimensional figure that shows the relationship of the customer requirements to the engineering characteristics of the product. It encapsulates the various aspects of the customers' requirements for the product and their opinion of the competitors' products, and conveys the IP^{2}D^2 team's *interpretation and judgment* of the relationship of each EC to each CR. The twelve regions of the house of quality are shown in Figure 4.1 and are described below. Several examples are given subsequently in Figures 4.2 and 4.3 and in Tables 4.3 and 4.4.

1. **Objective:** State the objective of the product, which is the goal of the IP^{2}D^2 team effort.
2. **List of characteristics:** Obtain a list of characteristics of the product as defined by the customers (CRs). When possible, the customer requirements should be grouped at their highest level by arranging them according to the applicable dimensions of Garvin's eight dimensions of quality given in Section 1.4.2. This helps to clearly delineate the differences between customer preferences for performance attributes and for features. These CRs are also used to evaluate the candidate concepts generated to satisfy each FR as shown in Section 6.4.

* Examples of applications of QFD can be found at http://www.qfdi.org/.

Customer Requirements		Customer Importance	Remove Material → Generate Sander Motion							Remove Material → Generate Rotatory Motion				Remove Material → Provide User Grip						Collect Dust			Connect to Power		
			Material removal rate	Brake life	Time to full speed	Random orbit offset	Off-work pad speed	Pad life	Pad size	Motor speed	Motor power	Motor brush life	No. thermal overloads	Unit weight	Body grip diameter	Center of gravity	Grip acceleration	Grip height	Impact resistance	Vacuum flow rate	Dust bag size	Dust collection efficiency	Cord length	Cord life	Switch life
Performance	Fast material removal	5	9			9			9	3	3									1					
Performance	Powerful	5	9		9	1				9	9						1			1		1			
Performance	No initial gouge	7		9			9			1															
Performance	Smooth finish	9				3	9																		
Performance	Low vibration	9				3	9		9	9							9								
Features	Pad brake	9		9																					
Features	Long cord	2																					9		
Features	Dust collection	7		1																9	1	9			
Ergonomics	Hand close to work	5																9							
Ergonomics	Controllable	9		9		9	9		9	9					9	3		3							
Ergonomics	Comfortable	7													9		9	9							
Ergonomics	Low fatigue	8												3	9	3	9								
Durability	Durable	9		3				9				9	1						9					9	9
Units of target values			cc/sec	hrs	sec	mm	rpm	rpm	area	cc	W	hrs		kg	cm	cm	g	cm	kg	cc/min	cc	%	m	hrs	
Absolute importance			90	259	45	185	306	81	207	229	60	81	9	24	216	51	221	144	81	73	7	72	18	81	81
Relative importance (%)			3.4	9.9	1.7	7.1	11.7	3.1	7.9	8.7	2.3	3.1	0.3	0.9	8.2	1.9	8.4	5.5	3.1	2.8	0.3	2.8	0.7	3.1	3.1

FIGURE 4.2 Partially completed QFD table for the redesign of a random orbital sander. (Target values have been omitted.)

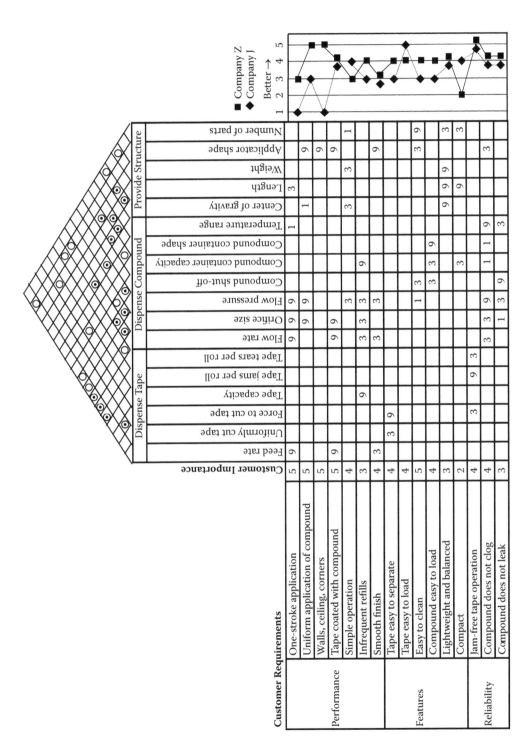

FIGURE 4.3 Partially completed QFD table for a drywall taping system.

TABLE 4.3
Customer Requirements Arranged by Garvin's
Quality Characteristics for the Steel Frame Joining
Tool of Section 5.2.6

Quality Characteristic	Customer Requirements
Performance	Short time to make connection
	Strong connection
	Makes connection every time (repeatable)
Features	Lightweight
	Accessible to hard-to-reach places
	Overload protection
	Easy to undo connection
Conformance	Safe
	Quiet
	Easy to use
	Impact resistant
	Low vibrations (if applicable)
Ergonomics	Not fatiguing to use
	Controllable/maneuverable
	Comfortable to hold
	Compact/balanced
Serviceability	Little or no maintenance
	Components easy to replace
	Replacement components available
Aesthetics	Rugged-looking

3. **Importance ratings:** Customer importance ratings are the weighted values obtained in the manner discussed at the end of Section 4.1. They are assigned to the product's characteristics indicating their relative importance as indicated by the customers.

4. **The quantities required to achieve the product's characteristics**: These are the engineering characteristics expressed in terms of measurable quantities. (See target values discussed in region 6 below.) Although it is not customary to do so, it is suggested that the engineering characteristics be grouped by their functional requirements in the manner discussed in Section 5.1. That is, the components of the product's functional decomposition should be used to organize the engineering characteristics in the house of quality. Furthermore, the engineering characteristics should contain a column for each of the constraints or qualifiers to each component of the functional grouping, such as waterproof to x meters, impact resistant to a y meter drop, volume less than z cm^3, power consumption less than w watts, etc. Sometimes it is desirable to place in front of each EC either an arrow ($\Uparrow$ or $\Downarrow$) or a "+" or "−" sign to indicate the direction of the magnitude of each characteristic. Thus, either a $\Downarrow$ or a "−" indicates that a lower value is better, and, conversely, either a $\Uparrow$ or a "+" indicates that a higher value is better.

5. **Correlation matrix:** A correlation matrix shows the degree of interaction among the product's engineering characteristics. It gives some idea of the degree of coupling that will exist when trying to satisfy the engineering characteristics, and indicates the degree of coupling frequently brought about by the physical laws governing the product's engineering characteristics. This matrix, then, indicates where design tradeoffs may have to take place. It is suggested that in the roof of the house of quality only two symbols be used, one to indicate

TABLE 4.4
Engineering Characteristics for Several Functional Requirements of the Steel Frame Joining Tool of Section 5.2.6

Functional Requirements	Engineering Characteristics (units)
Maintain position of stud and track flanges	Number of operations
	Elapsed time (s)
	Force (N)
	Weight of device (kg)
Provide structure for tool and its components	Weight (kg)
	Center of gravity with respect to hand (cm)
	Impact resistance—height of drop (m)
	Length (cm)
Generate energy	Power (W)
	Time to reach peak magnitude (s)
	Life expectancy of source (yr)
Convert energy to working element	Efficiency (%)
	Life (number of operations)
	Reliability (%)
	Force (N)
Make connection	Elapsed time to make connection (s)
	Joint strength (N)
	Number of operations to achieve connection
	Noise level (dBA)
Remove/disengage tool	Reliability of connection (%)
	Number of fasteners, if applicable
	Number of operations to remove tool
	Force (N)
	Time to disengage (s)

strong correlation and the other to indicate some correlation. This level of detail is often sufficient to identify where interactions occur.

6. **Target values:** Target values for each of a product's engineering characteristics are frequently determined from benchmarking data (see discussion of region 9 below) and from an independent assessment of how strongly the values impact the product's performance attributes and features. Sometimes a marketing strategy may require that certain values be met simply because the competition meets them.

7. **Relationship matrix:** A relationship matrix is a systematic means for identifying the levels of influence and effect between each engineering characteristic and the customers' requirements. A scale of 9-3-1 is used to weight disproportionately those engineering characteristics that strongly affect the customer requirements. This nonlinear scale aids in the identification of those quantities having the highest absolute importance described in the discussion of region 11 below. The assignment of the values relies on the interpretation and judgment of the IP^2D^2 team members.

8. **Customer competitive assessment:** Customer competitive assessment is a summary of the top two or three competitive products' characteristics in comparison with the product being developed. If the company does not have an existing product, then it indicates how the current products are viewed by the buyers. The higher the value in the chart, the better that requirement is perceived. This assessment is unrelated to the customer importance rating, to the values that are assigned to the relationship matrix, and to the target values.

9. **Technical competitive assessments:** Technical competitive assessments (benchmarks) compare the competitors' specifications for each of the product's engineering characteristics to those for the proposed specification.* Each of these specifications should meet or exceed each of the competitors' characteristics.

10. **Technical difficulty:** Technical difficulties are values that indicate the ease with which each of the product's characteristics can be achieved. The lower the number, the easier it is; that is, the risk of not meeting that characteristic is lower. This judgment is frequently based on both the IP^2D^2 team's experience and the company's experience with the product, or similar products, and with any techniques and technologies that are anticipated.

11. **Absolute importance:** Absolute importance is the sum of the product of the numerical value of each element in a column of the relationship matrix with its corresponding customer importance rating. This is the preliminary step in obtaining the final results that are displayed in region 12.

12. **Relative importance:** Relative importance is the determination of the percentage of the total numerical score that each EC has. The total numerical score is the sum of all the values appearing in the row comprising region 11. The percentage assigned to each EC is that EC's score multiplied by 100 and divided by the total score. Those ECs that have the highest rankings are the characteristics on which the IP^2D^2 team should concentrate its efforts, since they are clearly related to those CRs that are, in the aggregate, the most important to the customer. The satisfaction of these highest-ranked ECs provides the highest leverage for the allocation of the team's resources: time and money. Also, the relationship between the relative importance and the corresponding technical difficulty determined in region 9 provides additional guidance for the planning of resources during the product's development. Furthermore, the highest-ranked ECs indicate those functional requirements that are the most important to satisfy, which, in turn, indicate the attributes of the corresponding physical entities that have to receive the most attention.

There are several observations to be made about the QFD method, since it summarizes a large amount of information in a single diagram. First, the customers' competitive assessment is independent of the customer importance rating. Secondly, the ECs are first determined (selected) independently of the CRs, and then related to the CRs via the relationship matrix. Also, usually only a few of the CRs will relate to a group of ECs belonging to one FR, and to no other ECs belonging to any other FR, if the FRs are included in the QFD table. This aspect has important implications when it comes to selecting evaluation criteria for candidate concepts. For an example, see how the CRs in Figure 4.3 are used in Table 6.15. The ECs can also help decide the factors that should be selected for designed statistical experiments, which are discussed in Section 11.2.

The target values in the house of quality not only state the numerical goals for the corresponding engineering characteristic, but they also imply the ability to measure them. This is a very important part of the QFD process, for if one cannot measure it, then one cannot be sure that the engineering characteristic has been improved and, therefore, one cannot be certain that the customers' requirements have been met. Thus, an inherent part of completing the house of quality implicitly requires the knowledge and facilities to make the appropriate measurements. In addition, there are many standards that explicitly state the way in which certain tests should be conducted. Refer to the organizations cited in Section 1.4.2.

After the most important engineering characteristics have been identified by their rank in region 12 of the house of quality, the methods discussed in Chapter 6 can be used to obtain their solutions. Those rankings indicating high relative importance should be used as some of the concept evaluation criteria. See Table 6.15 for an illustration of this. It should be realized that one doesn't have to

---

* The technical competitive assessment is not very different from what has been published for years in *Consumer Reports*, that is, evaluation and comparison of competing brands of a consumer product.

wait for the house of quality to be completed before starting the concept generation and selection process. However, the evaluation of the concepts should wait until the first QFD iteration is completed. As shown in Figure 2.2, the concept selection process feeds back to the product definition stage of the product development cycle and is an important part of the team's refinement of their understanding of the product's definition. Furthermore, the QFD table has to be continually revisited as the understanding of the product and its relationship to the customer requirements become well understood. This is especially important if the increased understanding requires modifications to the original set of functional requirements as discussed in Chapter 5.

Some caution is needed when using the QFD method. In recording the voice of the customer, it must be realized that beyond a certain point the gathering of more information will only have incremental value. Also, there are no set rules for QFD implementation, and every product development undertaking should use as much or as little of it as is appropriate. The completion of the house of quality is not an end unto itself. Also, any reasonable rationale for organizing the house of quality is acceptable, as long as it leads to the satisfaction of the customers' requirements.

A partially completed house of quality[*] for the random orbital sander previously mentioned in Section 4.1 is given in Figure 4.2. This is an example of a redesigned product for which the concepts have already been determined. For another example of a QFD table, we consider the drywall taping system discussed in Section 5.2.5. The development of this product was preceded by a survey that resulted in the verbatims given in Table 4.2. A QFD table for this system is given in Figure 4.3, and is an example of a QFD table wherein the concepts are still to be determined (see Table 6.4). The functional requirements for this device are given in Section 5.2.5.

It is noted that in both examples the relationship matrix is sparse, which is common. The customer competitive assessment in Figure 4.3 has been displayed in graphical form; however, tabulated numerical values could have also been used to indicate the relative ranking of each customer requirement. For those products that are being redesigned, this portion of the house of quality should also include the customers' opinions of the company's present product.

As a last example, consider the steel frame joining tool that is discussed in Section 5.2.6. Here we will focus only on the customer requirements and the engineering characteristics, which are given in Tables 4.3 and 4.4, respectively. Table 4.4 is an example of how one arrives at the ECs prior to the generation and selection of the concepts that will satisfy their respective FRs. It also illustrates an advantage of organizing the ECs according to their FRs. It would be difficult to specify all the relevant ECs unless the FRs were first established. Notice, also, that at this stage the ECs have been determined independently of the concepts, which are given in Table 6.16.

The QFD method is also used as a guide through the product development cycle after the concepts and the embodiments have been decided. There are four phases: (1) design, (2) details, (3) process, and (4) production. These phases help communicate product requirements from the customers to the design teams to the production operators. Each phase has a matrix consisting of a row of customer requirements and a column of what is needed to achieve them, the ECs. At each stage, the ECs that are the most important, require new technology, or have high risk are carried to the next phase. A detailed example of these four phases for a copier's paper feed mechanism can be found in Clausing.[†]

4.3 PRODUCT DESIGN SPECIFICATION

During the process of completing the house of quality, the IP^2D^2 team should also be developing information that will form the product's design specification (PDS). The PDS contains all the facts relating to the product's outcome. It is a statement of what the product has to do and is the fundamental control mechanism and basic reference source for the entire product development activity.

---

[*] For additional examples, visit http://www.npd-solutions.com/.
[†] D. Clausing, ibid.

TABLE 4.5
Components of the Product Design Specification[a]

General	Constraints (continued)
Purpose of product	Political, social, and legal requirements
Target cost to customer	Maintenance and service requirements
Need for product	Packaging (including ability to re-use and recycle)
Benefits to user	Reliability
Time for product to first reach customer (scheduling)	Shelf life
Customers	Patents (search, apply for, obtain license)
Size	Environment: factory floor, packaged, stored, during
Weight	transportation, and in-use (see Table 4.6)
Quantity	Testing
Competition (benchmarking)	Safety: compliance issues
Service life (planned obsolescence)	Materials and components: recyclable, disposable,
Market evaluation: trends, growth, share	availability, suppliers
Trademark, brand name, logo	Ergonomics
Performance	Standards: U.S. and international
Functions	Interfaces: electrical and mechanical connections
Features	Aesthetics (appearance)
Constraints	End user requirements: professional installation,
Shipping modes and costs	preassembled, user-assembled
Disposal and recycling	Suppliers
Manufacturing facilities, processes, and capacities:	Energy consumption
in-house, in-country, out of country	Product operational costs
	User training and learning requirements: documentation

[a] Compiled, in part, from S. Pugh, ibid.; B. Hollins and S. Pugh, *Successful Product Design: What To Do and When,* Butterworths, London, 1990; and P. G. Smith and D. G. Reinertsen, *Developing Products in Half the Time,* Van Nostrand Reinhold, New York, 1991

The PDS is evolutionary; it may change as the design and development process proceeds. However, upon completion of the process, the PDS is a written document that matches the product itself, and it forms the basis for the specification of the product as designed and manufactured. In addition, the PDS must be comprehensive and written in terms that are understandable by all entities of the manufacturing enterprise. One of the most important things about the PDS is that the IP^{2}D^2 team members must agree to it, and by doing so take ownership.[*]

The product definition should address the items listed in Table 4.5 and should state the following:

- The product title.
- What purpose or function the product is to perform.
- Against what types of product it will be competing and who makes them.
- What market it will serve.
- Why there is a need.
- The anticipated demand and target price.
- Product identity.
- Relationship to the company's current product lines.

To illustrate the product design specification, consider the drywall taping system discussed in Section 5.2.5. A preliminary PDS for this device could be as follows.

[*] S. Pugh, *Total Design: Integrated Methods for Successful Product Engineering,* Addison-Wesley, Reading, MA, 1991.

General
Product Title
Drywall Taping System
Purpose
To dispense joint compound and tape simultaneously to fill gaps between adjacent drywall sheets
Need
Brand *A* is a relatively inexpensive tool ($150) but does not provide the customer with the level of performance they have requested.

Brand *B* is a more complete tool, but it is costly ($1650 plus accessories). Hence, there is a need to satisfy customer requirements with a reasonably priced tool.
Benefits
Reduces the number of separate operations to apply the tape coat

Reduces the time it takes to complete the tape coat

Simplifies the application of the tape coat
Competitors
Company *A*, model *X*

Company *B*, model *Y*
Customers/Market
Primary customers: professional drywall companies

Secondary customers: homeowners and tool rental centers
Quantity
Production will be 12,000 units in the first year and 20,000 units in each of the following two years
Product Cost
Retail: <$500

Company's selling price: <$200
Time Scale
In the marketplace by: (month/year)

Performance
Functions
Tape inside corners

Tape joints in any orientation

Tape joints without leaking joint compound

Taped joints will require no additional smoothing

Tape will not break prematurely

Compound will flow evenly across tape/wall

Will work with both paper and fiberglass tape, with tapes whose widths vary from 44.5 to 57 mm and whose thickness vary from 0.13 to 0.18 mm

Will work with thick or thin viscosity compounds

Dispense tape at the rate of 150 mm/s [see end of Section 6.1]

Dispense the joint compound at the rate of 32 cm^3/s [see end of Section 6.1]

Constraints
Size
Small enough to be easily transported by one person

Device attaching to end of handle will be smaller than $74 \times 30 \times 41$cm
Weight
Hand-held portion will weigh less than 2.3 kg empty

Will weigh less than 9 kg when fully loaded with joint compound and tape

Manufacturing Facilities and Processes

No new manufacturing facilities will be required

Manufacturing operations per component to be minimized

No components to require expensive or time consuming operations

Shipping

Will be shock, vibration, and weather resistant for shipping by any means:

Vibration environment from 4-33 Hz at 1.5 mm amplitude

Shock environment simulated with a 2.4 m drop test

Insensitive to temperature in the range 0 to 60°C and humidity to 100%

Will be packaged in a rectangular cardboard box that can be stacked to 2.5 m

Disposal and Recycling

Product will neither contain nor be manufactured with any environmentally hazardous materials

Easily disassembled for component recycling, reuse, and disposal

Political, Social, and Legal

Will conform to all applicable standards and local and national compliance regulations [List]

Product Environment [Table 4.6]

Product must operate properly in the following environment:

Temperatures between 15°C and 50°C

Atmospheric pressure from sea level to 2 km

Relative humidity to 100%

Testing

All purchased components will be obtained from suppliers with ISO 9000 certification

Vibration and shock testing as described in *Shipping*, and thermal testing as described in *Product Environment*

TABLE 4.6
Environmental Factors

Naturally Occurring	Man-Made
Weather	Mechanical
Temperature	Acceleration/deceleration
Humidity	Acoustic
Rain	Impact/shock
Snow, sleet, hail, ice	Vibration
Wind	Pressure
Dew point	Heating
Lightning	Cooling
Thunder	Radiation: electromagnetic, nuclear
Ultraviolet light	Radio/TV/cellular telephone waves
Atmospheric pressure	Chemicals/pesticides
Airborne	
Sand	
Dust	
Salt water spray	
Insects, fungus, pollen	
Cosmic radiation	
Air quality (pollution)	

Safety

All sharp edges, with the exception of the smoothing device, and any electrical, hot, and moving parts will be shielded to prevent contact with the operator

Product will be well-balanced in order to provide safe and comfortable operation

Documentation

A user's manual, maintenance manual, and guarantee will accompany product

All testing will be documented by the company or by outside company performing the test

Customer will be asked to complete a questionnaire to determine whether the product meets his/her expectations

Life Span

Product will be modular to allow component upgrades as they become available

Minimum operational life of 5 years

Materials

All materials will resist corrosion

Materials must be well suited to company's manufacturing methods

Ergonomics

Single operator

Two-handed use

Product must be comfortable to use for long periods of time without inducing excessive fatigue

Installation

Product will be shipped assembled except for handle

Aesthetics

Product will have a durable finish

Colors will be green and black

Product will convey ruggedness

Maintenance

All fasteners will be standard

All components subject to wear to be easily replaceable with relatively inexpensive parts and with standard tools and no special skills

Design will be modular for easy repair and cleaning

Joint compound must removed after each day's work and device devoid of any compound residue

No or very few lubrication points

Packaging

Will not require special handling during shipment

Container will have the name, function, critical performance specifications, and picture of the product on the outside

Must stack safely up to 2.4 m

Reliability

Will withstand repeated shipping and testing conditions without affecting performance

Components will have a service life of more than 5 years

Tape severing device will last for at least 10,000 tape separations

Shelf Life

Shelf life greater than 5 years

Patents

Product must not infringe on the following patents [list applicable patents]:

5 Product Functional Requirements and Functional Decomposition

The notions of product functional requirements and functional decomposition are discussed, and the axiomatic method is introduced as one means of clarifying and focusing the product's goals.

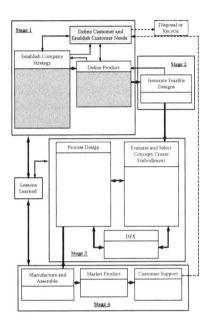

5.1 FUNCTIONAL MODELING

5.1.1 INTRODUCTION

The generation of a product's or a system's functional requirements (*FR*s) is done in the context of the overall goals for that product, which are directly related to the customer requirements. Accompanying the product's goals are constraints that the functions must satisfy. These constraints form an additional basis by which all solutions that are created to satisfy the *FR*s are judged. As indicated in Section 4.2.1, a function requirement specifies a mandatory action of a product or system. A constraint is an external factor or bound that the *FR* must satisfy. For example, a constraint may be bounds on size, weight, material, efficiency, strength, cost, geometric shape, laws of nature, or any of the life-cycle and social design aspects listed in Table 2.1. The same constraint may apply to more than one *FR* and one or more constraints may impact other constraints.

A product's overall function may be divided into a hierarchical set of subfunctions by decomposition, which is the process of dividing a system's functions into smaller, coherent, self-contained functional elements. The interrelationships among these functional elements will dictate the decisions by which a solution is obtained. Performing a functional decomposition has several advantages. It provides a means of transforming the overall complexity of the product into functional units that have lower complexity. It forces the design team to focus initially on *what* the product should do before determining *how* to do it.[*] It can provide the design team with a basis for organizing itself and the tasks that need to be accomplished. Lastly, if one or more of the functional units is independent of all other functional units, then tasks with respect to these units may be performed in parallel with the tasks for the other functional units.

---

[*] Einstein once said that if he had 20 days to solve a problem, he would take 19 days to define it.

5.1.2 Functional Decomposition and the Axiomatic Approach: Introduction

A method that is well-suited to decomposition is the axiomatic approach.[*,†,‡] This approach does not give a step-by-step procedure to generate a design, but rather it gives a means of clarifying and focusing both the product's functions and the objectives that the design should meet. The method is one means of attaining "clarity of the design."[§] As will be seen subsequently, the axiomatic approach provides a compact visual way of expressing the design intent and the overall design objective.

In the context of the axiomatic method, the following notation and definitions are introduced. Functional requirements are defined as the minimum nonunique set of independent mandatory requirements that completely characterize the design objectives for a specific need. If possible, they must be independent of each other at every level in the design hierarchy. Several examples of functional independence are given in Section 5.2.1. Design parameters (*DPs*) denote the physical entities that will be created by the design process to fulfill the *FRs*. In other words, the functional requirement describes what action or series of actions is required to satisfy the customer needs, and the design parameter is the physical entity (component/module/unit) that has to be created to satisfy its functional requirement. The creation of the *DP* requires that a concept (specific principle, method, or means) be selected. The generation, evaluation, and selection of candidate concepts are discussed in Chapter 6.

The *FRs* reside in the function domain and are presented in a *solution-neutral* manner. The term *solution-neutral* means the avoidance of any preconceived ideas of what would work best. The functional requirements, therefore, are stated in simple, unambiguous terms that neither refer to, nor imply, specific operations or processes. The functional requirements should have a noun and a verb, not use jargon, be stated in the affirmative rather than the negative, and be quantifiable. As stated in Section 4.2.1, each specific measurable attribute, or characteristic, of each *FR* is called an engineering characteristic (*EC*).

To illustrate what is meant by solution neutral, we shall examine the common two-knob water faucet. The design objective for this water faucet is to provide water continuously at a desired flow rate and at a desired temperature *when the flow rate of the hot and the cold water is controlled separately*. It is seen that the minimum number of functional requirements is two: obtain the desired flow rate and obtain the desired water temperature. In terms of the axiomatic design procedure, this system is modeled as follows:

$(FR)_1$ = Obtain water flow rate
$(FR)_2$ = Obtain water temperature

The design parameters are dictated by the fact that we are to control the flow rates of the hot and cold water separately. Thus,

$(DP)_1$ = Means to adjust cold water flow
$(DP)_2$ = Means to adjust hot water flow

It is seen, however, that the problem was not stated in a solution-neutral manner, for the *DPs* were stated as having two independent controls, one for adjusting the hot water flow rate and the other for adjusting the cold water flow rate. Thus, in the statement of the functional goals of the system,

---

[*] N. P. Suh, *Principles of Design*, Oxford University Press, New York, 1990.

[†] D. Dimarogonas, "On the Axiomatic Foundation of Design," in *ASME Design Theory and Methodology*, DE-Vol. 53, pp. 253–258, 1993.

[‡] N. P. Suh, *Axiomatic Design: Advances and Applications*, Oxford University Press, New York, 2001. Extensions of the axiomatic method to complex systems can be found in N. P. Suh, *Complexity*, Oxford University Press, New York, 2005.

[§] G. Pahl, W. Beitz, J. Feldhusen, and K.-H. Grote, ibid.

the means of *how* it was to be done was mandated. These two design parameters do not address the temperature adjustment directly. A better way to present this problem is given in Section 5.2.1.

As mentioned in the previous section, the functional requirements are subject to constraints. In the context of the axiomatic method, constraints differ from *FR*s in that they do not have to be independent of other *FR*s or other constraints. The constraints in the two-knob water faucet problem could be the maximum flow rates of the hot and cold water and their respective temperature ranges.

For an example of constraints on a system, consider a few of one manufacturer's requirements that must be satisfied by its all-wheel drive (AWD) system:[*]

- Torque transfer up to 2,400 Nm.
- Built-in torque transfer limitation.
- Full function in reverse.
- Instant activation on differential speed.
- Fully integrates with brake systems and stability systems.
- Can be deactivated in less than 60 ms.
- Provide limited/full AWD function with a flat tire.
- Requires no additional sensors.
- No wind-up during tight cornering and parking.
- Optimal traction during acceleration.
- No functional problems with tires having uneven wear pressure or size.
- No functional problems when towing with one axle lifted.
- Transparent actuation.

For another example of constraints, consider the Palm Pilot personal digital assistant, which was originally conceived in 1994. It had just four requirements (constraints): (1) fit in a shirt pocket, (2) seamlessly synchronize with a PC, (3) be fast and easy to use, and (4) sell for no more that $299. When this list was first made, it was unclear how to satisfy these conditions, but it turned out that these four conditions were the proper ones to focus on.

The effect of constraints on the outcome of the design process cannot be overstated. We shall illustrate this with three examples of commonly used products. Consider first two bicycles, one designed to be a touring/racing bicycle and one designed as a mountain bicycle. The functional requirements for both bicycles are the same: support rider, manually propel bicycle, steer, and stop. The touring bicycle constraints will be directed to high speed on smooth riding surfaces. For the mountain bicycle, the constraints will be directed toward maneuverability and climbing ability on a wide variety of unpaved surfaces. An example of the resulting products is shown in Figure 5.1, where it is seen that the differences are virtually everywhere: wheels and tires, frame and wheel yokes, seats and handle bars, and pedals. These striking differences are because the constraints are different, even though the functional requirements for both bicycles are the same.

For a second example of how constraints affect the final outcome of a product, consider a door hinge. The function of a hinge is to connect two solid objects so that one object can rotate about an axis with respect to the other object. Referring to Figure 5.2, we see examples of the devices that satisfy different constraints indicated in the figure caption. As a last example, consider the two paper clips shown in Figure 5.3. The one on the right satisfies the constraint that the paper clip does not substantially extend beyond the border of the paper. It is noted that both clips have the same operating principle.

An example of how the overall goal of a product can change the final outcome, consider the following different goals for automobiles and representative products that illustrate the attainment of that goal. It should be realized that each goal has been attained by making tradeoffs with respect to

---

[*] Haldex Traction AB [http://www.haldex-traction.com].

FIGURE 5.1 Different final products due to different constraints: (a) mountain bicycle and (b) touring/racing bicycle. (© 2008 Trek Bicycle Corporation, Waterloo, Wisconsin. Reprinted with permission.)

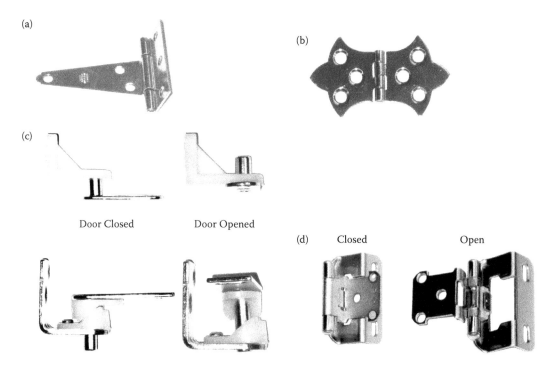

FIGURE 5.2 Examples of hinges that satisfy different constraints: (a) gate tee hinge, (b) decorative furniture hinge, (c) swinging café door hinge with gravity return, and (d) self-closing overlay cabinet hinge.

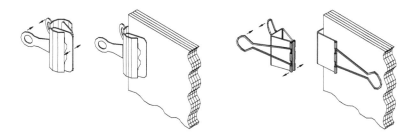

FIGURE 5.3 Two paper clips that operate with the same principle but satisfy different constraints.

vehicle size, weight, fuel economy, safety, conveniences, ride comfort, cost of ownership, to name a few.

Goal	Representative Product
High acceleration	Corvette
High fuel economy	Prius
No emissions	Tesla roadster
Luxury	Maybach
Sporty	BMW Z4
Family friendly	Odyssey
Lowest purchase price	Aveo
Small size	Smart, Polo

We see that in the context presented, design is the creation of synthesized solutions (products, processes, systems) that satisfy perceived needs through the mapping between the *FR*s (in function space) and the *DP*s (in the physical domain) through the proper selection of the *DP*s that satisfy the *FR*s. If the *FR*s change, then the solution changes; that is, a new solution must be found. Simply modifying the previous solution is unacceptable. Thus, design involves the continuous interplay between what we want to achieve (the *FR*s) and how we want to achieve it (the *DP*s) so that the final design cannot be better than the set of *FR*s and their constraints that it was created to satisfy.

5.1.3 FUNCTIONAL DECOMPOSITION AND THE AXIOMATIC APPROACH: TWO AXIOMS

In practice, the functional requirements are generated to satisfy a set of customer requirements (*CR*s). These *CR*s are then related into the *FR*s by some method, an effective one being the QFD method presented in Section 4.2. (In the QFD method, the quantities termed engineering character-istics were used to quantify the attributes of the individual *FR*s.) Additionally, after the *DP*s have been obtained, they become the inputs to the process domain, that is, the domain that determines how the *DP*s will be made. These four domains: the customer, functional, physical, and process domains—comprise the four distinct domains in the product realization process. Once the customer needs have been formalized and the set of functional requirements determined as being those that will satisfy these needs, then ideas are generated to create a product. This product is then analyzed and compared with the original set of *FR*s using the methods suggested in Chapter 6. When the product does not fully satisfy the specified *FR*s, one must either come up with a new idea or change the *FR*s to make the original needs more amenable to solution. This iterative process continues until an acceptable result is produced.

There are two design axioms that have been proposed as a means of evaluating a good design.[*,†]

Axiom 1. The Independence Axiom
   Maintain the independence of the *FR*s.

Axiom 2. The Information Axiom
   Minimize the information content of the design.

---

[*] N. P. Suh, 1990, ibid.

[†] In general, most commercial organizations are interested in optimizing the expected net present value of the future profits resulting from their products. Posing this problem formally at the design stage is quite involved for most real-life products because it involves accounting for customer preferences, competitor responses, and the influence of all possible design parameters on a product's performance and cost. Solving this optimization problem is even more challenging. Hence, to keep the design tasks tractable, most product development teams in day-to-day tasks prefer to use simple rules, such as Axiom 2 or heuristics, which are easier to utilize in comparing design options. These rules and heuristics should indi-rectly help to maximize profits, but it is by no means guaranteed that they will.

Axiom 1 deals with the relationship between functions and physical variables. It states that during the design process, as one goes from *DP*s to *FR*s the mapping must be such that a perturbation in a particular *DP* only affects its referent *FR*. When we say that a *DP* satisfies a certain *FR*, it is understood that the *FR* is satisfied subject to any specified constraints. This approach is more formal, but analogous to that employed by Pahl and Beitz[*] who state that a clear and unambiguous relationship must exist between the input and output (or cause and effect).

Axiom 2 deals with the complexity of the design. It states that among all designs that satisfy Axiom 1, the one with the minimum information content is the best. Thus, designs that integrate parts while preserving their functional independence, designs that use standard and interchangeable parts, and designs that use symmetry as much as possible will result in designs that have reduced information content. In its simplified form, information content can be considered the number of instructions that must be given in order to produce the design. Consider a standard piece of cylindrical steel bar stock of length L and diameter d. If one accepts the tolerances of d, then one only has to give instructions as to what the length should be. On the other hand, if one specifies slightly smaller dimension than d, then more instructions must be given, which may require the specification of fixturing, tooling, machine type, etc.

The notion of information content seems to be in agreement with the concept of complexity, which has been defined[†] by a complexity factor C_f expressed as

$$C_f = \left(\frac{K}{f} \right) (N_p N_t N_i)^{1/3}$$

where K is a constant of convenience, N_p is the number of parts, N_i is the number of interconnections and interfaces, N_t is the number of types of parts, and f is the number of functions that the product is expected to perform. Thus, for a given number of functions f, the complexity decreases as the number and the types of parts decrease, and the number of interfaces decreases.

Since there are potentially a very large number of designs that can satisfy a given set of FRs, Axioms 1 and 2 may be restated as follows:[‡]

Axiom 1. The Independence Axiom
   Alternate Statement 1
   An optimal design always maintains the independence of the *FR*s.
   Alternate Statement 2
   In an acceptable design, the *DP*s and the *FR*s are related in such a way that a specific *DP* can be adjusted to satisfy its corresponding *FR* without affecting other functional requirements.

Axiom 2. The Information Axiom
   Alternate Statement
   The best design is a functionally uncoupled design that has minimum information content.

Functional coupling is different than physical coupling, which is often desirable as a consequence of Axiom 2. Integration of more than one function in a single part, as long as the functions remain independent, often reduces complexity. As an example, consider a garlic press whose functional requirements are

$(FR)_1$ = Press garlic
$(FR)_2$ = Clean press

[*] G. Pahl, W. Beitz, J. Feldhusen, and K.-H. Grote, ibid.
[†] S. Pugh, ibid.
[‡] N. P. Suh, 1990, ibid.

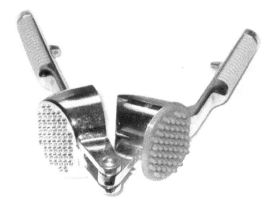

FIGURE 5.4 Example of integrating functions: a garlic press in which the means to clean the press is a part of the handle.

One device that satisfies these requirements is shown in Figure 5.4, where the means to clean (dislodge) the garlic from the press is an integral part of one handle. Notice that, because the functions are independent, the user of this device can apply either function in any order, although as a practical matter this does not happen.

5.1.4 FUNCTIONAL DECOMPOSITION AND THE AXIOMATIC APPROACH: MATHEMATICAL REPRESENTATION

We now give a brief mathematical representation of Axiom 1. We define a functional requirement vector $\{FR\}$ and a design parameter vector $\{DP\}$ as[*]

$$\{FR\} = \begin{Bmatrix} (FR)_1 \\ \vdots \\ (FR)_n \end{Bmatrix} \quad \text{and} \quad \{DP\} = \begin{Bmatrix} (DP)_1 \\ \vdots \\ (DP)_n \end{Bmatrix}$$

Then the design process involves choosing the right set of DPs such that the design equation

$$\{FR\} = [A]\{DP\}$$

is satisfied, where $[A]$ is the *design matrix* given by

$$[A] = \begin{bmatrix} A_{11} & A_{12} & \cdots & A_{1n} \\ A_{21} & A_{22} & \cdots & A_{2n} \\ \cdot & \cdot & \cdots & \cdot \\ A_{n1} & A_{n2} & \cdots & A_{nn} \end{bmatrix}$$

The vector $\{FR\}$ is what we want in terms of design goals, and $[A]\{DP\}$ is how we hope to satisfy the design's functional requirements. It has been shown[†] that the number of FRs must equal the number of DPs. The A_{ij} have two forms: (1) they take on the value of either x or 0 to indicate that a relationship (dependency) exists (x) or does not exist (0) between the FRs and the DPs, but the

---

[*] N. P. Suh, 1990, ibid.
[†] N. P. Suh, 1990, ibid.

specific relationship is not of interest; or (2) they take on a specific relationship because the governing equations between the FRs and the DPs are known and are of interest. Examples of the former type are given in Section 5.2.2 and Sections 5.2.4 to 5.2.6. An example of the latter is given in Section 5.2.3.

There are three types of solutions to the problem. The first type of solution is the one that satisfies Axiom 1 and is attained when $[A]$ is a diagonal matrix. This is called the *uncoupled* solution. For the case of three functional requirements, $n = 3$ and we have

$$\begin{Bmatrix} (FR)_1 \\ (FR)_2 \\ (FR)_3 \end{Bmatrix} = \begin{bmatrix} A_{11} & 0 & 0 \\ 0 & A_{22} & 0 \\ 0 & 0 & A_{33} \end{bmatrix} \begin{Bmatrix} (DP)_1 \\ (DP)_2 \\ (DP)_3 \end{Bmatrix}$$

The second type of solution always violates Axiom 1. In this case

$$\begin{Bmatrix} (FR)_1 \\ (FR)_2 \\ (FR)_3 \end{Bmatrix} = \begin{bmatrix} A_{11} & A_{12} & A_{13} \\ A_{21} & A_{22} & A_{23} \\ A_{31} & A_{32} & A_{33} \end{bmatrix} \begin{Bmatrix} (DP)_1 \\ (DP)_2 \\ (DP)_3 \end{Bmatrix}$$

and the solution is called *coupled*. The third type of solution is when

$$\begin{Bmatrix} (FR)_1 \\ (FR)_2 \\ (FR)_3 \end{Bmatrix} = \begin{bmatrix} A_{11} & 0 & 0 \\ A_{21} & A_{22} & 0 \\ A_{31} & A_{32} & A_{33} \end{bmatrix} \begin{Bmatrix} (DP)_1 \\ (DP)_2 \\ (DP)_3 \end{Bmatrix}$$

This solution is called a *decoupled* solution, and the independence of the FRs can be assured if we arrange the DPs in a certain order to arrive at the design matrix as shown. In this case, Axiom 1 is satisfied. These three types of solutions are shown pictorially in Figure 5.5. Examples of uncoupled, decoupled, and coupled systems are given in Section 5.2.

The design equation must be satisfied at each level in the hierarchy. However, as one proceeds deeper into the hierarchy not all FRs have to have a next lower level. This is illustrated in Section 5.2.5. Although in an uncoupled design the order in which the FRs are applied is in principle arbitrary, in many cases the satisfaction of the next higher level FR may require that a certain order be used, as illustrated in Section 5.2.4.

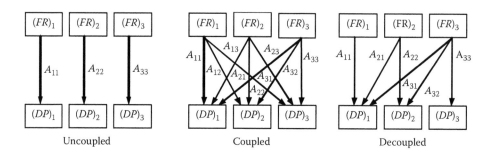

FIGURE 5.5 Graphical representations of the three types of design equations.

5.2　EXAMPLES OF FUNCTIONAL DECOMPOSITION

5.2.1　INTRODUCTION

There are two important points to make about the use of functional decomposition and the axiomatic method. The first point is that the power of the method is in its structure and overall goal: uncoupled and decoupled designs. The explicit hierarchical nature of the design matrix helps the IP^2D^2 team clarify, focus, and properly pose the product's goals (*FR*s), that is, satisfy the functional requirements that, in turn, will satisfy the customers' requirements. The second point is that the *DP*s are to specify what the physical embodiment is supposed to do, and eventually how it is to be done. It is suggested that the *FR*s and the *DP*s be formulated, for as many levels in the hierarchy as practical, without regard to how the *FR*s will be satisfied. This will have the effect of ensuring that the *FR*s will indeed be stated in a solution-neutral manner. Eventually, how the *DP*s are to be satisfied will have to be addressed. At that point, the *FR*s in the next level in the hierarchy should again be stated in a solution-neutral manner before proposing how the corresponding *DP*s will satisfy them.

　　Before proceeding with several examples, the concept of coupled/decoupled/uncoupled is briefly elaborated on using a few simple examples. Consider first the utilities provided to multi-family and single-family dwellings in large urban areas: electricity, gas, water, and telephone. Each of these is uncoupled from the other. For example, during a power outage, we still have gas, telephone, and water. If the water main breaks, then one will have gas, telephone, and electricity. However, within the dwellings these utilities are frequently coupled. If the electricity is unavailable and the hot water heater is electric, then one is unable to generate hot water. However, cold water is still available. If the electricity goes out in the winter, one is unable to provide heat to a home with either a heat pump system or a gas/oil forced air heating system, because the blower cannot work.

　　We return to the example of the common two-knob water faucet. It was stated that the design objective of a water faucet is to provide water continuously at a desired flow rate and at a desired temperature when the hot and cold water are controlled separately. In terms of the axiomatic design procedure, the *FR*s for the two knob system as stated previously are

$(FR)_1$ = Obtain water flow rate
$(FR)_2$ = Obtain water temperature

and the corresponding *DP*s are

$(DP)_1$ = Means to adjust cold water flow
$(DP)_2$ = Means to adjust hot water flow

The design equation is then given by

$$\left\{ \begin{array}{c} (FR)_1 \\ (FR)_2 \end{array} \right\} = \left[\begin{array}{cc} x & x \\ x & x \end{array} \right] \left\{ \begin{array}{c} (DP)_1 \\ (DP)_2 \end{array} \right\}$$

This equation illustrates what everyone knows, that to obtain the desired water flow rate and temperature one must adjust the hot and cold water amounts simultaneously (or iteratively). In other words, the two-handed faucet is a coupled system. According to the axiomatic approach, a better solution is obtained by re-visiting the situation without dictating *how* it is done. The functional requirements remain the same. However, we reformulate the *DP*s without specifying how it is to be done as follows:

$(DP)_1$ = Water flow regulating device
$(DP)_2$ = Water temperature regulating device

The solution in the physical domain must satisfy the conditions (constraints) that the water is to flow continuously and, second, that the hot and cold water are provided separately. The design equation becomes

$$\left\{ \begin{array}{c} (FR)_1 \\ (FR)_2 \end{array} \right\} = \left[\begin{array}{cc} x & 0 \\ 0 & x \end{array} \right] \left\{ \begin{array}{c} (DP)_1 \\ (DP)_2 \end{array} \right\}$$

That is, the design goal is to be able to control the water temperature independently of the water volume. This is the design equation for all single-handle faucets.

A third example of how one goes about separating (uncoupling) functions is to consider the design of Hewlett Packard's first low-cost plotter. Most plotter designs at that time used a mechanism to move the plotter pen in both the x and y directions; that is, the x-axis motor, say, had also to move the y-axis frame, motor, and pen assembly. Because this mechanism was fairly massive, it required large motors to move pen quickly over the paper. Hewlett Packard recognized, however, that the function of movement did not necessarily have to be concentrated in the pen assembly. Instead, they decided to move the paper in one axis and the pen in the other. Now the pen only had to move, say, from left to right, while the paper moved up and down. Since the paper was light in weight it required smaller motors to move it, and it could move quickly. The overall result was a fast-drawing, inexpensive plotter.

Another example of uncoupling is to consider the character manipulation ability of word processor programs. Here, on a character-by-character basis one can independently change its typeface (Arial, Times Roman, etc.), its size (4–48 points), and its style with any combination of bold, italic, and underline.

As a final example, we consider the functional requirements for a system that safely removes a crew member from an in-flight fighter aircraft. The requirements and sequence of events that are required are as follows:

(FR) = Safely remove an aircrew member from an aircraft in flight
    $(FR)_1$ = Initiate egress
        $(FR)_{11}$ = Initiate ejection
        $(FR)_{12}$ = Disconnect aircrew member from services ($< t_o$)
        $(FR)_{13}$ = Obtain proper body position ($< t_y$)
    $(FR)_2$ = Aircrew member extraction ($< t_{esc}$)
        $(FR)_{21}$ = Create escape path
        $(FR)_{22}$ = Separate aircrew member from aircraft
        $(FR)_{23}$ = Propel aircrew member through escape path
    $(FR)_3$ = Recover aircrew member
        $(FR)_{31}$ = Orient/stabilize aircrew member
        $(FR)_{32}$ = Reduce horizontal airspeed
        $(FR)_{33}$ = Reduce vertical airspeed

The corresponding design parameters are

(DP) = System to safely remove an aircrew member from an aircraft in flight
    $(DP)_1$ = Means to initiate egress
        $(DP)_{11}$ = Means to initiate egress
        $(DP)_{12}$ = Means to disconnect services
        $(DP)_{13}$ = Means to orient body position
    $(DP)_2$ = Means to extract aircrew member ($< t_{esc}$)
        $(DP)_{21}$ = Means to create escape path

$(DP)_{22}$ = Means to separate aircrew member from aircraft
$(DP)_{23}$ = Means to propel aircrew member through escape path
$(DP)_3$ = Means to recover aircrew member
$(DP)_{31}$ = Means to orient/stabilize aircrew member
$(DP)_{32}$ = Means to reduce horizontal airspeed
$(DP)_{33}$ = Means to reduce vertical airspeed

We now clarify several aspects of the use of this method.

5.2.1.1 Functional Independence versus Integration versus Modularity

To distinguish between functional independence (separateness) and integration and modularity, consider an automobile door. A typical door contains a means of opening, closing and locking it, a window with its opening and closing mechanism, a loudspeaker, and an arm rest. Each of these systems is essentially a module that can be used independently of each other, irrespective of whether the door is open or closed. In addition, the door itself provides structural support and protection from weather and noise, and its visible surfaces are aesthetically pleasing.

5.2.1.2 Phrasing of the Functional Requirements

One should be careful of the phrasing of the functional requirements so that the *FR* is not stated in a way that inadvertently drives the solution. For example, consider the following three versions of a *FR*: (1) "Tow a disabled automobile from one location to another," (2) "Transport a disabled automobile from one location to another," and (3) "Move a disabled automobile from one location to another." The word *tow* in the first version seems to imply a pulling or towing operation. The second version seems to imply a carrying of the automobile. The third version is the most neutral; it does not connote how one should do it. The phrasing of the functions in a solution neutral manner also has the advantage that it doesn't bias the search for concepts that can satisfy the functional requirement. Consider the phrasing of the task: how to raise a bridge in order for boats of all heights to pass through. A better way to state this is to ask how one creates a structure to let through boats of all heights. Maybe a better solution will be to rotate the bridge, or to build a tunnel, or even to submerge the bridge below the boats.

5.2.1.3 Physical Coupling

There are situations in which the *FR*s may not be able to be met because of competing physical laws; that is, the functional requirements are coupled by physical laws. For example, a *FR* that asks for a transducer that has good sensitivity over a wide frequency range may be difficult to attain. Good sensitivity usually requires a high degree of response to the stimulus. Suppose that one idea is to attach a strain gage to a cantilever beam-like structure to measure a dynamic force. In this case, the beam's stiffness k and its first natural frequency f_n are, respectively, proportional to

$$ k \propto \left(\frac{h}{L}\right)^2 \quad \text{and} \quad f_n \propto \frac{h}{L^2} $$

where h is the height of the beam and L its length. It is seen that choosing h and L to give a low stiffness (high sensitivity) also lowers f_n, making it difficult to simultaneously achieve a wide frequency range.

5.2.2 EXAMPLE 1—CARTON TAPING SYSTEM

Consider a family of cartons whose sizes vary over a modest range. A typical carton is shown in Figure 5.6. Prior to the cartons arriving at the taping system their orientation may not be correctly

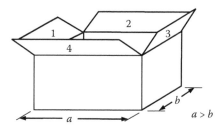

FIGURE 5.6 Carton orientation and nomenclature.

aligned in the horizontal plane. However, it can be assumed that the side (top) to be sealed is always oriented as shown in the figure and that one of the two flaps (1 or 3) that is to be covered by the other two flaps perpendicular to them (2 and 4) is always first to enter the system. The cartons arrive at the sealing system on a horizontal conveyor system. The bottom of the carton has been previously sealed. The cartons arrive at the taping station at irregular intervals.

The objective is to determine the design equation from which a family of automatic carton taping systems can be developed. The functional statement at the highest level of the hierarchy is to develop an automatic carton taping system subject to the constraints implied in the previous paragraph. The corresponding design parameter is a means to tape the carton. The functional requirements at the next level are now given. Functional requirements $(FR)_{12}$, $(FR)_{15}$, and $(FR)_{16}$ do not require another level.

$(FR)_{11}$ = Place one carton into the system
$(FR)_{12}$ = Maintain position of carton
$(FR)_{13}$ = Close the carton's flaps
$(FR)_{14}$ = Tape carton
$(FR)_{15}$ = Release carton
$(FR)_{16}$ = Remove sealed box from system

The corresponding design parameters are:

$(DP)_{11}$ = means to place carton into system
$(DP)_{12}$ = means to maintain position of carton
$(DP)_{13}$ = device to close the carton's flaps
$(DP)_{14}$ = taping mechanism
$(DP)_{15}$ = means to release carton
$(DP)_{16}$ = carton removal device

The design equation is

$$
\begin{Bmatrix}
(FR)_{11} \\
(FR)_{12} \\
(FR)_{13} \\
(FR)_{14} \\
(FR)_{15} \\
(FR)_{16}
\end{Bmatrix}
=
\begin{bmatrix}
x & 0 & 0 & 0 & 0 & 0 \\
x & x & 0 & 0 & 0 & 0 \\
x & x & x & 0 & 0 & 0 \\
x & x & x & x & 0 & 0 \\
x & x & x & x & x & 0 \\
x & x & x & x & x & x
\end{bmatrix}
\begin{Bmatrix}
(DP)_{11} \\
(DP)_{12} \\
(DP)_{13} \\
(DP)_{14} \\
(DP)_{15} \\
(DP)_{16}
\end{Bmatrix}
$$

The next level in the hierarchy for $(FR)_{11}$ is

$(FR)_{111}$ = Orient carton
$(FR)_{112}$ = Propel one carton into the system

with the corresponding design parameters

$(DP)_{111}$ = Carton orienting device
$(DP)_{112}$ = Carton insertion (forward motion) device

The design equation is

$$\left\{ \begin{array}{c} (FR)_{111} \\ (FR)_{112} \end{array} \right\} = \left[\begin{array}{cc} x & 0 \\ x & x \end{array} \right] \left\{ \begin{array}{c} (DP)_{111} \\ (DP)_{112} \end{array} \right\}$$

The next level in the hierarchy for $(FR)_{13}$ is

$(FR)_{131}$ = Close flap 1 and hold
$(FR)_{132}$ = Close flap 3 and hold until either flap 2 or 4 starts to close
$(FR)_{133}$ = Close flap 4 and hold until taping is completed
$(FR)_{134}$ = Close flap 2 and hold until taping is completed

One of the objectives, however, in defining the FRs is to define a minimum number of them. Consequently, in this situation, we can close flaps 1 and 3 simultaneously and close flaps 4 and 2 simultaneously. Also, from Axiom 2, the devices for the satisfaction of $(FR)_{132}$ and $(FR)_{133}$ should be the same. Thus, it seems reasonable that the previous functional requirements can be restated as

$(FR)_{131}$ = Close flap 1 and 3 simultaneously and hold until flaps 2 and 4 start to close
$(FR)_{132}$ = Close flap 2 and 4 simultaneously and hold until taping is completed

The corresponding design parameters are

$(DP)_{131}$ = Flap closing system
$(DP)_{132}$ = Flap closing system

The design equation is

$$\left\{ \begin{array}{c} (FR)_{131} \\ (FR)_{132} \end{array} \right\} = \left[\begin{array}{cc} x & 0 \\ x & x \end{array} \right] \left\{ \begin{array}{c} (DP)_{131} \\ (DP)_{132} \end{array} \right\}$$

The next level in the hierarchy for $(FR)_{14}$ is

$(FR)_{141}$ = Tape box
$(FR)_{142}$ = Sever tape

and the corresponding design parameters are

$(DP)_{141}$ = Taping mechanism
$(DP)_{142}$ = Tape severing device

The design equation is

$$\left\{ \begin{array}{c} (FR)_{141} \\ (FR)_{142} \end{array} \right\} = \left[\begin{array}{cc} x & 0 \\ x & x \end{array} \right] \left\{ \begin{array}{c} (DP)_{141} \\ (DP)_{142} \end{array} \right\}$$

It is seen that all the design equations are decoupled. In examining the functional requirements, we see that the phrases "one carton into…," "hold flaps in place…," and "hold until taping…" are constraints imposed on the various functional requirements. Notice that the design parameters neither specify nor indicate how the functional requirements will be satisfied; they merely state what type of device or system will be required.

From Axiom 1, we note that the overall taping system should be independent of the system moving the boxes to the taping system. Finally, it may be possible to eliminate $(FR)_{14}$ and combine it with $(FR)_{13}$ depending on whether the taping mechanism moves with respect to box or the taping mechanism is stationary and the box moves relative to it. In view of Axiom 1, which states that we should have a minimum number of independent FRs, the latter approach is better.

The decomposition and mapping for the carton taping system are shown in Figure 5.7.

5.2.3   Example 2—Intelligent V-Bending Machine[*]

The objective is to develop a procedure that produces a curved metal part of constant thickness from a thin, flat sheet of metal. The generation of the means to satisfy this objective is governed by certain physical laws, and this example has been chosen for this reason. The previously stated objective is the FR at the highest level. The corresponding DP is a procedure that produces the curved part. The process that we shall use for sheet metal bending is the V-bending process. Then, at the next level in the hierarchy, the FR is

FR = Produce a bend angle $\theta_f \pm \Delta\theta_f$ using sheet metal bending,
        regardless of how the material and thickness properties vary

The corresponding DP is

DP = System to generate and control the bend angle

The V-bending process is schematically illustrated in Figure 5.8a and its corresponding typical bending moment M versus bending angle θ is shown in Figure 5.8b. As seen in Figure 5.8b, the moment M_o is sufficiently high so that it causes the plate to undergo permanent deformation at the corresponding bend angle θ_o. Corresponding to θ_o is a displacement X_a under the applied force F_o. When M_o is released, however, there is a certain amount of spring-back to a bend angle $\theta_f < \theta_o$. Corresponding to θ_f is a displacement ΔX_a, which is the amount of permanent deformation under the point where the force was applied. From classical beam/plate theory, it is known that $X_a \sim \theta \sim F/EI$, $M \sim F$ and, therefore, $M/\theta \sim EI$, where E is the Young's modulus and I is the moment of inertia of the cross section.

Thus, for a fixed method of supporting the beam/plate, we have that

$$M_o = F_o d/2$$

$$\theta_o = \tan^{-1}(X_a/d)$$

$$\theta_f = \tan^{-1}(\Delta X_a/d) \tag{5.1}$$

---

[*] N. P. Suh, 1990, ibid.

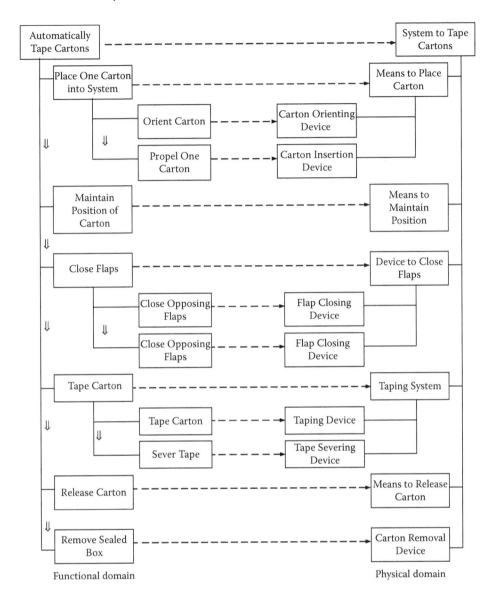

FIGURE 5.7 Mapping of the carton taping functional requirements to the physical domain. The arrows (⇓) indicate the order in which the *FR*s must be performed, since the design equation is decoupled.

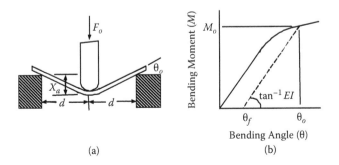

FIGURE 5.8 (a) Geometry of a V-bending device and (b) bending angle versus bending moment.

If we are able to measure F_o and X_a (and, consequently, ΔX_a), then we have a means of controlling the process.

Referring to Figure 5.8b, it is seen that

$$\tan^{-1} EI = \frac{M_o}{\theta_o - \theta_f} \tag{5.2}$$

From Equations 5.1 and 5.2, we obtain

$$\theta_f = \theta_o - \frac{M_o}{\tan^{-1} EI} = \tan^{-1}(X_a/d) - \frac{F_o d/2}{\tan^{-1} EI} \tag{5.3}$$

It is seen from these equations that we have three independent parameters: M_o, which is proportional to the applied force F_o; EI, which is a function of a physical property of the material and the cross-sectional dimensions of the plate; and θ_f, which is the resulting bend angle of the plate after the release of M_o. Therefore, in order to obtain θ_f, we have the following three *FRs*:

$$(FR)_1 = M_o \quad \text{(Generate moment)}$$

$$(FR)_2 = \theta_o \quad \text{(Bend and deform metal)}$$

$$(FR)_3 = \theta_f \quad \text{(Release to final bend angle)}$$

In this case, the *DPs* are not explicitly chosen since they are governed by Equation 5.1. Thus, the design equation becomes

$$\begin{Bmatrix} M_o \\ \theta_o \\ \theta_f \end{Bmatrix} = \begin{bmatrix} 1 & 0 & 0 \\ 0 & 1 & 0 \\ 0 & 1 & -1 \end{bmatrix} \begin{Bmatrix} F_o d/2 \\ \tan^{-1}(X_a/d) \\ F_o d/(2\tan^{-1} EI) \end{Bmatrix}$$

In order to implement this design equation, the following procedure is employed. The punch is brought down and the plate is subjected to a force F_o', which results in a displacement under it of X_a'. The punch is removed and $\Delta X_a'$ is measured. From these three measurements, we use Equations 5.1 and 5.2 to determine $\tan^{-1} EI$. We now apply a slowly increasing force F_o and continuously monitor F_o and X_a until their values produce the desired θ_f as computed from Equation 5.3.

5.2.4 EXAMPLE 3—HIGH-SPEED IN-PRESS TRANSFER MECHANISM[*]

There are products that are, or can be, produced by a sequence of stamping operations within a press bed. On the press bed, there is placed a set of progressively shaped dies and punches that form the part to its final shape. At each cycle of the press, each part at each die location is simultaneously subjected to the press action. After each cycle of operation of the press, the part has to be removed from the die and placed into the adjacent die. Each removal and placement occurs simultaneously

[*] L. W. Tsai, E. B. Magrab, and Y. J. Ou, "Just-In-Time Transfer Mechanism," Department of Mechanical Engineering, University of Maryland, Final report to M. S. Willett, Co., Cockeysville, MD, October 1991.

at each die location. The devices that remove the parts from each die after each operation are called transfer mechanisms. The mechanism lifts each part out of each die to the same height and moves it forward the same distance, which is called the *pitch distance* and is the distance between the centers of adjacent die. Typically, the transfer mechanism straddles the die and operates with synchronized pairs of opposing grippers at each die. Only the shape of the gripping mechanism and its offset (or standoff, the distance from the gripper to the part) may be different at each die, depending on the shape of the part at the end of each operation. The raw material, usually a roll of metal stock, is fed in at one end and a formed (or partially formed) product leaves at the other end. Typical products that are made this way are the tops of pull-tab soda cans, door hinges, and door knobs.

A company that specializes in making transfer mechanisms wants to broaden its product line by greatly increasing the pitch to handle a larger family of products, such as automotive engine oil pans. At the highest level in the hierarchy, we have that the functional requirement is to pick and place a part from a set of die within a press under the following constraints:

- All timing is with respect to the motion of the press.
- Mechanism must be comprised of mechanical elements—no hydraulics, servomechanisms, and pneumatics.
- Must work on press beds up to 4 m long.
- Small enough to fit on press bed alongside the die sets.

The corresponding *DP* is a device that grips the parts from each die simultaneously and places them in the adjacent die.

At the next level in the hierarchy, the functional requirements are the following:

$(FR)_1$ = Grip part with a stand-off of either 10, 15, or 20 cm
$(FR)_2$ = Lift part over a continuously adjustable range of 6 to 10 cm
$(FR)_3$ = Move each part forward over a continuously adjustable range of 20 to 50 cm ±0.5 mm
        at a rate of at least 10 per minute
$(FR)_4$ = Release the part into the die
$(FR)_5$ = Return gripper to its original position

The corresponding design parameters are the following:

$(DP)_1$ = Gripping mechanism
$(DP)_2$ = Lifting mechanism
$(DP)_3$ = Pitch mechanism
$(DP)_4$ = Release mechanism
$(DP)_5$ = Return mechanism

The design equation is

$$
\begin{Bmatrix} (FR)_1 \\ (FR)_2 \\ (FR)_3 \\ (FR)_4 \\ (FR)_5 \end{Bmatrix} = \begin{bmatrix} x & 0 & 0 & 0 & 0 \\ x & x & 0 & 0 & 0 \\ x & x & x & 0 & 0 \\ x & x & x & x & 0 \\ x & x & x & x & x \end{bmatrix} \begin{Bmatrix} (DP)_1 \\ (DP)_2 \\ (DP)_3 \\ (DP)_4 \\ (DP)_5 \end{Bmatrix}
$$

Note that the gripping, lifting, and pitching motions must be independent of each other. However, they also must occur in the order given; consequently, we have a decoupled design and these three functions cannot be considered independently of the remaining two functional requirements.

5.2.5   EXAMPLE 4—DRYWALL TAPING SYSTEM

The three preceding examples illustrated uncoupled and decoupled designs. The example considered in this section is chosen to illustrate that a coupled design can sometimes be the design objective.

In the construction of interior walls of dwellings, the most frequently used method is to fabricate them from sheets of gypsum board, which is called *drywall*. After these sheets are attached to the wall, there is a seam formed wherever the sheets abut each other. After the drywall has been put in place, the next phase is to fill the seams with a joint compound so that after the wall has been painted the wall appears as one smooth surface.

The drywall plastering process is usually done in three steps: tape coat, block (or bed) coat, and the skim coat. Typically, each step after the first step is performed after a minimum of 24 hours have elapsed. The traditional installation method is to apply a layer of joint compound in the drywall joint, and then to apply the paper tape. The tape prevents the compound from cracking after it has dried. After the tape has been placed over the joint compound a blade is swept over the length of the seam to simultaneously smooth it and to imbed the tape into the compound. The tape is typically 5 cm in width and comes in rolls 60 m in length. Once the joint is dry, the block coat is applied. The block coat is the major joint filling portion of the process. Most of the time, after it has dried, the block coat is sanded to a semismooth finish. A skim coat is applied next using a somewhat diluted compound. The skim coat is applied to level the wall surface, since the compound shrinks as it dries, and to fill in any holes caused by air bubbles. The skim coat is spread over a much wider area than the immediate joint region to make the joint smooth and continuous with the rest of the drywall.

The objective is to develop a device that permits a drywall finisher to perform the first step of the process in one motion: to apply the joint compound and the tape simultaneously to the drywall seam leaving a smooth surface. The key word in this functional statement is *simultaneously*, for it demands that as the tape is dispensed, the compound is applied in the right amount.

The FRs at the next level of the design process are

$(FR)_1$ = Load joint compound
$(FR)_2$ = Load tape
$(FR)_3$ = Dispense tape and joint compound uniformly and smoothly to wall
$(FR)_4$ = Unload tape
$(FR)_5$ = Unload joint compound

The corresponding DPs are

$(DP)_1$ = Joint compound loading system
$(DP)_2$ = Tape holding device
$(DP)_3$ = Tape and joint compound dispensing device
$(DP)_4$ = Tape unloading device
$(DP)_5$ = Joint compound removal system

The functional requirement $(FR)_3$ has a next level, which is decomposed to

$(FR)_{31}$ = Dispense tape simultaneously with joint compound
$(FR)_{32}$ = Provide joint compound to dispenser
$(FR)_{33}$ = Dispense joint compound uniformly

$(FR)_{34}$ = Apply tape and joint compound smoothly to wall
$(FR)_{35}$ = Separate tape from roll

The corresponding design parameters are

$(DP)_{31}$ = Tape dispensing device
$(DP)_{32}$ = Means to move joint compound through system
$(DP)_{33}$ = Joint compound dispenser
$(DP)_{34}$ = Tape and joint compound wall application device
$(DP)_{35}$ = Tape severing device

The design equation is

$$
\begin{Bmatrix} (FR)_{31} \\ (FR)_{32} \\ (FR)_{33} \\ (FR)_{34} \\ (FR)_{35} \end{Bmatrix} = \begin{bmatrix} x & x & 0 & 0 & 0 \\ x & x & 0 & 0 & 0 \\ 0 & x & x & 0 & 0 \\ x & 0 & 0 & x & 0 \\ x & 0 & 0 & x & x \end{bmatrix} \begin{Bmatrix} (DP)_{31} \\ (DP)_{32} \\ (DP)_{33} \\ (DP)_{34} \\ (DP)_{35} \end{Bmatrix}
$$

Several concepts for the $(DP)_{ij}$ that satisfy this equation are given in Table 6.4.

It is instructive to reformulate the design equation when the requirement that the tape and joint compound be dispensed simultaneously is removed. In this case, $(FR)_{31}$ is no longer coupled to $(DP)_{32}$, and we can rearrange the above design equation to obtain

$$
\begin{Bmatrix} (FR)_{32} \\ (FR)_{33} \\ (FR)_{31} \\ (FR)_{34} \\ (FR)_{35} \end{Bmatrix} = \begin{bmatrix} x & 0 & 0 & 0 & 0 \\ x & x & 0 & 0 & 0 \\ 0 & 0 & x & 0 & 0 \\ 0 & 0 & x & x & 0 \\ 0 & 0 & x & x & x \end{bmatrix} \begin{Bmatrix} (DP)_{32} \\ (DP)_{33} \\ (DP)_{31} \\ (DP)_{34} \\ (DP)_{35} \end{Bmatrix}
$$

which results in the following independent design equations:

$$
\begin{Bmatrix} (FR)_{32} \\ (FR)_{33} \end{Bmatrix} = \begin{bmatrix} x & 0 \\ x & x \end{bmatrix} \begin{Bmatrix} (DP)_{32} \\ (DP)_{33} \end{Bmatrix}
$$

and

$$
\begin{Bmatrix} (FR)_{31} \\ (FR)_{34} \\ (FR)_{35} \end{Bmatrix} = \begin{bmatrix} x & 0 & 0 \\ x & x & 0 \\ x & x & x \end{bmatrix} \begin{Bmatrix} (DP)_{31} \\ (DP)_{34} \\ (DP)_{35} \end{Bmatrix}
$$

It is seen that the two functions—to dispense the joint compound and to dispense the tape—are indeed uncoupled from each other. Hence, two separate and uncoupled devices will satisfy this equation: one device to apply the joint compound and one to apply the tape.

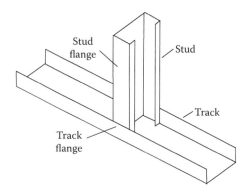

FIGURE 5.9 Steel stud and track framing system.

5.2.6 EXAMPLE 5—STEEL FRAME JOINING TOOL

Wood frame construction has dominated the home building industry for decades. However, the increasing cost of wood has the home building industry looking for alternative materials and methods in an attempt to reduce costs. Since the early 1990s, the use of thin gauge steel tracks and studs has been introduced as a replacement for wood. The track and stud are C-section beams made of cold-rolled steel as shown in Figure 5.9. Their thickness varies from 0.45 to 0.68 mm for partition wall members (nonload bearing), from 0.83 to 1.09 mm for structural studs, and from 1.09 to 1.37 mm for joists and headers. The advantages of steel over wood are its consistent strength and its termite and fire resistance. One of its disadvantages is its high thermal conductivity.

The current method of securing the track to the stud is by means of self-tapping screws, using standard portable electric screwdrivers. Since there is a tendency of the screw to push the stud flange away from the track flange after the screw has passed through the track flange, the mechanic has to first clamp the two elements together. The clamp placement, screwing, and clamp removal take around 25 seconds to complete. What is desired is a portable, manually-operated joining system that is as fast as a electric/pneumatic nail gun, not more fatiguing to operate, and provides connections as strong as the screws currently used. A secondary consideration is to be able to disassemble the joint when an error has been made in the location of the stud.

The functional requirement at the highest level is to join a steel stud to a steel track such that it results in a joint able to withstand 2000 N of separation force and can be joined in less than 5 seconds. The corresponding design parameter is a tool to join a steel stud to a steel track. The functional requirements at the next level are

$(FR)_1$ = Maintain position of stud and track flanges
$(FR)_2$ = Manually position/orient tool
$(FR)_3$ = Execute joining operation
$(FR)_4$ = Manually remove/disengage tool

The corresponding design parameters are

$(DP)_1$ = Means to maintain track and stud flange position
$(DP)_2$ = Means to give tool structure
$(DP)_3$ = Means to join steel stud to track
$(DP)_4$ = Means to separate/disengage tool from track and stud

The first functional requirement $(FR)_1$ may or may not be required depending on the fastening method chosen. In the present method of screwing the track and stud together, a clamp is used. The fourth functional requirement $(FR)_4$ reminds us that if the fastening technique incorporates a clamping type action, then this action must be disengaged prior to removing the tool.

The design equation is

$$
\begin{Bmatrix} (FR)_1 \\ (FR)_2 \\ (FR)_3 \\ (FR)_4 \end{Bmatrix} = \begin{bmatrix} x & 0 & 0 & 0 \\ x & x & 0 & 0 \\ x & x & x & 0 \\ x & x & x & x \end{bmatrix} \begin{Bmatrix} (DP)_1 \\ (DP)_2 \\ (DP)_3 \\ (DP)_4 \end{Bmatrix}
$$

The functional requirements $(FR)_2$ and $(FR)_3$ are decomposed to their next level. For $(FR)_2$, the decomposition is

$(FR)_{21}$ = Provide structure for tool's components
$(FR)_{22}$ = Provide user grip
$(FR)_{23}$ = Provide user protection

The corresponding design parameters are

$(DP)_{21}$ = Tool structure
$(DP)_{22}$ = Means to grip tool
$(DP)_{23}$ = Means to protect user

For $(FR)_3$, the functional requirements are

$(FR)_{31}$ = Generate energy
$(FR)_{32}$ = Convert energy to working element
$(FR)_{33}$ = Make connection
$(FR)_{34}$ = Disconnect/release energy

The corresponding design parameters are

$(DP)_{31}$ = Means to generate energy
$(DP)_{32}$ = Means to convert energy to fastening system
$(DP)_{33}$ = Means to fasten steel stud and track
$(DP)_{34}$ = Means to disconnect/release/remove energy

For $(FR)_3$, the design equation is

$$
\begin{Bmatrix} (FR)_{31} \\ (FR)_{32} \\ (FR)_{33} \\ (FR)_{34} \end{Bmatrix} = \begin{bmatrix} x & 0 & 0 & 0 \\ x & x & 0 & 0 \\ x & x & x & 0 \\ x & x & x & x \end{bmatrix} \begin{Bmatrix} (DP)_{31} \\ (DP)_{32} \\ (DP)_{33} \\ (DP)_{34} \end{Bmatrix}
$$

Further discussion of this example appears in Section 6.4.

6 Product Concepts and Embodiments

Several techniques are introduced that can be used to obtain and evaluate candidate product concepts, which are then converted into physical entities that satisfy the customers' requirements.

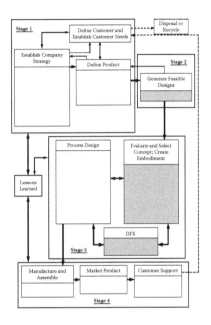

6.1 INTRODUCTION

In this chapter, techniques are introduced that can be used to generate the widest possible set of solutions on how the various functional requirements at each level in the hierarchy are going to be satisfied. Then, one means is described that can be used to evaluate the various solutions according to how well they satisfy the important items in Table 2.2, and the most important customer requirements that are recorded in the house of quality. The purpose of the evaluation is to ensure that the proposed means of satisfying the FRs also satisfy the customer requirements and the constraints on the FRs.

In Chapter 5, the axiomatic method was introduced as one means of performing functional decomposition. In this approach, the product is the physical entity (DP) that satisfies the highest level FR. At the lower levels in the hierarchy, the DPs can be thought of as modules, components, or units whose purpose is to satisfy their corresponding FR. We recall from Section 5.1.2 that corresponding to each DP there is a concept (specific principle, method, or means) that will be used to satisfy the referent FR. There are three stages that one goes through to arrive at the product's final form: (1) concept generation and evaluation; (2) determination of the configurations (spatial relationships) of the functional units with respect to each other and the configurations of the components comprising each unit; and (3) determination of the physical form (embodiment) that embraces the concept(s), as employed in their units, and their most suitable configuration. The second and third stages are usually implemented in an overlapping, integrated, and iterative manner.

The last stages of the process, which determine the physical form that each DP will take at each level in the hierarchy, involves the consideration of many aspects and can influence the configurations previously explored. Examples of what aspects may influence the final embodiment are keeping a temperature-sensitive module from a heat-producing one; meeting mandated safety requirements; satisfying aesthetics and ergonomic requirements such as balance and gripping surface; meeting certain restrictions inherent in a specific technology, such as size and shape; meeting strength requirements; providing accessibility for assembly, disassembly, and service; meeting producibility requirements in order to use a certain manufacturing process; and obtaining operational efficiency.

To illustrate the distinctions between concept, configuration, and embodiment and their impact on each other, consider the common portable hair dryer. The functional requirements are straightforward: draw in air, heat the air, and discharge the air in a directed stream. The design equation is

$$\left\{ \begin{array}{c} \text{Draw air} \\ \text{Heat air} \\ \text{Discharge air} \end{array} \right\} = \left[\begin{array}{ccc} x & 0 & 0 \\ x & x & 0 \\ x & x & x \end{array} \right] \left\{ \begin{array}{c} \text{Means to draw air} \\ \text{Means to heat air} \\ \text{Means to discharge air} \end{array} \right\}$$

The various systems that can perform these tasks are familiar to most everyone. In Figure 6.1, several different configurations and embodiments from several different manufacturers are shown. The overall shapes vary, the types of fan vary, the locations and shapes of the intake ports vary, and the shapes and the sizes of the heating elements vary. Each of the five embodiments has the same functional modules: an electrically operated heating element, an electric motor, a fan, a case (structure) to contain and support the modules, and openings in the case for the air intake and for the exhaust (heated) air. However, the embodiments of each heating element are different, from the shape of the electric element itself (either a bent wire or coiled like a spring), to how the element is spatially oriented, and to the way the element is supported within the heating module.

There are three hair dryers that have centrifugal fans (Figure 6.1a,b,c) and two that have axial fans (Figure 6.1d,e). The dryers with centrifugal fans each have a fan with a different shape. In addition, they have their air intakes on the side of the case, whereas those with axial fans have their intakes in the rear. The configuration of the motors is different in three of the five dryers. The dryers with the axial fans have their motors in-line with the flow, whereas in the dryers with centrifugal fans the motors are perpendicular to the flow. In addition, for one of the dryers with a centrifugal fan (Figure 6.1b) the motor is exterior to the fan blades.

6.1.1 INITIAL FEASIBILITY ANALYSIS

During the design's concept evaluation process, one must ascertain that the proposed solution (concept) is feasible by determining that the important performance attributes and the major constraints that have been imposed on that functional requirement can be met by that concept. The ability to quickly obtain the order of magnitude of the required performance metrics is needed to prevent the IP²D² team from pursuing a solution path that is either not feasible or impractical. However, at this stage of the evaluation process, a back-of-the-envelope type of analysis is usually sufficient. There are several approaches that one can use to perform this type of calculation. The first is when one or more members of the IP²D² team have sufficient knowledge to do this. The second way is for the team to refer to an appropriate handbook to find the fundamental formulas that permit quick calculations. (There seems to be one handbook for almost every aspect of engineering.) A third approach that can be used is attributed to Enrico Fermi.[*] In general, a Fermi problem requires knowledge of facts not mentioned in the statement of the problem. However, when the problem is broken down into sub problems, each one answerable without the help of experts or reference books, an estimate can be made that comes remarkably close to the exact solution. Suppose, for example, that one wants to determine the Earth's circumference without looking it up. It is known that New York and Los Angeles are separated by about 5,000 kilometers and that the time difference between the two coasts is 3 hours. These 3 hours correspond to one eighth of a day, and a day is the time it takes earth to complete one revolution. Thus, its circumference must be $8 \times 5,000 = 40,000$ km, an answer that differs from the true value at the equator of 39,843 km by less than 0.4%.

Fermi's intent was to show that, although at the outset the answer's order of magnitude is unknown, one can proceed on the basis of different assumptions and still arrive at estimates that fall within range of the answer. The reason is that, in any string of calculations, errors tend to cancel one another so that the final results will converge toward the right number.

---

[*] H. C. Von Baeyer, *The Fermi Solution*, Random House, Inc., New York, 1993.

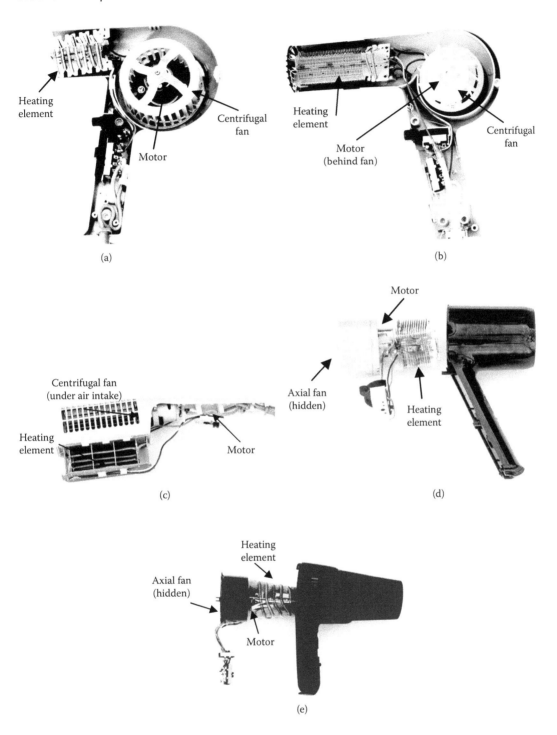

FIGURE 6.1 Five embodiments of portable hair dryers: embodiments (a) to (c) employ centrifugal fans and (d) and (e) axial fans.

TABLE 6.1
Well-to-Wheel Comparisons of Different Automotive Engines with Respect to Efficiency, CO_2 Emissions, and Acceleration[a]

Engine Type	CO_2 Emissions (g/km)	Efficiency (km/MJ)	Acceleration 0–96 km/h (s)
Electric (Tesla prototype)	46.1	1.14	3.9
Natural gas (Honda CNG)	166.0	0.32	12.0
Diesel (VW Jetta)	152.7	0.48	11.0
Gasoline (Honda Civic VX)	141.7	0.51	9.4
Hybrid (Toyota Prius)	130.4	0.56	10.3
Hydrogen fuel cell (Honda FCX)	151.7	0.35	15.8

[a] Data taken from Eberhard and M. Tarpenning, ibid.

6.1.2 ESTIMATION EXAMPLE 1

We now illustrate what we mean by approximating several performance parameters by estimating the amount of joint compound that is required by the drywall taping system described in Section 5.2.5, the rate at which it has to be dispensed, and the force required to dispense it. If we assume that an average size room is 3.6 m × 4.8 m × 2.4 m and each sheet of drywall is 1.2 m × 2.4 m, then about 20 sheets are required. If each sheet has the joint compound applied to one of its short edges and one of its long edges, then each sheet requires 3.6 m of tape and joint compound, and therefore, the room requires 72 m. If the tape is 63 mm wide and the joint compound will be applied in a layer 3.2 mm thick, then the total amount needed to do one room is 0.015 m³, or about 15 L. Furthermore, if it takes 15 s to tape a 2.4 m seam, then the joint compound must be able to flow at the rate of 32.3 cm³/s (= 240 × 6.3 × 0.32/15). The force on the joint compound to maintain this flow rate through an orifice 3.2 × 63 mm is about 1.6 N/cm. This value was obtained by determining that the velocity is 16 cm/s, and by assuming that the viscosity of the joint compound is 1000 Ns/m².

6.1.3 ESTIMATION EXAMPLE 2

In some cases, it is possible to use available data to estimate if it is worthwhile to pursue a candidate concept. Let us suppose that we would like to create an electric sports car* that is very efficient, produces very low emissions, and has very high acceleration. We need to compare this idea with competing technologies. The metrics that we shall use are the well-to-wheel energy efficiency and the well-to-wheel CO_2 emissions. The well-to-wheel energy efficiency is taken as the ratio of the EPA mileage for a given automobile to the amount of energy needed to produce 1 L of gasoline at the pump. This value for the energy must take into account the energy that is expended to produce the fuel and to transport the fuel to its distribution point (e.g., a gas station). When this is taken into account, this reduces the available energy by 18.3%. For an all-electric car, the energy is determined from the number of watt hours it takes to charge the batteries in order for the car to go 1 km. This value must be reduced by taking into account the charging-discharging losses, which are about 14%. In addition, the best well-to-electric-outlet efficiency is about 52.5%. The CO_2 emissions for the electric car are those generated by electric power plants. Similar estimates are made for diesel engines, hybrid systems, natural gas engines, and hydrogen fuel cell systems. The results are presented in Table 6.1. It is seen from these metrics that the electric sports car looks very promising.

---

* M. Eberhard and M. Tarpenning, *The 21st Century Electric Car*, Tesla Motors Inc., 2006, http://www.teslamotors.com/.

6.2 CONCEPT GENERATION AND THE SEARCH FOR SOLUTIONS

6.2.1 INTRODUCTION

There are several formal and not-so-formal methods that have been proposed to facilitate the generation of ideas. Several of these methods were introduced in Section 2.4.2: transformation, random input, why? why? why?, and counter planning. Other methods are synectics and the concept map.[*] The major ones that we shall discuss are brainstorming, which was introduced in Section 2.4.2; morphological analysis, which is introduced in Section 6.2.2; TRIZ, which is introduced in Section 6.2.3; and bio-inspired ideas, which is introduced in Section 6.2.4.

6.2.1.1 General Activities That Can Generate Ideas

Ideas may occur to the team members during the information-gathering phase and during the benchmarking phase of the process, where various competing products have been tested, analyzed, and torn down. Additional ideas also can sometimes be obtained from a search of the U.S. patent office's database at http://www.uspto.gov/. Their databases can be searched by a wide range of attributes including patent title, number, inventor, assignee, issue date, and keywords. Other sources for ideas can be found on the Web site of *Design News* at http://www.designnews.com/ and at the Web site of *Machine Design* at http://machinedesign.com/. A Web search for "design awards" will turn up the annually selected results of Cooper-Hewitt, *Fortune, Business Week*, and *R&D Magazine*. These will be the year's best as determined by the judges from the respective organizations. Lastly, *R&D Magazine* annually issues their research and development 100 awards, which can be found at http://www.rdmag.com/awards.html.

6.2.1.2 Ideas That Can Come from a Brainstorming Session

In a brainstorming session, one might use some phrases that can trigger an idea. We shall show, in an expo facto sense, some examples that can result from the use of the following words: *magnify, adapt, combine, modify, put to another use*, and the combination of some of these. Consider first the suggestion to magnify. A self-cleaning oven works because the highest cooking temperature, around 260°C, is greatly increased (magnified) to around 400°C, a temperature at which food particles and drippings are reduced to ash. Another example of magnification is the cart shown in Figure 6.2. Here the cart's wheels are very large, thick-tube bicycle wheels. Their large (magnified) size creates two advantages: (1) the position of the wheels is such that the loads in the cart tend to be very close to the wheel's axle, requiring very little effort on the part of the user to pivot the support legs off the ground; and (2) the large wheels make the cart very stable and easy to navigate over irregular and soft terrain.

To illustrate *adaptation*, consider the use of laminated safety glass in residential home windows. After an analysis of the damage from Hurricane Andrew in South Florida, it was found that a substantial amount of the damage was caused by flying debris. South Florida is now considering a change in its building code to require that homes be built with windows (or shutters) that can withstand the impact of the end of a 2×4 in. (5×10 cm) wood stud traveling at more than 160 km/h. Some laminated glass, similar to that used for windshields of cars, has been able to meet this condition. An added benefit will be improved home security.

To illustrate the *combining* (integrating) of functions, consider the products shown in Figure 5.4 and Figures 6.3 to 6.6. In Figure 5.4, we showed a garlic press in which the means to clean (dislodge) the garlic is an integral part of one handle. In Figure 6.3, we show a nondisassemblable

---

[*] For a discussion of these methods see K. Otto and K. Wood, *Product Design: Techniques in Reverse Engineering and New Product Development*, Prentice Hall, Upper Saddle River, NJ, 2001, Chapter 10; and G. E. Dieter and L. C. Schmidt, *Engineering Design*, 4th ed., McGraw Hill, New York, 2009, Chapter 6.

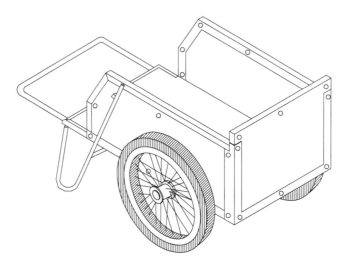

FIGURE 6.2 Example of magnification: a cart with very large wheels.

child's play yard in which the play yard's floor also serves as its carrying case. This combination of functions is possible because each of these individual functions is not required to be performed simultaneously. In Figure 6.4, we show the integration of an acoustic muffler into the casting of the structure that generates a vacuum using the Bernoulli principle. When the open end as shown is sealed with a flat plate, the volumes indicated act as acoustic filters. This product is further analyzed in Section 11.5.2. For another example of combining, consider a walker frequently used by elderly people. These can be purchased with a seat and a basket, which have been added to the

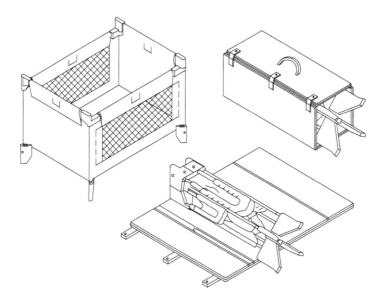

FIGURE 6.3 Example of combining functions: a child's portable play yard wherein the floor is also the play yard's carrying case.

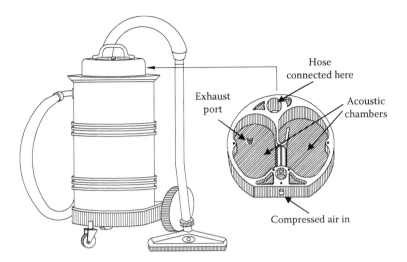

Hose
connected here

Exhaust
port

Acoustic
chambers

Compressed air in

FIGURE 6.4 Example of combining functions: integration of an acoustic muffler into the structure of an air-driven vacuum cleaner. See also Section 11.5.2.

walker.[*] However, the walker's envelop remains the same. For our last example, consider a piece of wheeled luggage. One can purchase a collapsible baby seat that attaches to it.[†] An example of putting a device to *another use* is the hand-held battery-operated laser pointer, which is frequently used by lecturers to highlight objects on an optically projected image.

In some cases, products are created by rethinking the application entirely differently. Consider the creeper, which is a low-wheeled platform for supporting the body when working under vehicles. The most expensive parts on these devices are the wheels, which, when the creeper is in use, tend to fall into cracks in the floor or to get diverted when encountering a bolt or tool. A new design was created that eliminated the wheels and instead used compressed air, which is always plentiful in automotive repair shops, to create a hovercraft-like platform.[‡] It has the added advantage that when the creeper is at the desired location the air is cut off and the device falls to the floor, where it offers a stationary platform to work from.

In recent years, several tire manufacturers have offered pneumatic tires that can "run flat" for 100 km or so at reduced speeds. Michelin also offers tires that perform this function; however, they have also taken a different approach by seeking a solution that does not require air in the first place. The result is an airless, molded tire called the TWEEL.[§] It consists of four pieces that are bonded together: the hub, a polyurethane spoke section, a "shear band" surrounding the spokes, and the tread band. The tire is virtually maintenance-free, is puncture-proof, has longer service life than radial tires, has a reusable structure for retreading, and has improved shock and road hazard resistance. Its current disadvantages are that it is noisy, has about 5% higher rolling friction than a radial tire, and it is as heavy as a conventional tire. Initially, the tire will be marketed for military and construction vehicles.

---

[*] See, for example, http://www.hugoanywhere.com/products.asp.

[†] http://www.rideoncarryon.com/.

[‡] http://www.davisoninternational.com/successes/hover_creeper.php.

[§] See, for example, http://www.gizmag.com/go/3603/.

Other examples of novel solutions are the self-erecting tent,[*] the Segway personal transporter,[†] the use of two elevator cabs in one elevator shaft,[‡] and the Paslode cordless framing nail tool that uses a replaceable fuel and battery cartridge for the energy source.[§]

6.2.1.3 Ideas That Can Come from Thinking about Simplifying Things

Making simplicity a goal in the concept generation process for the functional and physical attributes of a product can sometimes lead to a different way in which the product is viewed and may provide additional ideas. In some sense, simplicity is about removing the obvious and adding the meaningful.[¶] Some ways to achieve simplicity are (1) seek a balance between simplicity and complexity—the more complex something is the greater the need for it to be simpler to use; (2) organize its elements to make it appear that less is there; and (3) knowledge can make things appear simpler than they are. It should be realized, however, that some things cannot be made simpler.

6.2.1.4 Crowdsourcing: Consumers as a Source of Ideas[**]

In some cases, it has been found that consumers can be a source of product ideas. This type of consumer participation has been called "user-centered innovation" and a subset of this is called "crowdsourcing." Crowdsourcing is an Internet-enabled business trend where companies get amateurs to design products in their spare time. User-centered innovation on the other hand typically involves ad hoc groups of customers who develop or improve products of interest to that group such as open software and wind surfing gear. These groups feed manufacturers their needs, their desired products, and their enhancements to these products. The manufacturers then take these results and create products for their markets.

Both user-centered innovation and crowdsourcing are part of a trend whereby the Internet is used to get large numbers of people to help with a task.[††] *Innocentive* has successfully solved 250 challenges this way by offering prizes in the $10,000 to $25,000 range. Some of their successful achievements include a compound for skin tanning, a method for preventing snack chip breakage, and a mini extruder for brick-making. In early 2008, the $10,000,000 X Prize was created[‡‡] for the group that could bring to market a vehicle that got the equivalent of 100 mpg (42.5 km/l) gasoline engine, one that people wanted to buy, and met the market needs for price, size, capability, safety, and performance.

6.2.2 MORPHOLOGICAL METHOD

Rather than just generate ideas haphazardly, it is sometimes better to structure the concept generation process by first organizing the general search space with the aim of finding all conceivable solutions that satisfy each functional requirement. One method of doing this is the *morphological method*, whose goal is to find for each functional requirement all conceivable solutions that theoretically satisfy them. The word *theoretical* is used to convey the sense that these solutions satisfy the functional requirements in the absence of evaluation and comparison with other solutions. As in brainstorming, the evaluation occurs after it is determined that no more solutions can be generated; that is, after as many solutions as possible are generated and categorized. The "theoretical" solution set is composed of any combination of solutions taken from those that satisfy each functional requirement.

---

[*] http://seconds.quechua.com/index.php5#/home/.

[†] www.segway.com.

[‡] http://www.thyssenkruppelevator.com/.

[§] http://www.paslode-cordless.com/.

[¶] J. Maeda, *The Laws of Simplicity*: http://lawsofsimplicity.com/category/laws?order=ASC.

[**] P. Boutin, "Crowdsourcing: Consumers and Creators," *BusinessWeek*, July 13, 2008; http://www.businessweek.com/innovate/content/jul2006/id20060713_755844.htm.

[††] C. Dean, "If You Have a Problem, Ask Everyone," *New York Times*, July 22, 2008; http://www.nytimes.com/2008/07/22/science/22inno.html?scp=1&sq=ask%20everyone&st=cse.

[‡‡] http://www.progressiveautoxprize.org/

TABLE 6.2
Metal Removal Concepts and Examples of the Tools That Utilize Them

Concept	Tools That Utilize Concept
Mechanical	
Rigid	Milling machine, lathe
Flexible	Band saw
Abrasive	Grinder
Liquid	Water jet
Gas	Acetylene torch, plasma
Chemical	Etching
Optical	Laser
Electrical	Electric discharge machining

In generating solutions from a morphological exercise, it should be realized that the combination selected can have many different embodiments. Suppose, for example, that the task is to create a mode of transportation for one person. One possible combination is to have the person stand while the system is in motion, an electric motor for the energy source, and a cable for the means of causing movement to the system. Such a combination could be describing either a cable car in San Francisco, a tow lift at a ski resort, or an elevator.

The morphological method can be used several ways. One way is to classify the solutions by physical principles, and to then determine the ways in which these principles can be implemented. Consider the function of storing energy. Energy can be stored either mechanically, electrically, thermally, or hydraulically. In each of the categories, the energy can be stored in different ways. For example, mechanical energy can be stored in a spring or in a wheel (cylinder) rotating either about a fixed axis or rolling down an incline. Electrical energy can be stored in batteries and in capacitors. Thermal energy can be stored in a heated solid, liquid, or gas. Hydraulic energy can be stored in a liquid at an elevated height (potential energy). Another example of exploring concepts can be illustrated for the case of metal removal. In Table 6.2, we have listed several concepts along with examples of tools that utilize these concepts.

Although different solution principles will produce differently functioning products to satisfy the *FR*s, they may share a very large number of similar components and outwardly appear the same. For example, consider automotive engines. Except for the type of fuel and the absence of a spark plug, the gasoline and diesel engines are very similar in appearance, structure, major components, and fuel injection and lubrication systems. However, in terms of weight, safety, reliability, and certain performance characteristics, differences exist.

To further illustrate the various ways in which the morphological method can be used, consider the following examples. For the first example, we consider all the ways that one can apply glue to the back of a carpet. In this case, what is of interest is the manner in which the glue applicator should traverse the carpet or the carpet should traverse the applicator while the applicator dispenses the glue. The possible individual motions for the applicator and the carpet are stationary (no motion), translating, rotating, oscillating, rotating and translating, rotating and oscillating, and oscillating and translating. This gives 49 different combinations of carpet and applicator motions. In addition to these combinations, the number, locations, and the shape of the applicators still have to be identified.

Suppose that one wants to use a motor to generate a linear motion with a specified stroke, force, and/or velocity. Representative types of mechanical devices that are governed by different principles and that can be used to satisfy these requirements are shown in Figure 6.5.

(a)

(b)

(c)

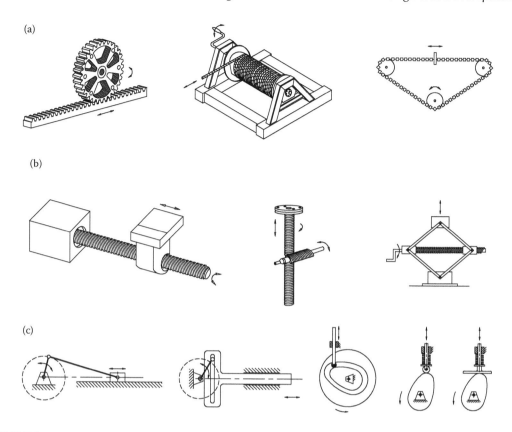

FIGURE 6.5 Means whereby rotational motion is converted to linear motion.

For another example, consider a riding lawn mower that is to be modified to prevent the backing over of small children and to reduce the severity of the injuries in these types of accidents. Any solution that is proposed to alleviate or minimize this problem is subject to the following conditions: the safety features must fit on the existing platforms of the product line; they must not adversely affect the grass cutting properties of the mower; the addition of such a device must not significantly increase the cost of the mower; the modifications themselves must not adversely affect the overall safety of the mower; and the means that will be employed must be reliable and not easily disabled by the operator. The morphology of possible solutions or combination of solutions is summarized in Table 6.3. When it comes to evaluating these concepts, the conditions stated can also serve as the evaluation criteria.

We consider as our final examples the drywall taping system described in Section 5.2.5 and the steel stud wall joining system described in Section 5.2.6. A brainstorming session led to the candidate concepts shown in Tables 6.4 and 6.5, which also show two ways in which to represent the information. The presentation given in Table 6.5 is used when it is neither easy nor practical to generate sketches. It should be realized that, although the various concepts have been listed as independent quantities, in attempting to attain candidate concepts many of these concepts may be combined. For example, in $(FR)_{32}$ in Table 6.5 it is reasonable to consider a piston that is activated by a cam and is driven by an electric motor. The solution for each of these products can be obtained, in principle, by taking in their respective tables one concept or combination of concepts from each row. Since there are about 2,700 combinations in Table 6.4 and 400,000 in Table 6.5, it is impractical to examine even a small percentage of them. A method is presented in Section 6.4 that permits one to perform a systematic evaluation efficiently.

TABLE 6.3
Concept Morphology for the Redesign of a Riding Lawnmower to Prevent Backing Over a Child and to Reduce the Severity of Injury

Increase Operator and Victim Awareness of Each Other			
Reduce Noise and Vibration	**Warn Victim**	**Warn Operator**	**Reverse Interlocks**
Electric motor	Light	Mirror	Pivot seat to engage reverse
Damp panel vibrations	Sound	Video display of rear	Very slow reverse speed
Quieter muffler		Light	Lever behind seat to keep blade engaged
Quieter air intake		Sound	in reverse

Prevent Blade Contact		
Restrict Access	**Sense Victim's Presence**	**Use Different Method to Cut Grass**
Grill/grating parallel to blade	Radar	Like hedge trimmers
Skirt around perimeter	Infrared	Like electric razor
	Sonar	Laser
	Light beam	Water jet
	Tilt sensor	Heated filament
	Mechanical trip-switch	Chain/cord/string
		Like grater
		Rubber/plastic blade

Prevent Backing Over Victim	
Physical Barrier	**Brake**
Hinged bumper	Mechanical
Hinged "feelers"	Electromagnetic
"Scoop" to lift victim	Engine
Lift rear wheels off ground	Anchor/stake
Deck independent of blade height	Block blade
Electric fence	

6.2.3 TRIZ[*,†,‡]

A very useful method called TRIZ has been recently developed to help one obtain solutions to engineering design problems at the conceptual stage. TRIZ is a Russian acronym that expands in translation to the Theory of Inventive Problem Solving. It is a qualitative means developed in the former U.S.S.R. by Genrich S. Altshuller (1926–1998) to obtain solutions to technical problems. Altshuller started the project in the early 1940s and by the end of the 1980s he and a loosely formed group of volunteer engineers and academicians analyzed over one million patents. From this analysis, he determined that the key to solving technical problems is the elimination of contradiction (conflict) from competing aspects of the solution; that is, an ideal solution was one that could be obtained without making any tradeoffs. In the study of these patents, Altshuller was able to deduce the following. First, he determined that there were 39 standard technical characteristics that could describe the causes of conflict. He called these *engineering parameters*. Second, he determined that there were 40 principles that could be used to resolve these conflicts. They are listed in Table 6.6

---

[*] K. Rantanen and E. Domb, *Simplified TRIZ*, 2nd ed., Auerbach Publications, Boca Raton, FL, 2008, Chapter 10.

[†] M. A. Orloff, *Inventive Thinking through TRIZ: A Practical Guide*, Springer, Berlin, 2003, Appendix 4.

[‡] "TRIZ 40 Principles," http://www.triz40.com/aff_Principles.htm.

TABLE 6.4
Concepts for Several of the *DP*s for the Drywall Taping System of Section 5.2.5

Design parameter	Concept						
Means to dispense tape [(DP)$_{31}$]	11 Tape	12 Tape	13 Tape	14 Tape			
Means to move joint compound [(DP)$_{32}$]	21 Gear	22 Screw	23 Bellows	24 Friction	25 Piston	26 Air pressure	
Means to dispense joint compound [(DP)$_{33}$]	31 Spray	32 Bead	33 Slot orifice	34 Viscous shear			
Means to apply tape and joint compound to wall [(DP)$_{34}$]	41 Plate	42 Brush	43 Roller	44 Blade			
Means to sever tape from roll [(DP)$_{35}$]	51 Rotating saw	52 Rotating blade	53 Translating blade	54 Tear	55 Hot wire	56 Chemical	57 Laser

TABLE 6.5
Concepts for Several of the *DP*s for the Steel Stud and Track Joining System of Section 5.2.6 That Satisfy Their Referent *FR*s

$(FR)_1$ (Maintain Position of Stud and Track)	$(FR)_{22}$ (Provide User Grip)	$(FR)_{31}$ (Generate Energy)	$(FR)_{32}$ (Convert Energy to Working Element)	$(FR)_{33}$ (Make Connection)

Mechanical grips
  C-clamp
  Vise grip
  Eccentric (cam)
  Wedge
  Lever
Spring-loaded clip
Tongs
Vise
Hand
Magnet
Adhesive
Insert wedge to spread stud flange
Press-fit stud between track flanges
	One-handed
	  Pistol
	  Knob
	  Screwdriver-like
	  Hand saw/snow shovel
	Two-handed
	  Bicycle
	  Plunger (two hands on one rod)
	Parallel
	Perpendicular (e.g., impact drill)
		Electrical
		  AC
		  Battery
		Thermal
		  Gas expansion
		  Bimetallic effect
		  Flame
		  Solar
		Mechanical
		  Compressed spring
		  Rotating flywheel
		Magnetic
		Pneumatic
		Hydraulic
		Explosive
			Mechanical
			  Mechanical linkages
			  Slider crank
			  Chain/belt drive
			  Gears and shaft
			  Cams
			  Screw drives
			  Lever
			Electromechanical
			  Magnetostrictive
			  Piezoelectric
			  Solenoid
			Piston
			  Pneumatic
			  Hydraulic
			  Mechanical
			Motors
			  AC/DC
			  Linear
			Electric arc
			Heat
			Pneumatic pressure
			Hydraulic pressure
				Mechanical deformation
				  Dimpling
				  Clinch
				  Spot
				  "V"
				  Metal stitching (perforation and bending)
				Mechanical fastener
				  Clip
				Self-piercing
				  Screw
				  Nail
				Self-piercing and then deformed
				  Staple
				  Nail
				  Hollow metal piercing rivet
				Pre-drill holes
				  Bolt and nut
				  Snap fitting
				  Rivet
				Adhesive
				  Glue
				  Tape
				Thermal
				  Solder
				  Braze
				  Weld (arc, resistance, gas)

TABLE 6.6
General TRIZ Principles to Resolve Conflicts in the Quest for Good Engineering Solutions[a]

1. **Segment (fragment).** Divide a part into independent parts.
   Examples: laptop computer, multi-stage rocket, sectional sofa, cupcake versus a cake, mulching lawn mower versus bagging, modular furniture, cloud computing, window shades versus blinds, truck versus tractor-trailer, chain versus cable.
2. **Separate (remove, extract).** Separate or remove the necessary part from an object or separate or remove the unnecessary part from the object.
   Examples: air bearing, hovercraft, single-family home central vacuum cleaner (noise reduction), diving bell versus submarine.
3. **Local quality.** Change object's structure or local environment so that each part of the object will function more suitably in that environment or more suitably satisfies a function.
   Examples: quenching or annealing metal, pencil with an eraser, socket wrench set, tires with dual layer treads for longer wear life and reduced noise.
4. **Change symmetry.** Move from a symmetrical shape to an asymmetrical one or increase existing asymmetry.
   Examples: asymmetry in eyeglasses to correct astigmatism, main deck of an aircraft carrier to permit takeoff and landing simultaneously, buoy (weighted base to recover upright position), paint can mixer (unbalanced mass to generate vibration amplitude), cement mixer, keyed shaft attached to a gear.
5. **Merge (join, unite, combine, integrate).** Merge or temporarily merge identical or similar objects to perform parallel or contiguous tasks.
   Examples: combine electronic devices for centralized cooling, integrated circuits, read/write heads on tape recorders, DVD player as part of a television set, rocket motor composed of several smaller rocket motors, large window with smaller framed panes of glass, Internet, mass spectrometer.
6. **Universality (multifunctional).** Perform several functions and eliminate redundant objects.
   Examples: adjustable wrench, door knob, AM/FM radio, toothpaste cap containing floss, luggage with wheels and telescoping handle, cell phone with camera and display.
7. **Nesting.** Place one or more objects inside each other or pass one object through a cavity of another object.
   Examples: telescope, radio antenna, pointer, double-hull ship, measuring cups, pencil, pen, packaging, lift mechanism for hydraulically actuated elevators, adjustable bicycle seat, retractable seat belt, airplane landing gear.
8. **Weight compensation (counterweight).** Compensate for the weight of an object with an externally applied weight or an externally applied force, such as aerodynamic lift or buoyancy.
   Examples: blimp, life jacket, tower cranes, oil field pumps, high "g" flight simulator centrifuge, tire balancing, internal combustion engine valve springs, boats using hydrofoils to lift boat out of water.
9. **Preliminary counteraction.** In anticipation of a harmful action by an object, create an initial opposite action.
   Examples: banking highways and railroad tracks on turns, stretching reinforcing rods prior to pouring the concrete in forming reinforced concrete beams, kitchen exhaust fan, self-cleaning oven, using a lead apron while taking x-rays, using masking tape prior to painting.
10. **Preliminary action.** Perform a required action before it is needed or arrange objects so that a required action can be performed immediately.
    Examples: prefabricated houses, machining of net shape castings, perforated edges of a roll of postage stamps, pressure treated lumber, sterilization of medical instruments, placement of material in a just-in-time manufacturing environment.
11. **Prior compensation (advance protection).** Provide safety measures in advance for low reliability objects.
    Examples: airbags, electrical fuses and circuit breakers, fire extinguishers, emergency exits, automotive antilock brake systems, life jackets, lifeboats, water heater pressure relief valves, ground fault detector electrical outlets, parachutes.
12. **Equipotentiality.** Change conditions so that an object does not work against a potential field such as gravity.
    Examples: automotive repair pit, loading dock, electrical grounding of machinery, factory supplies available at tabletop level, elevated walkways connecting buildings, canal locks.

TABLE 6.6 (CONTINUED)
General TRIZ Principles to Resolve Conflicts in the Quest for Good Engineering Solutions[a]

13. **Inverse action (opposite solution, other way around).** Take prescribed action and reverse it; from heating to cooling, from movable to stationary, rotate orientation, turn upside down, expand instead of contract, from horizontal to vertical, etc.

     Examples: track versus treadmill, mouse versus touchpad, hinged door versus rotating door, agitator (top-loading) washing machine versus tumbler (front-loading) washing machine, milling machine (tool rotates) versus lathe (work piece rotates), moving sidewalk versus stationary sidewalk.

14. **Change curvature (spheriodality, increase curvature).** Change linear lines to curved ones, flat surfaces to spherical ones, cubes or parallelepipeds to spherical ones, linear motion to curved or rotary motion. Use rollers, balls, spirals, and domes.

     Examples: roller casters versus spherical casters, flat roof construction versus geodesic dome, flat surfaces versus corrugated surfaces, glass plate photography versus a roll of film, circular saw versus chain saw, hydraulic jack versus screw jack.

15. **Dynamism (dynamic parts).** Allow an object flexibility or relative movement to find an optimum operating condition.

     Examples: aircraft wing with adjustable angle of attack, lighted message traffic signs, pneumatic tires (replacing solid tires), diving board versus fixed platform, fixed focal length camera lens versus variable focal length camera lens, skis, adjustable steering column, plumber's snake.

16. **Partial or excessive action.** When 100% of a goal is too difficult to achieve, then try for a little less or a little more.

     Examples: stenciling, sketching versus engineering drawing, perforations in packaging, overfilling and then leveling.

17. **Dimensional change (transition to another dimension).** Change the dimensions of an object, reposition or reorient the object, use the reverse side of an object, layer the object, or transform from a relatively flat object to a full three-dimensional object.

     Examples: drawing tables that are oriented at an angle, high-rise garages, highway overpasses and underpasses, dump truck, three-axis milling machine to a four- or five-axis one, two-sided circuit boards.

18. **Mechanical vibrations (oscillations).** Cause an object to oscillate by mechanical, aerodynamic, hydrodynamic, magnetic, or piezoelectric means.

     Examples: ultrasonic dental cleaning, fire alarms, air horns, paint mixers, vibratory conveyor belts and parts feeders, active noise cancellation, jack hammer.

19. **Periodic action.** Go from continuous action to intermittent action, where the intermittent action can be periodic; if already periodic, alter period or duty cycle.

     Examples: continuous light versus blinking light, impact drill, siren with continuous sound versus one that warbles, pile drivers, spot welders using variable frequency current.

20. **Uninterrupted useful action (continuity of useful action).** Complete action at full capacity or maximum capability without interruption; avoid idling.

     Examples: power generation turbines, paper mills, petroleum refineries, inkjet printers that print on return stroke, production run in a just-in-time manufacturing facility, nonstop airplane trip.

21. **Hurrying (rushing through, skipping, quick jump).** Complete some or all dangerous, hazardous, or damaging actions as fast as possible.

     Examples: cut plastics quickly before they distort from heat buildup, use short exposure x-ray film, accelerate quickly through a system resonance to get to an operating speed.

22. **Turn bad to good (transform damage into use; "blessings in disguise," "turn lemons into lemonade").** Use harmful effects, including effects to the environment or immediate surroundings, to achieve a useful effect, combine the harmful effect with other harmful effects to neutralize them, or amplify the effect until it is no longer harmful.

     Examples: use waste heat to generate electricity, turn newspaper into kitty litter, create a backfire to control a forest fire.

23. **Feedback.** Introduce feedback or change existing feedback to improve an action.

     Examples: stability and antilock braking systems in automobiles, statistical process control, thermostat-controlled heating/cooling system, noise-canceling headset.

(continued)

TABLE 6.6 (CONTINUED)
General TRIZ Principles to Resolve Conflicts in the Quest for Good Engineering Solutions[a]

24. **Intermediary (mediator, go-between).** Use another object to transfer or transmit an action to another object or temporarily attach an easily separable object to another object.

   Examples: power steering on an automobile, masking in the production of printed circuit boards, gear trains, pot holder, jigs and fixtures, brake pads.

25. **Self-service.** Have object service itself with auxiliary and repair functions and reuse waste energy and materials.

   Examples: heating automobile interior with heat from engine cooling system, self-sealing pneumatic tires, urinals that automatically flush after user is finished, vehicle-actuated traffic lights.

26. **Copying.** Use inexpensive replicas, copies, or simplified versions of expensive, complicated, inaccessible, and fragile objects.

   Examples: finite element programs to simulate crash testing, rapid prototyping, measurements from aerial photographs, crash test dummies, medical imaging devices, CD recordings of performances, fight simulators.

27. **Inexpensive disposables (cheap short-life objects to replace expensive long-life ones).** Replace an expensive long-life object with one that is inexpensive and disposable.

   Examples: medical supplies, cell phones, fuses, paper and plastic tableware, diapers, bushing, paper coffee filters.

28. **Substitution of means (replacement of mechanical matter, mechanics substitution).** Replace mechanical schemes with acoustic, optical, and olfactory ones; use electrical, magnetic, or electromagnetic for interactions; and replace static fields with dynamic ones.

   Examples: magneto-rheological fluid damper, hybrid automotive engines, microwave oven, laser printers (electrostatic charge), electronic pet fence, sulfur-based compound to give natural gas an odor, magnetic levitation trains, CD player, laser machining, electrostatic air filters.

29. **Pneumatics and hydraulics.** Replace solid parts with gases or liquids.

   Examples: hydraulic lifts, air bearings, inflatable rafts, indoor tennis structures, waterbeds, shoe cushions, air springs, water jet cutting.

30. **Flexible membranes and thin films (flexible shells/covers and thin films).** Use flexible membranes instead of rigid structures and isolate objects from external environment with thin films.

   Examples: tarps over swimming pools, polyurethane on floors, hot houses, tents, plastic wrap for food, balloons, layer comprising safety glass, moisture and water resistant home wrap, rubber membrane for flat roofs, removable panels on new automobile exteriors during transport.

31. **Porous materials.** Make an object porous or use porous inserts; if already porous, fill pores with a useful substance.

   Examples: oil-filled powder metallurgy part, sponge, filters, weight reduction using honeycomb structures.

32. **Optical property changes (color changes).** Change color, lighting, or transparency of object or its surroundings.

   Examples: litmus paper, "smoke" introduced to air in wind tunnel tests, red lights in photographic darkrooms, self-darkening sunglasses and windows, food packaging, signs, flesh-colored bandages, camouflage clothing, mood lighting, bulbs that simulate natural light.

33. **Homogeneity.** Make objects interacting with a given object of the same material or of a material having substantially the same properties.

   Examples: meshing gears, electrical contacts, edible food containers (e.g., ice cream cone, ravioli), liquid Band-Aid, bioactive glass to aid in bone regeneration.

34. **Discarding and recovering (discard and renewal of parts, discarding and regenerating parts).** After a part has fulfilled it purpose and is no longer needed, it should be discarded (modified, evaporated, dissolved, deformed) and then replaced during the operations.

   Examples: separation of booster rockets after their fuel is expended, capsule containing medicine, self-cleaning electric razor, biodegradable packaging, recycling and purification of water used in industrial manufacturing processes.

35. **Change aggregate state (parameter changes).** Change object's physical state (solid, liquid, gas), consistency (viscosity), concentration, elasticity (pseudo-state), temperature, chemical reaction.

   Examples: transport gases in liquid state, baking cakes, cryogenic storage of biomaterials, liquid soap versus soap bar, hand warmers (exothermic reactions), freezing candy centers prior to application of chocolate coating.

TABLE 6.6 (CONTINUED)
General TRIZ Principles to Resolve Conflicts in the Quest for Good Engineering Solutions[a]

36. **Phase transitions.** Use attribute associated with a phase transition such as volume and loss or absorption of heat.
    Examples: heat pumps, expansion of frozen water, fire extinguisher, dynamite, spray foam insulation, plasma cutters.

37. **Thermal expansion.** Use the thermal expansion or contraction of an object or use materials with different coefficients of thermal expansion.
    Examples: bimetallic elements in thermostats, thermometers, compression fitting of concentric cylinders (cool inner cylinder and heat outer cylinder).

38. **Strong oxidation agents (strong oxidants).** Replace air with oxygen-enriched air or pure oxygen or use ionized oxygen or ozone.
    Examples: oxygen-nitrogen mixture for scuba diving tanks, oxy-acetylene torch, oxygen steelmaking.

39. **Inert media (inert atmosphere).** Replace a normal environment with an inert one, use a vacuum, or use inert additives.
    Examples: helium blimp, argon welding, light bulbs, medicines (most of tablet/capsule consists of inert ingredients), insect repellent (most of solution consists of inert ingredients).

40. **Composite materials.** Replace homogeneous materials with composite materials.
    Examples: airplane exterior surfaces, yachts, golf clubs and tennis racquets, reinforced concrete.

[a] The terms in parentheses are alternate expressions that are also used.

along with several examples of where these principles can be used.[*] The 39 engineering parameters are listed in Table 6.7. The numbers associated with each of the principles and with each of the parameters are integral parts of the TRIZ method, as will be shown subsequently.

The 40 principles and the 39 engineering parameters comprise the major part of TRIZ, and the method combines these two tables to form a contradiction matrix. (The contradiction matrix has not been given here; it can be found in the previously cited references.) The contradiction matrix is an array of 39 columns numbered from 1 to 39 and 39 rows numbered for 1 to 39. Each row number and column number corresponds to the numbered parameter shown in Table 6.7. The column numbers refer to a worsening aspect of the conflict and the row numbers refer to an improving aspect of the conflict. In each cell, there appears one number, two numbers, three numbers, four numbers, or no number. Each of these numbers corresponds to a principle listed in Table 6.6, which then can be used to generate ideas to resolve the conflict. These principles can be used to resolve conflicts between competing requirements, that is, to lessen the impact of the tradeoffs that inevitably have to be made. However, not all of these principles will be applicable and, additionally, principles not appearing in the cell may prove more useful. This is because, historically, these are the principles that were used the most frequently to resolve the conflict. In addition, when there are no numbers appearing in the cell, all 40 principles should be explored.

We will now outline how the TRIZ procedure is used.[†] One first selects a desired parameter from Table 6.7 that most closely describes the attribute that is to be improved and records the number associated with it, say N_1. Then one selects from Table 6.7 a parameter that most closely describes an attribute that will worsen when parameter N_1 is improved. We will denote this parameter as N_2; N_2 is the conflicting parameter. One then goes to the cell in the contradiction matrix defined by

---

[*] Additional examples of the general TRIZ principles can be found at http://www.mazur.net/triz/.
[†] The TRIZ method also appears in software programs, descriptions of which can be found at http://www.creaxinnovation suite.com/index.htm and http://www.ideationtriz.com/software.asp.

TABLE 6.7
Engineering Parameters That Are Used to Identify Contradictions

1. Weight of moving object	21. Power
2. Weight of nonmoving object	22. Waste of energy
3. Length of moving object	23. Waste of substance
4. Length of nonmoving object	24. Loss of information
5. Area of moving object	25. Waste of time
6. Area of nonmoving object	26. Amount of substance
7. Volume of moving object	27. Reliability
8. Volume of nonmoving object	28. Accuracy of measurement
9. Speed	29. Accuracy of manufacturing
10. Force	30. Harmful factors acting on object
11. Tension, pressure	31. Harmful side effects
12. Shape	32. Manufacturability
13. Stability of object	33. Convenience of use
14. Strength	34. Repairability
15. Durability of moving object	35. Adaptability
16. Durability of nonmoving object	36. Complexity of device
17. Temperature	37. Complexity of control
18. Brightness	38. Level of automation
19. Energy spent by moving object	39. Productivity
20. Energy spent by nonmoving object	

row N_1 and column N_2. For example, if one wants to make a cylindrical wall of a container thinner, then one selects parameter number 4 from Table 6.7; that is, $N_1 = 4$. However, when the thickness of the wall decreases, the stresses increase. Thus, one selects parameter number 11 from the table, that is, $N_2 = 11$. From the contradiction matrix, we would find that the (4, 11) cell would contain the numbers 1, 14, and 35. These numbers refer to the principles that would then be interpreted in the context of this particular example in order to arrive at a solution so that the conflicting parameters are simultaneously satisfied. Interpretation of these principles may lead one to examine changing the cylindrical wall to a corrugated one (inspired by principle 14) or to using a stronger material (inspired by principle 35).

In many cases, these principles can be grouped to resolve four major types of conflicts.[*] To help resolve those conflicts involving *spatial properties*, we can explore the following principles:

(2) separate; (4) change symmetry; (6) universality; (7) nesting; (14) change curvature; (17) dimensional change; (18) mechanical vibrations; (26) copying; (29) pneumatics and hydraulics; and (30) flexible membranes and thin films

The principles that can be used to resolve those conflicts involving *temporal properties* are

(5) merge; (9) preliminary counteraction; (10) preliminary action; (11) prior compensation; (15) dynamism; (19) periodic action; (20) uninterrupted useful action; (21) hurrying; and (24) intermediary

---

[*] M. A. Orloff, ibid.

TABLE 6.8
Examples of Nature's Unique Solutions

Geckos can walk upside down.

Hummingbirds can hover and can fly straight up, down, and backwards.

Migratory locusts can fly continuously for 9 hr.

A 32 kg octopus can fit through a 40 mm diameter hole.

The sperm whale can dive to a depth of 1.5 km.

Leaves 110 m from the base of a redwood tree receive water from its roots.

Termites build columns 2,500 times their height.

A golden eagle can see an 18 cm object from a distance of 3.2 km.

A dog's nose can detect more than 10,000 distinct odors.

The shape of the boxfish provides extremely low drag coefficient (<0.06).

Squid, octopus, and cuttlefish can camouflage themselves to the point of being virtually invisible.[a]

Penguin feathers interlock to provide excellent thermal insulation against cold and wind.

Spider silk requires three times the energy to break than Kevlar and 100 times more than steel.

[a] J. C. Anderson, R. J. Baddeley, D. Osorio, N. Shashar, C. W. Tyler, V. S. Ramachandran, A. C. Crook, and R. T. Hanlon, Modular organization of adaptive colouration in flounder and cuttlefish revealed by independent component analysis, *Computation in Neural Systems*, Vol. 14, Issue 2, 2003, pp. 321–333.

The principles that can be used to resolve those conflicts involving *property of structure* are

(1) segment; (3) local quality; (5) merge; (13) inverse action; (24) intermediary; and (34) discarding and recovering

The principles that can be used to resolve those conflicts involving *property of material* are

(25) self-service; (27) inexpensive disposables; (28) substitution of means; (31) porous materials; (33) homogeneity; (35) change aggregate state; (36) phase transitions; (37) thermal expansion; (38) strong oxidation agents; (39) inert media; and (40) composite materials

6.2.4 BIO-INSPIRED CONCEPTS[*]

In the previous sections, we have suggested various ways in which one could obtain ideas for arriving at solutions that satisfy the functional requirements: benchmarking, patent search, morphological methods, and TRIZ. Each of these methods has indirectly limited the search space to the physical world. In this section, we will show that another large source of ideas can be found in nature. We will see that nature's solutions are often optimal ones for a given environment, are energy efficient, are high performing, and are multifunctional. We will show by example that in examining a wide range of biological entities we will find ideas for materials, geometry, sensors, signal processing, adaptation, and kinematics.

Nature has many remarkable animals, insects, reptiles, birds, and plants that were developed through millions of years of evolution. These living systems are able to generate energy, sense the environment, perform complex motion, transport material, dissipate heat, and communicate. Research over the years has shown that in some cases nature's way of performing a function can be quite remarkable as indicated in Table 6.8. Thus, engineers are starting to look to nature for ideas for

[*] See bibliography at end of the chapter.

TABLE 6.9
Commercial Products Inspired by Nature

Inspiration	Description of Product
Plant burr seed sacs	Velcro
Wax crystals on lotus leaves[a]	Lotus Effect aerosol spray manufactured by BASF to make surfaces water and dirt repellant
Timber beetle larvae (ergastes spiculatus)[b]	Chain saw teeth
Ants[c]	Bimorph piezoelectric motors
Nature's flow efficiencies translated into streamlined geometries[d]	PAX fans and mixers with significantly improved performance and efficiencies

[a] Lotus: http://nanotechweb.org/cws/article/tech/16392
[b] Asserted by Joe Cox, who in 1947 formed Oregon Chain Saw Mfg. Co.
[c] http://www.piezomotor.se/pages/PLtechnology.html
[d] http://www.paxscientific.com/Index.html

solutions. In fact, over the years, several successful products have been developed using inspiration from the nature, a few of which are listed in Table 6.9.

However, there are important differences between entities created by nature and artifacts manufactured by humans.[*] Nature, of course, doesn't use metals but it uses materials whose properties vary spatially. Nature uses curved surfaces whereas man uses mostly flat surfaces that are at right angles to other flat surfaces. Many of nature's entities bend, twist, and stretch at predetermined locations whereas man's artifacts are stiff and rotate and slide with respect to each other. In nature, drag is reduced by flexible, reconfigurable bodies whereas man's artifacts are streamlined fixed shapes. In nature, mechanical advantage is used to amplify distance at the expense of force whereas man amplifies force at the expense of distance. Nature's creatures produce artifacts that are larger than they are whereas man's factories dwarf the artifacts.

The method to examine the solutions found in nature can range from a straightforward approach to a quite involved one depending upon the functional requirement to be satisfied. In Table 6.10, we have listed nature's solutions in terms of several functional requirements. As an example, let us consider the case of a robot design. Robots are usually designed to perform navigation, locomotion, and manipulation tasks in certain environments. These tasks are typically characterized by the required number of degrees of freedom and expected motions. So a biological source of inspiration can be located by identifying an animal that operates in the similar environment and performs the similar types of motions. A list of a variety of candidate biological sources is given in Table 6.11.

In more general applications, locating a biological source of inspiration is much more challenging. One approach is to break down the design task into the following three areas[†]: (1) sensing, (2) actuation, and (3) structural. For each area, a different search for the biological inspiration sources needs to be carried out. Representative results of such a search for sensors is given in Table 6.12 and for structures in Table 6.13.

---

[*] S. Vogel, *Cats' Paws and Catapults*, Norton, New York, 1998, pp. 289–291.
[†] See the journal titled *Bioinspiration and Biomimetics* for advances in these areas.

TABLE 6.10
Nature's Solution to Different Functional Requirements

Functional Requirement	Nature's Solution
Move fast	
On land	Cheetah (sprints: 115 km/hr)
In the air	Swift (flying: 170 km/hr)
	Peregrine falcon (diving: 160–320 km/hr)
	Dragonfly (flying insect: 58 km/hr)
In water	Sailfish (110 km/hr: combination of in and out of water)
Lift object	Elephant (1,000 kg)
	Rhinoceros beetle (850 times its weight)
Jump far	Kangaroo (10 m)
	Flea (200 times its length)
Leap high	Puma (5.4 m)
	Flea (150 times its length)
Glide far	Flying squirrel (195 m)
Crush	Snake
	Lobster claws

TABLE 6.11
Nature's Variety of Locomotion Techniques That Have Been Used in the Development of New Robots

Biological Inspiration[a]	Type of Motion or Feature
Legged creatures	
Four-footed animals, spiders, cockroach, crabs, lobsters	Stable and rapid forward motion over irregular terrain
Snakes	Lateral undulation, concertina progression, rectilinear motion, side-winding
Inchworms	Reaching and pulling from opposite ends
Earthworms	Stretching and contracting
Cuttlefish, squirting cucumbers, mussels	Jet propulsion
Dolphin skin (compliant) and shape, shark skin (rough) and shape	High speed underwater due to shape and less drag
Tuna	High underwater speed and maneuverability

[a] Some examples of prototypes that have resulted from these biological inspirations can be found at

[1] Cockroach-inspired—http://biorobots.cwru.edu/projects/onrprojects.htm

[2] Six-legged-creature-inspired—http://www-robotics.jpl.nasa.gov/tasks/taskImage.cfm?TaskID=30&tdaID=2585&Image=144

[3] Cockroach-inspired—http://www-cdr.stanford.edu/biomimetics/sprawlmedia/sprawl-media.html

[4] Lobster-inspired—http://www.biology.neu.edu/faculty03/initiatives/neurobiology.html

[5] Snake-and-spider-inspired—http://www.isi.edu/robots/conro/

[6] Snake-inspired—http://download.srv.cs.cmu.edu/~biorobotics//robots_quad.html

[7] Tuna-inspired—http://web.mit.edu/towtank/www/Tuna/tuna.html

[8] Crab-inspired—http://www.bath.ac.uk/news/2008/3/18/crabrobot.html

TABLE 6.12
Examples of Nature's Use of Sensors

Homing pigeons have a magnetic material between their brain and skull that may act as a magnetic field detector.[a]

Neurological elements embedded in the last joint of each of a scorpion's eight legs are used to sense vibration caused by surface waves in sand. Time-of-arrival sensed at each joint is used to determine the direction of the source.[b]

High spatial resolution sonar echo location in bats and dolphins.

The shape of a barn owl's ears and head along with special neurological functions provide exceptional external sound localization capabilities.

Hair cells in the lateral line of some fish detect flow.[c]

Crickets have external hairs that can detect very minute changes in air currents to sense the approach of predators.

[a] C. Walcott, J. L. Gould, and J. L. Kirschvink, Pigeons have magnets, *Science*, Vol. 205, September 1979, pp. 1027–1029.

[b] P. H. Brownell and J. L. van Hemmen, Vibration sensitivity and a computational theory for prey-localizing behavior in sand scorpions, *American Zoologist*, Vol. 41, 2001, pp. 1229–1240.

[c] Z. Fan, J. Chen, J. Zou, D. Bullen, C. Liu, and F. Delcomyn, Design and fabrication of artificial lateral line flow sensors, *J. Micromechanics and Microengineering*, Vol. 12, No. 5, 2002, pp. 655–661.

6.3 PRODUCT MODULARITY AND ARCHITECTURE

Product modularity has developed in different domains and has, therefore, evolved to mean slightly different things, depending on one's point of view. In an attempt to resolve these shades of meaning, the following five definitions have been proposed in the context of a perspective.[*]

1. **Component commonality** refers to modularity as a standard kit of components to be used, several at a time, in a number of applications.
2. **Component combinability** refers to the mixing and matching of components taken from a given set.
3. **Function binding** refers to the design of machines, assemblies, and components that fulfill various overall functions through the combination of distinct building blocks or modules.
4. **Interface standardization** refers to a set of design parameters describing how two objects mutually interact; however, much freedom is left to the definition of a module with regard to what is connected by the interface.
5. **Loose coupling** refers to a way to partition a system effectively and implies that some degree of complexity should be imbedded in the modules; that is, too many simple modules may result in a complex or inefficient system.

Modularity, when determined by the axiomatic approach, tends to shorten development times because most, if not all, of the modules can be developed more or less independently of each other, provided that the FRs are either uncoupled or decoupled. There are occasions, however, when costs can increase because the modules must be interfaced with each other. The effect of these costs is often insignificant when compared to the benefits of reducing the time to market. In addition, modularity makes it possible to change, maintain, and improve the overall functionality without having to completely redo a design. These changes can be brought about by product upgrades, product add-ons (e.g., computer mass storage devices), adaptation (e.g., gasoline engine conversion to a propane engine), wear (e.g., tires, brake pads), consumption (e.g., replacement of batteries and copier toner cartridges), and product flexibility (e.g., lens changes in cameras). Consequently, the use of modular manufacture can result in simplified schedules and shortened delivery dates, easier

---

[*] F. Salvador, "Toward a Product System Modularity Construct: Literature Review and Reconceptualization," *IEEE Transactions on Engineering Management*, Vol. 54, No. 2, May 2007, pp. 219–240.

TABLE 6.13
Variety of Structures Appearing in Nature

Bio-Inspiration	Description of Important Feature
Bamboo	Ring reinforced tubes
Bird bone	Very porous, with high strength-to-weight ratio
Seaweed	Sandwich structure with honeycomb center
Water lily (Victoria amazonia)	Rib-reinforced flat structure (typically 2 m in diameter)
Daffodil	Increased structural integrity due to the bending and torsion flexibility of stem
Abalone shells[a]	Very hard and difficult to break
Plant leaves and flowers	Deployable structures
Venus fly trap[b]	Hydroelastically created curvature creates a bi-stable system
Angler fish mouth	4-bar-mechanism-like jaw that permits fish to swallow prey twice its size
Moths	Sound absorbing hairs and soft layers thwart echolocation from bats

[a] Z. Tang, N. A. Kotov, S. Magonov, and B. Ozturk, Nanostructured artificial nacre, *Nature Materials*, Vol. 2, 2003, pp. 413–418.

[b] A. G. Volkov, T. Adesina, V. S. Markin, and E. Jananov, Kinetics and mechanism of *Dionaea muscipula* trap closing, *Plant Physiology*, February 2008, Vol. 146, pp. 694–702.

assembly, continued use of certain modules in new products, easier repair and maintenance, and improved availability of spare parts.

In the process of decomposing the functional requirements, the architecture (layout) of the product is also indirectly being created, since the functional requirements will be met by the modules/ units created to satisfy them. In other words, the spatial distribution of the various modules determines the product's architecture. With each module, there is an associated interface that may be required in order to connect it to another module. In addition, each module may have a technical risk associated with its development. (Recall region 10 of the house of quality in Figure 4.1.) Thus, there are several issues that must be addressed: concentration and location of the modules, interfaces, and technical risk. Each of these decisions has important implications for the product's development speed, and will impact the IP^2D^2 team's latitude during the product realization process.

Modularity also permits a variety of product models to be generated without increasing the product's complexity and manufacturing costs; that is, it makes mass customization possible.* There are six types of modularity for the mass customization of products and services.

1. **Component-sharing modularity** is using the same component across multiple products to provide economics of scope. This type of modularity permits low-cost production of a great variety of products and services. Its net effect is to reduce the number of different parts and, hence, the cost of a product line that has a wide variety of products. The use of the same automobile engine in different models is an example. Another example is the Black & Decker rechargeable battery pack, which has been designed to fit a wide range of their portable electric hand tools: multifunction saw, detail sander, hand vacuum, variable-speed drill, flashlight, and screwdriver.

2. **Component-swapping modularity** is the complement of component-sharing modularity. Here, different components are paired with the same basic product. Bowling balls, where holes are drilled to fit the purchaser's hand, T-shirts with individually selected ironed-on decals, and eyeglass lenses are examples.

* B. J. Pine II, *Mass Customization: The New Frontier in Business Competition*, Harvard Business Press, Boston, MA, 1993.

3. **Cut-to-fit modularity** is where one or more components are continuously variable within prescribed limits. Clothing and salad bars are examples.
4. **Mix modularity** can use any of the above three types of modularity such that when the modules are put together, they form something different. Typical examples are those that are governed by recipes, such as foods, chemicals, paints, and fertilizer.
5. **Bus modularity** uses standard structures to which different kinds of components (modules) can be attached. Examples are computer I/O buses, track lighting, and bicycle frames.
6. **Sectional modularity** allows the configuration of any number of different types of components in arbitrary ways, as long as each component is connected to another at standard interfaces. Lego building blocks (toys) and modular office partitioning panels are examples.

Examples of product architecture are given in Figure 6.1, where various hair dryers are shown. It is seen in these pictures that the placement of the various components is a function of the type of fan chosen, where the placement of the intake and exhaust ports are located, and how the device is to be gripped with respect to the flow direction of the heated air.

Modules may require interfaces to connect them with other modules, for example, the crankshaft of an engine with a transmission, a terminal to a computer, etc. Using the right approach to the interfaces can shorten development cycles. To do so means designing stable, robust, standard, and simple interfaces. Since the various modules should be designed concurrently, the interfaces must be defined early in the product realization process and be prevented from changing.

Another principle of good interface selection is to use standard interfaces whenever possible. Standard interfaces have the advantage that designers and suppliers already understand them. As an example of the standardization of interfaces, consider bicycles, in which the seats, handle bars, wheels, tires, pedals, and pedal crank arms have a standard method of attachment, so that a bicycle can be customized, piece by piece, from products built by many different manufacturers.

In the system design process, a decision must be made on the degree of technical risk to be taken with each module. It is a best practice to concentrate technical risk in as few modules as practical. Also, modules with risky technology tend to be communication intensive, and their designers must communicate frequently with the designers of other modules. Confining the risk to only one or two modules minimizes the amount of external communication, and its associated risk of miscommunication.

6.4 CONCEPT EVALUATION AND SELECTION

The various concepts are evaluated after many different concepts and their variations have been generated. In the linear motion example shown in Figure 6.5, the constraints and the operating environment would be itemized and a rating given for each item. For example, the device may be required to work in any orientation, generate the force in either direction (in/out), work in a contaminated environment, and weigh less than a certain value. Each of the concepts shown in Figure 6.5 would be given a rating as to how well it performed in each evaluation category. In addition, one would have to analyze these various concepts to estimate their feasibility as a solution. This may include the determination of the capacity of the motor, the maximum buckling loads, an estimate of fatigue life, the temperature rise of the components, etc. At this stage in the design process, these calculations should be performed approximately and quickly.

There are several evaluation schemes that can be used, such as utility theory, analytic hierarchy process, and the Pugh selection method. Utility theory is an attempt to infer subjective value, or utility, from choices. It can be used in both decision making under risk where the probabilities are explicitly given and in decision making under uncertainty where the probabilities are not explicitly given. The analytic hierarchy process is a structured technique whereby the first step in the process is to decompose the decision problem into a hierarchy of more easily comprehended sub problems,

TABLE 6.14
Rating Scales and Their Meaning[a]

Eleven-Point Scale	Meaning of Scale Value	Five-Point Scale	Meaning of Scale Value	Three-Point Scale	Meaning of Scale Value
0	Useless				
1	Inadequate	0	Inadequate		
2	Very poor			0 (or −1 or "−")	Poor (worse)
3	Poor	1	Weak		
4	Tolerable				
5	Adequate	2	Satisfactory	1 (or 0 or S)	Satisfactory (same)
6	Satisfactory				
7	Good	3	Good		
8	Very good			2 (or +1 or "+")	Good (better)
9	Excellent	4	Excellent		
10	Perfect				

[a] Based, in part, on N. Cross, *Engineering Design Methods*, John Wiley & Sons, Chichester, 1989.

each of which can be analyzed independently. Once the hierarchy is built, one systematically evaluates its various elements, comparing them to one another in pairs. Both of these techniques require some training to use them correctly. The third technique, the Pugh method, is relatively straightforward to use and is the one that we will introduce and employ here.

The Pugh evaluation rating scheme can use any of several rating scales, which are summarized in Table 6.14. For the concept evaluation stage, it is often sufficient to use the three-point scale. The way in which the various concepts can be evaluated is with a matrix, in which the columns of the matrix denote the concepts and the rows the evaluation criteria. The evaluation criteria should be some or all of the customer requirements recorded in the QFD diagram that pertain to the particular functional requirement one is attempting to satisfy. If only a portion of the customer requirements is used, then those with the highest customer importance ratings should be selected.

There are two ways in which to fill the matrix. The first way is to select one of the concepts as the reference or datum concept, and then compare all the other concepts to it by simply recording for each evaluation criteria the judgment that the other concept is either better than, the same as, or worse than the datum concept. At the end of this process, a tally is made of each concept (column). Those with the highest numerical score or with the greatest number of pluses (+) that exceed the number of minuses (−) are kept for further consideration.

Another way to evaluate the concepts is to give each evaluation criterion a weighting, or value, equivalent to the values recorded in the QFD method for the customer importance. In this approach, no datum concept is required, and instead, for each evaluation criterion a numerical score is recorded: 0, 1, 2 or −1, 0, +1. The overall score for each concept is the sum of the product of each numerical value and its corresponding weight. Pugh[*] recommends the use of the datum when evaluating the concepts. Pahl and Beitz[†] use the weighted method. Either method will work since the main purpose of this type of matrix scoring is to identify candidate concepts for further examination, refinement, and possible combination, and not to identify a single concept and then stop.

The differences in the evaluation scores provide a means of identifying those ideas that seem to satisfy a large number of the relevant customer requirements. When one or two concepts have scores

[*] S. Pugh, ibid.
[†] G. Pahl, W. Beitz, J. Feldhusen, and K.-H. Grote, ibid.

that greatly exceed the others, then only those concepts should be kept. When the differences in the highest scores are very small for several of the concepts, then these several are retained. In either case, this reduced set of candidate concepts for each *DP* is examined to see if they can be combined or modified in some way to improve their applicability. Then each of the concepts is reevaluated using additional criteria, which are usually comprised of the most important constraints and several appropriate requirements from Table 2.2, such as those pertaining to producibility, to the environment, and to social issues. The objective of the reevaluation is to see if the number of concepts can be reduced further.

To illustrate this evaluation procedure, the concepts shown in Table 6.4 are evaluated, and the results are shown in Table 6.15. The evaluation criteria have been grouped so that only those criteria that relate to a particular *FR* are used to evaluate it. The three criteria under the heading "System" were deemed important enough to be used to evaluate all the concepts for each of the five functional modules. The evaluation (+, 0, −) in each of the columns in Table 6.15 is arrived at by team consensus. However, it is suggested that before placing a concept in an evaluation table, like Table 6.15, it should be subjected to an analysis using the appropriate physical criteria in order to determine, for example, whether or not the concept can meet the size requirements, can meet the power requirements, can generate sufficient force, etc. In addition, each concept should be further scrutinized to determine whether it can satisfy the imposed constraints or requirements, such as being nontoxic, battery-operated, portable, etc. If a concept does not appear to satisfy one or more of the constraints, then before eliminating it from further consideration the concept should be reexamined to determine whether a variation, modification, or combination with another concept would remove the limitation, thereby making it suitable as a candidate for comparison with the other concepts.

The advantages of having originally performed the functional decomposition and having determined whether the design matrix was coupled, decoupled, or uncoupled should now be apparent. When the design matrix is either uncoupled or decoupled, it is possible for the IP^{2}D^2 team to evaluate the concepts for each *DP* independently. Thus, since only the concepts that scored the highest in their relative comparisons are retained, the total number of combinations that can be used as a candidate solution for the product is greatly reduced. This makes it possible to evaluate a small group of combinations that will form the product and determine the best one. Also, since each remaining concept satisfies the customer requirements, any combination of them will most likely satisfy the customer requirements.

We now illustrate the next step in the concept evaluation stage by considering further the candidate concepts listed in Table 6.5. The concepts shown in Table 6.5 can be evaluated using the criteria presented in Table 6.16 in the same manner as used in Table 6.15. Suppose that after an evaluation of these concepts using the criteria in Table 6.16, the following concepts ranked the highest (those for $[FR]_{22}$ have been excluded):

$(FR)_1$
    Mechanical C-clamp
$(FR)_{31}$
    Electric
    Explosive
$(FR)_{32}$
    Directly through (no transfer to working element required)
    Mechanical
    Piston
$(FR)_{33}$
    Weld
    Mechanical fastener
    Metal deformation

TABLE 6.15

Evaluation of the Concepts for the Drywall Taping System of Table 6.4 (The Concept Numbers Correspond to the Numbers in Table 6.4; the Evaluation Criteria are from Figure 4.3.)

Functional Requirement →	Dispense Tape				Dispense Compound						Apply Compound				Applicator				Separate Tape						
Evaluation Criteria | Concept →	11	12	13	14	21	22	23	24	25	26	31	32	33	34	41	42	43	44	51	52	53	54	55	56	57
System																									
Low weight	−	D	−	+	+	+	0	D	+	+	0	D	0	−	0	D	−	+	−	0	D	−	−	0	−
Low cost	−	A	−	+	0	+	0	A	+	0	0	A	0	−	+	A	−	+	−	0	A	+	+	−	−
Can be made compact	−	T	−	+	+	+	−	T	+	0	0	T	0	−	0	T	−	+	−	0	T	+	−	−	−
Dispense tape																									
Doesn't jam	+	U	+	−																					
Doesn't tear/fold	0	M	+	−																					
Synchronized with compound	0		0	+																					
Dispense compound																									
Supply at specified rate					+	+	−	U	+	−															
Easy to clean					−	−	0	M	+	+															
Simple operation					+	+	−		+	0															
Doesn't clog					+	+	−		+	−															
Doesn't leak					+	+	−		+	+															
Infrequent refills					+	+	−		−	−															
Apply compound																									
Compound uniformly distributed											−	U	+	−											
Doesn't clog											−	M	0												
Applicator																									
Walls, corners, ceiling															−	U	−	+							
One-stroke															+	M	+	+							
Smooth finish															+		+	+							
Separate tape																									
Easy to separate tape																			+	0	U	−	−	−	+
Straight line, not jagged																			−	0	M	−	+	−	+
Works when taping corners																			−	0		+	−	−	+
Works every time																			0	0	0	0	0	0	−
Totals	−2	0	−1	2	6	7	−6	0	7	0	−2	0	1	−3	2	0	−2	6	−4	0	2	2	−3	−6	−1
Rank	4	2	3	1	3	1	6	4	1	4	3	2	1	4	2	3	4	1	6	2	2	1	5	7	4

TABLE 6.16

Customer Requirements and Other Criteria for the Evaluation of Four Functional Requirements of the Steel Frame Joining Tool in Table 6.5

Provide user grip [$(FR)_{22}$]
    Controllable/maneuverable
    Comfortable
    Aesthetics
    Nonfatiguing
    Weight
    Size/shape
    Balance
    Safe
    Impact resistant
    One-hand/two-hand operation[a]
Generate energy[$(FR)_{31}$]
    Reliable
    Safe
    Easy to maintain
    Lightweight
    Controllable
    Availability (of energy source)
    Quiet
    Size[a]
    Energy density[a]
    Power consumption[a]
    Environmentally nonhazardous[a]

Convert energy to working element [$(FR)_{32}$]
    Fast conversion
    Lightweight
    Reliable
    Quiet
    Easy to maintain
    Safe
    Controllable
    High efficiency (of transfer to working element)[a]
    Overload protection[a]
    Complexity/simplicity[a]
    Size[a]
Make connection [$(FR)_{33}$]
    Fast
    Results in strong joint
    Makes connection every time
    Non-fatiguing
    Safe
    Number of operations per application[a]
    Works over a range of track and stud thickness[a]
    Complexity/simplicity[a]

[a] Additional criteria and constraints that are not part of the *CR*s.

It is seen that all the concepts are distinct except for two of the concepts for $(FR)_{32}$: mechanical means and a piston. Since at this stage these two concepts are not yet explicitly defined, they will be temporarily combined and denoted mechanical/piston. Thus, the *practical* combinations of these candidate concepts are the following:

Combination	$(FR)_1$	$(FR)_{31}$	$(FR)_{32}$	$(FR)_{33}$
1	C-clamp	Electric	Direct	Weld
2	C-clamp	Electric	Mechanical/piston	Mechanical fastener
3	C-clamp	Electric	Mechanical/piston	Mechanical deformation
4	C-clamp	Explosive	Direct	Mechanical fastener

When these four combinations are evaluated using the customer requirements given in Table 4.4, it is found that combinations 2 and 4 ranked the highest. Returning to Table 6.5 with configurations 2 and 4 in mind, the following decisions can be made. Since there is no requirement that the tool be battery operated, an AC electric source is selected. In addition, one of the means for actuating a piston is mechanically. Since a piston can be used to compress air, and the sudden release of compressed air would result in an explosive-type motion, it appears that concept configurations 2 and 4 can be combined in the following manner. The electric source can be used to run an electric motor, which through a mechanical means actuates a piston to compress air. The sudden release of the compressed air can be used to propel the mechanical fastener through the track and stud flanges.

TABLE 6.17
Evaluation Criteria Used to Determine the Best
Concepts to Satisfy $(FR)_{32}$ and $(FR)_{33}$ in Table 6.5

$(FR)_{32}$	$(FR)_{33}$
Ease of manufacture	Joint strength
Ease of assembly	Maturity of fastener technology
Maximum force	Loading ease
Speed	Easy to deform
Length of stroke	Cost
Size	Consistency of penetration
Weight	Number of fasteners required per joint
Cost	Damage to flanges
Reliability	Penetration force
Complexity	

The mechanical fasteners that are amenable to an explosive-type force are (see the last column of Table 6.5) staples and nails that are bent after piercing the flanges and hollow metal piercing rivets that puncture both flanges, and are deformed after they exit the second flange. The mechanical energy transfer concepts that can have an electric motor to drive the piston to compress the air are (1) linear motor, (2) screw-type drive, (3) slider crank, and (4) cam. When each of these concepts is evaluated using the criteria given in Table 6.17, it is found that the hollow rivet and the motor/slider-crank/piston combination rank the highest in their respective categories. Notice that the evaluation criteria in Table 6.17 now include engineering and manufacturing criteria.

Thus, the final combination for a candidate solution for the steel frame joining tool is to use an AC electric motor to drive a slider-crank mechanism to move a piston to generate compressed air, which is then released suddenly to propel, by some means, a hollow metal piercing rivet into the stud and track flanges.

Having tentatively decided on these concepts, it is seen that the design hierarchy for the functional requirements and the design parameters can be expanded. Thus,

$(FR)_{31}$ = Generate energy	$(DP)_{31}$ = Electric motor
$(FR)_{311}$ = Generate air pressure	$(DP)_{311}$ = Slider-crank and piston
$(FR)_{32}$ = Convert energy to working element	$(DP)_{32}$ = System to convert energy
$(FR)_{321}$ = Store/contain pressurized air	$(DP)_{321}$ = Means to store air
$(FR)_{322}$ = Rapidly release pressurized air to impel working element	$(DP)_{322}$ = Means to release air
$(FR)_{323}$ = Impact working element with fastener	$(DP)_{323}$ = Means to impact working element
$(FR)_{324}$ = Retract working element	$(DP)_{324}$ = Means to retract working element
$(FR)_{33}$ = Make connection	$(DP)_{33}$ = Impulsively actuated hollow rivet

It is seen that at the second level of the hierarchy we have specified how the referent FRs will be satisfied. However, we start with solution neutral statements at the third level.

We now explore in a little more detail the means to attain the $(DP)_{32j}$, $j = 1, 2, 3, 4$. One possible solution is to use the compressed air to impel another piston at sufficient speed into the rivet. As shown in Figure 6.6, one idea for the impact system module requires that three chambers be pressured in an appropriate order. At the start of the cycle, chamber 1 is pressurized to the desired

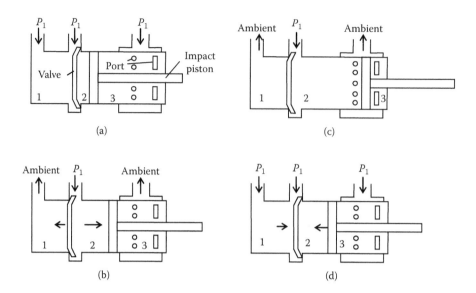

FIGURE 6.6 A means to propel a piston to impact a rivet: (a) at start of cycle impact piston is stationary; (b) chambers 1 and 3 are vented to ambient, valve diverts P_1 to chamber 2, and piston moves toward rivet; (c) impact piston at end of stroke; (d) valve moves to seal chamber 2 so that impact piston can return to the start position.

pressure P_1 as shown in Figure 6.6a. This pressure is generated by the motor/slider-crank/piston shown in Figure 6.7. For each revolution of the motor, the pressurizing piston compresses (pushes) the air in its chamber through the one-way valve into chambers 1 and 3 of the impact piston module. The movement of the impact piston to its return position causes chamber 2 to be slightly pressurized, which causes the spring-loaded valve to move backwards (left). Upon actuation of the tool's trigger mechanism, the pressure in chambers 1 and 3 is vented to ambient conditions, at which time the valve moves backwards (left), opening the port to allow the pressurized air at P_1 into chamber 2 as indicated in Figure 6.6b. The injection of the pressurized air propels the impact piston forward (right), where its end eventually makes contact with the rivet. In Figure 6.6c, it is seen that at the end of the impact piston's stroke, the piston exposes vents to the ambient conditions thereby further reducing the pressure in chamber 2. Chambers 1 and 3 are again pressurized. The impact piston is then forced back to its starting position. The total area of the vent holes is considerably less than the entrance ports on the forward (right) side of the piston, thus permitting more air to flow into chamber 3 than into chamber 2 and forcing the impact piston back to its starting position as indicated in Figure 6.6d.

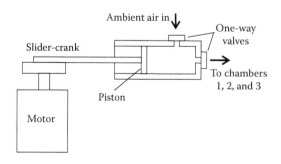

FIGURE 6.7 Means of generating pressure in impact piston chambers of Figure 6.6.

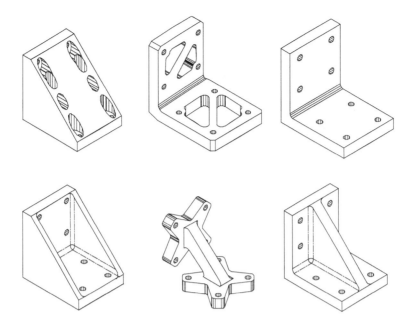

FIGURE 6.8 Six possible embodiments for a right-angled bracket. (Adapted, in part, from S. R. Burgett, R. T. Bush, S. S. Sastry, and C. H. Sequin, "Mechanical Design Synthesis from Sparse, Feature-Based Input," in *SPIE Proceedings, Mathematics and Control in Smart Structures*, V. V. Varadan, Ed., Vol. 2442, pp. 280–291, 1995.)

6.5 PRODUCT EMBODIMENTS

The conversion of a concept to a physical form should also involve a morphological-type of analysis. The idea is to consider as many possible forms as practical that a component or a module can have, which also may be dependent on their arrangements (configuration). The various forms for the component and the various arrangements of the component within the product are then evaluated using the most important criteria listed in Table 2.2 in the manner described in Section 6.4. These evaluation criteria usually relate to one or more aspects of producibility. (It is mentioned that in many patents where more than one embodiment is given the preferred one is identified by the inventor.)

It is during the embodiment stage of the product realization process that one constructs one or more prototypes. In order to do this, preliminary engineering drawings are generated. Then the prototype* is built and operated to verify and confirm that (1) the performance criteria and the customer requirements are satisfied, (2) the reliability and environmental criteria are met, (3) the manufacturing and assembly operations meet their appropriate criteria, (4) safety and legal issues are satisfied, (5) raw materials and purchased components will meet performance and delivery requirements, and (6) cost and time-to-market will be within specified limits. Once these factors have been verified, the final details are completed: the engineering drawings, manufacturing and assembly process plans, and production scheduling. The product then enters production, and shortly thereafter, the marketplace.

To illustrate the various aspects of the conversion of a concept to an embodiment, several different products are examined. Consider first the goal of creating a rigid coupling to connect two plates at right angles to each other. The connection is to be made to each plate by four bolts. Six possible forms that such a bracket could have are shown in Figure 6.8. Which form is selected would

---

* See Section 9.10 for one set of techniques called *layered manufacturing* that can be used to generate prototypes.

be based on its relationship to the overall product, and such evaluation criteria as cost, materials, manufacturing processes, loading, weight, ease of attachment and disassembly, and so on.

It is observed that the gulf between generating a concept and generating its embodiment is large. In fact, there are a very large number of ways the final embodiment can end up, very few of which can be anticipated in the concept development stage. This is why the concept generation, evaluation, configuration, and embodiment aspects of the product realization process have to be performed in a highly overlapping and iterative manner. The iterative nature of this process may also be necessary because one may not always be able to convert the highest ranked concept into an embodiment that can be produced easily and for the targeted cost. Consequently, previously rejected concepts may have to be reexamined and tradeoffs made.

BIBLIOGRAPHY FOR BIO-INSPIRED CONCEPTS

R. M. Alexander, *Size and Shape*, Edward Arnold, Great Britain, 1971.

R. M. Alexander, *Biomechanics*, Chapman and Hall, London, 1975.

R. M. Alexander, *Optima for Animals*, Princeton University Press, Princeton, NJ, 1996.

Y. Bar-Cohen and C. L. Breazeal, Eds., *Biologically-Inspired Intelligent Robots*, SPIE Publications, Bellingham, WA, 2003.

A. Bejan, *Shape and Structure, from Engineering to Nature*, Cambridge University Press, Cambridge, 2000.

J. M. Benyus, *Biomimicry: Innovation Inspired by Nature*, Harper Perennial, New York, 2002.

C. G. Gebelein, *Biomimetic Polymers*, Springer, Berlin, 1990.

S. Hirose, P. Cave, and C. Goulden, *Biologically Inspired Robots: Snake-Like Locomotors and Manipulators*, Oxford University Press, Cambridge, 1993.

E. Laithwaite, *An Inventor in the Garden of Eden*, Cambridge University Press, Cambridge, 1994.

C. Mattheck, *Design in Nature: Learning from Trees*, Springer, Berlin, 1998.

J. McKittrick, J. Aizenberg, C. Orme, and P Vekilov, Eds., *Biological and Biomimetic Materials: Properties to Function*, Material Research Society, Warrendale, PA, 2002.

S. Nolfi and D. Floreano, *Evolutionary Robotics: The Biology, Intelligence, and Technology of Self-Organizing Machines*, MIT Press, Cambridge, MA, 2004.

I. Overington, *Computer Vision: A Unified, Biologically-Inspired Approach*, Elsevier Publishing Company, 1992.

F. R. Paturi, *Nature, Mother of Invention: The Engineering of Plant Life*, Harper & Row, New York, 1976.

H. Tennekes, *The Simple Science of Flight: From Insects to Jumbo Jets*, MIT Press, Cambridge, MA, 1996.

K. Toko, *Biomimetic Sensor Technology*, Cambridge University Press, Cambridge, 2005.

H. Tributsch, *How Life Learned to Live: Adaptation in Nature*, MIT Press, Cambridge, MA, 1982

J. F. V. Vincent, *Structural Biomaterials* (revised edition), Princeton University Press, Princeton, NJ, 1990.

S. Vogel, *Cat's Paws and Catapults: Mechanical Worlds of Nature and People*, Norton, New York, 1998.

S. A. Wainwright, W. D. Biggs, J. D. Currey, and J. M. Gosline, *Mechanical Design in Organisms*, John Wiley & Sons, New York, 1976.

K. Wunderlich and W. Gloede, *Nature as Constructor*, Arco Publishing, New York, 1979.

7 Design for Assembly and Disassembly

Suggestions and examples are given to show how components and products can be created for simplified assembly and disassembly operations.

7.1 INTRODUCTION

There are three very important and inextricably linked elements in the product development cycle that greatly affect the product's cost, time to market, plant productivity, degree of manufacturing automation, producibility, and reliability. These are assembly methods, manufacturing processes, and material selection. As will be discussed subsequently, reducing the number of parts that comprise the component or product is one means of improving its ease of assembly and reducing cost. This reduction is often accomplished by combining several of the individual parts into a single piece, which, in turn, may decrease the number of different materials used in the product. Some manufacturing processes that can be used to accomplish this are casting, injection molding, and sheet-metal bending. Each of these processes works

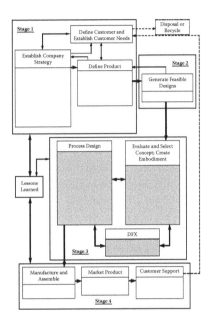

better with certain materials than with others, and with certain shapes, sizes, and geometric attributes. Therefore, one must consider these three elements more or less concurrently. Furthermore, assembly considerations greatly impact, and are impacted by, maintenance (serviceability), testability, and inspection requirements, as discussed in Section 10.3. In addition, how assembly is to be performed is also influenced by, and will influence, the product's architecture, degree of integration, and the degree of modularity employed as discussed in Section 6.3.

This more or less simultaneous consideration of these three elements as the design concept progresses acts to facilitate and integrate the product development process. Furthermore, when using the concept and embodiment evaluation procedures suggested in Chapter 6, these three elements are frequently the most critical, and the ones for which many of the tradeoffs with respect to the satisfaction of the customer requirements are made.

There are also many products that are assembled by the end user, such as bicycles, strollers, desks, bookcases, etc. These products must be designed so that assembly can be performed by the customer in such a manner that they can be assembled easily, quickly, and without mistakes, and that the assembly can be performed with a minimum number of tools, or with no tools.

In this chapter, guidelines for improving product assembly operations are given. Material selection and manufacturing processes are given in Chapters 8 and 9, respectively. Several formal methods are available for the evaluation of a given design to determine whether that design needs to be improved so that it is easier and less costly to assemble. These methods are the Hitachi assemblability evaluation method,[*] the Lucas design for assembly method,[†] and the Boothroyd–Dewhurst

---

[*] U.S. Patent 6223092: Automatic manufacturing evaluation method and system.
[†] For a brief description of the method see http://deed.ryerson.ca/~fil/t/dfmlucas.html.

design for assembly method.[*] These methods, to varying degrees, determine the relative difficulty of an assembly operation, the relative cost of an assembly operation, the number of assembly operations, the appropriate assembly method, and the assembly sequences. From these determinations, the user of one of these methods is guided in how to alter the design to create cost-effective and easy-to-assemble components and products.

Assembly consists of many activities that are more than simply joining parts. In addition, assembly itself may be hierarchical, in which assemblies are joined to other assemblies. The main activities of assembly are[†,‡]

- **Marshaling**, which is a logistic function that may be performed according to strategies that are based on estimates of work schedules, the planned production of various product types, and the parts needed for each type of assembly. Two types of strategies are generally used: (1) the push type, which operates on the basis of a planned production schedule of anticipated final needs for finished assemblies; and (2) the pull strategies, which is another term for just-in-time methods.
- **Transport**, which is the short-term logistic implementation of marshaling and accomplishes the actual carrying of parts or assemblies between stations or work areas. The distance between stations should be kept as close to each other as possible.
- **Part presentation**, which takes parts from the delivery system and places and orients them so that assembly can occur. To facilitate these operations, one should avoid visual obstructions and potentially dangerous arrangements of the assembly components.
- **Part mating**, which is the actual process of fitting parts together.
- **Joining**, which accompanies mating and usually involves fastening in some way: with screws and bolts; adhesive bonding; welding, brazing, and soldering; and press and snap fittings.
- **Inspection**, which determines incorrect assembly operations.
- **Documentation**, which is an indirect operation that records the test results to provide the ability to trace problems back to their causes, to maintain control of the processes, and to permit improvement of the assembly process.

7.2 DESIGN FOR ASSEMBLY

7.2.1 WHY ASSEMBLE?

There are eight reasons why assembly is required[§]:

1. **Relative movement of components:** Various elements must provide a degree of mobility to achieve the desired function, such as hinged elements. Although relative motion requires two or more pieces, these pieces do not have to be made of different materials.
2. **Material differentiation:** The functional realization depends on particular material characteristics, such as the electrical insulation properties of a circuit board.
3. **Production considerations:** Some parts are easier to produce by division into subparts, such as a pipe and a flange that are assembled by welding and some parts can have their manufacturing yield improved if the integration is not excessive, such as combining several smaller integrated circuits versus making one very large-scale integrated circuit.

---

[*] http://www.dfma.com/.

[†] M. M. Andreasen, S. Kahler, T. Lund, and K. Swift, *Design for Assembly*, 2nd ed., IFS Publications, Springer-Verlag, Berlin, 1988.

[‡] D. E. Whitney, *Mechanical Assemblies: Their Design, Manufacture, and Role in Product Development*, Oxford University Press, New York, 2004.

[§] M. M. Andreasen et al., ibid.

4. **Replacement and upgrade:** Service may require disassembly and replacement, such as automobile brake pads and upgrading a PC may require adding more memory.
5. **Differentiation of functions:** A function can be carried out by a single element or a combination of such in the form of more elements, such as the use of a roller bearing to support radial loads and, simultaneously, the use of an axial bearing to support axial loads.
6. **Particular functional conditions:** The requirements of accessibility, demounting, cleaning, inspection, etc., can necessitate a division into elements, such as the hood of a car.
7. **Design considerations:** Aesthetics require a division of the form, which will consequently require assembly, such as the trim on an automobile.
8. **Cost:** It may be less expensive to assemble than to integrate, such as placing components on a printed circuit board versus creating a custom integrated circuit.

An important concept embodied in these seven reasons is the idea of a theoretical minimum number of parts.[*] Thus, by using these eight reasons as a guide, the IP^2D^2 team should attempt to eliminate as many separate parts as possible.

7.2.2 Assembly Principles and Guidelines

The basic idea in the design for assembly is to first reduce the number of components (parts, pieces) that must be assembled, and then to ensure that the remaining components are easy to assemble, are easy to manufacture, reduce the total cost of the assembly, and, of course, satisfy the functional requirements.

The principles governing the design for assembly are as follows.

1. **Simplify and reduce the number of parts** because for each part there is an opportunity for a defective part and an assembly error. Fewer parts mean less of everything that is needed to manufacture a product: engineering time, drawings and part numbers; production control records and inventory; number of purchase orders and vendors; number of bins, containers, stock locations, and buffers; amount of material-handling equipment, containers, and number of moves; amount of accounting details and calculations; service parts and catalogues; number of items to inspect and type of inspections required; and amount and complexity of part production equipment and facilities, assembly, and training. Note that these suggestions are in accordance with satisfaction of Axiom 2, the information axiom, in Section 5.1.3.
2. **Standardize and use common parts and materials** to facilitate design activities, to minimize the amount of inventory in the system, and to standardize handling and assembly operations. The use of common parts will result in lower inventories, reduced costs, and simplified operator learning. Note that these suggestions are in accordance with satisfaction of Axiom 2, the information axiom, in Section 5.1.3.
3. **Mistake-proof product design and assembly** so that the assembly process is unambiguous. Components should be designed so that they can only be assembled in one way: they cannot be reversed. Notches, asymmetrical holes, and stops can be used to mistake-proof the assembly process. Design verification can be achieved with simple go-no-go tools in the form of notches or natural stopping points. Products should be designed to avoid adjustments. These ideas are discussed further in Section 10.2.
4. **Design for parts orientation and handling** to minimize the effort and ambiguity in orienting and merging parts. Parts should be designed to orient themselves when fed into a process. Product design must avoid parts that can become tangled, wedged, or disoriented. Part design should incorporate symmetry, low centers of gravity, easily identifiable

[*] G. Boothroyd and L. Alting, "Design for Assembly and Disassembly," *Annals of CIRP*, Vol. 4/2, pp. 625–636, 1992.

features, guide surfaces, and points for easy pick-up and handling. This type of design may allow the use of automation in parts handling and assembly such as vibratory bowls, tubes, magazines, pick and place robots, and vision systems. When purchasing components, consider acquiring materials already oriented in magazines, bands, tape, or strips. Note that these suggestions are in accordance with satisfaction of Axiom 2, the information axiom, in Section 5.1.3.

5. **Minimize flexible parts and interconnections** by avoiding flexible and flimsy parts such as belts, gaskets, tubing, cables, and wire harnesses. Their flexibility makes material handling and assembly more difficult, and these parts are more susceptible to damage. Use plug-in boards and back planes to minimize wire harnesses. Where harnesses are used, consider mistake-proofing electrical connectors by using unique connectors to avoid incorrectly connecting them. Interconnections such as wire harnesses, hydraulic lines, and piping are expensive to fabricate, assemble, and service. Partition the product to minimize interconnections between modules and place related modules adjacent to each other to minimize the routing of interconnections.

6. **Design for ease of assembly by utilizing simple patterns of movement and minimizing the number of axes of assembly**. Complex orientation and assembly movements in various directions should be avoided. Parts should include such features as chamfers and tapers. The product's design should enable assembly to begin with a base component with a large relative mass and a low center of gravity upon which other parts are added. Assembly should proceed vertically with other parts added on top and positioned with the aid of gravity. This minimizes the need to reorient the assembly and reduces the need for temporary fastening and additional fixturing. A product that is easy to assemble manually frequently will be easily assembled with automation.

7. **Design for efficient joining and fastening**. Threaded fasteners (screws, bolts, nuts, and washers) are time-consuming to use in assembly and difficult to automate. Where they must be used, standardize to minimize variety, and use fasteners such as self-threading screws and captured washers. Consider the use of adhesives and connectors that snap together. Match fastening techniques to materials and product requirements.

8. **Design modular products to facilitate assembly**. A modular design should minimize the number of part and assembly variations and manufacturing processes, while allowing for greater product variation during final assembly. This approach minimizes the total number of items to be manufactured, thereby reducing inventory and improving quality. Modules can be manufactured and assembled in parallel to reduce the product's overall production time, and are more easily tested before final assembly. Short final assembly lead-time can result in a wide variety of products being made in a short period of time without having to stock a significant level of inventory. These suggestions are important to the implementation of mass customization discussed in Section 1.7.

7.2.3 SUMMARY OF DESIGN-FOR-ASSEMBLY GUIDELINES

We shall now introduce in summary format several design-for-assembly guidelines.

1. **Overall component count should be minimized.**
   - Modify the design to make the number of parts as low as possible by reconsidering alternative product concepts.
   - Seek innovative ways to eliminate the reasons why a component must be comprised of separate parts.
   - Modify the design to reduce the number of extra parts needed to provide a desired range of product and model variations.
   - Check all parts for function and modify the design to eliminate redundant parts.

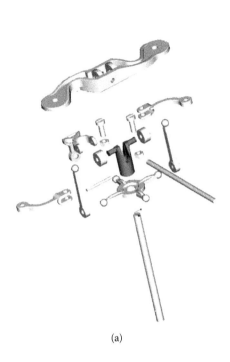

(a)

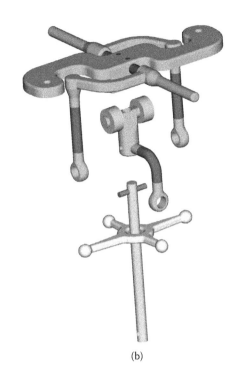

(b)

FIGURE 7.1 Examples of ways to reduce the overall number of components: (a) a component with many pieces; (b) part redesigned and manufactured by a multimold manufacturing process to substantially reduce the number of individual pieces and to provide some of the elements with elasticity. (From R.M. Gouker, S.K. Gupta, H.A. Bruck, and T. Holzschuh, "Manufacturing of multi-material compliant mechanisms using multi-material molding," *International Journal of Advanced Manufacturing Technology*, 30(11–12): 1049–1075, 2006. © Springer 2006. With permission from Springer Science+Business Media.)

To illustrate some of these guidelines,[*] consider the part shown in Figure 7.1. Figure 7.1a shows how a part that was originally composed of numerous pieces was redesigned and manufactured by a multimold process (see Section 9.2.6) to greatly minimize the number of pieces. The redesigned and manufactured system is shown in Figure 7.1b.

2. Use a minimum number of separate fasteners.
- Use fewer large fasteners rather than many small fasteners.
- Use a minimum number of types of fasteners.
- Design screw assembly for downward motion.
- Minimize the use of separate nuts and consider captive fasteners when applicable.
- Avoid separate washers.
- Use self-tapping screws when possible.
- Minimize electrical cables; plug electrical subassemblies directly together.
- Minimize the number of types of cables.

To illustrate some of these guidelines, consider the part shown in Figure 7.2 where it is shown that it is better to keep the holes, and therefore, the bolts or screws, the same size and to reduce the overall number of them.

---

[*] Additional examples can be found in G. Boothroyd, *Assembly Automation and Product Design*, Marcel Dekker, New York, 1992; G. Pahl, W. Beitz, J. Feldhusen, and K.-H. Grote, ibid.; M. M. Andreasen et al., ibid.; R. Bakerjian, Ed., *Tool and Manufacturing Engineers Handbook*, Vol. 6, *Design for Manufacturability*, SME, Dearborn, MI, 1993; and D. G. Ullman, ibid.

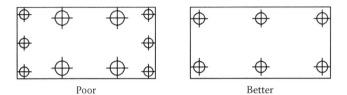

FIGURE 7.2 Reduce the number and types of fasteners.

3. Design the product with a base component for locating other components quickly and accurately.
- Don't require the base to be repositioned (reoriented) during assembly.
- Fastened parts should be located before a fastener is applied.
- Provide registration and fixturing locations.
- Provide self-locating assemblies.
- Design products so that they can be laid on top of each other, that is, stacked.

To illustrate some of these guidelines, consider the parts shown in Figure 7.3, which shows that a recess ensures alignment with a blind hole.

4. Make the assembly sequence efficient.
- Minimize handling.
- Use modular design.
- Avoid simultaneous operations.
- Provide for easy handling.
- Avoid flexible materials.
- Minimize part variation.
- Shape parts unambiguously so that they can't be assembled incorrectly.
- Use subassemblies, especially if different processes are used to produce different parts.
- Purchase subassemblies assembled and tested.

To illustrate some of these guidelines, consider the part shown in Figure 7.4. Here it is shown that by simply making one of the elements to be inserted longer than the other element that is to be inserted avoids having to perform two assembly operations simultaneously.

5. Avoid component characteristics that complicate retrieval.
- Avoid designs that cause tangling and nesting of identical parts.
- Employ features to provide positive holding of components.

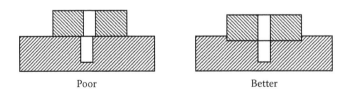

FIGURE 7.3 Provide guides for self-adjustment and alignments.

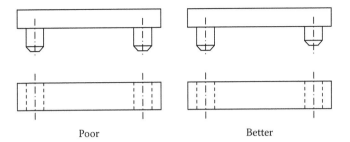

FIGURE 7.4 Avoid simultaneous operations.

To illustrate these guidelines, consider the parts shown in Figure 7.5. The part on the left side has its slot narrower than the thickness of the part, thereby preventing it from entering the slot of another part. The part on the right side was converted from an open part to one that is closed, thereby preventing them from tangling.

6. **Design components for a specific type of retrieval, handling, and insertion.**
   - Use assembly motions that are simple.
   - Standardize features and use standard parts where possible.
7. **Design components to mate through straight-line assembly, all from the same direction.**
8. **Make use of chamfer and compliance to facilitate insertion and alignment.**
   - Design for assembly motions that can be done with one hand.
   - Design for assembly motions that do not require skill or judgment.
   - Design products that do not need any mechanical or electrical adjustments.
   - Use the widest tolerances that are consistent with functional, quality, and safety objectives.

9. **Maximize component accessibility and visibility to provide unobstructed access to parts and tools.**
   - Make parts independently replaceable, that is, without having to remove other parts first.
   - Order the assembly so that the most reliable ones go in first and the most likely to fail, last.
   - Anticipate future options so that they can be added easily.
   - Ensure that the product's life can be extended with future upgrades; for example, leave room for future replacements.

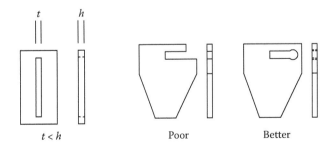

FIGURE 7.5 Create parts that can't tangle with themselves.

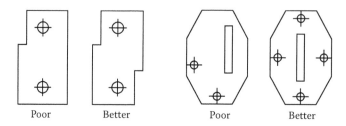

FIGURE 7.6 Use symmetrical features.

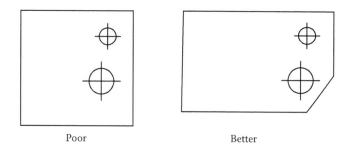

FIGURE 7.7 Increase asymmetry when symmetry can't be used.

10. Design components for either symmetry or asymmetry.
  - Design components for symmetry if no preferred orientation is required.
  - Design components for symmetry about their axes of insertion.
  - Design components that are not symmetric to be clearly asymmetric.
  - Avoid right and left hand parts.

Illustrations of how to increase symmetry and asymmetry are shown in Figures 7.6 and 7.7, respectively.

7.2.4 Manual Assembly versus Automatic Assembly[*]

One can generally assume that products designed for automatic assembly can easily be assembled manually, except for those operations demanding high power and/or accuracy. A good guideline for ease of automatic assembly is, therefore, to design the product so that it is easy to assemble manually; then there will be a high probability that it will be easily assembled automatically. Manual assembly has the great advantage of flexibility with respect to various product types, product variation, component variation, faulty components, and unforeseen assembly problems. It also requires lower investment in equipment and often provides greater job satisfaction than with mechanical and automatic assembly. Manual assembly is best, therefore, when conversion ability is required, either because of frequently changing tasks or complexity.

Design products for mechanized and partial or fully automated assembly in order to achieve assemblies of uniformly high quality and for large production runs. Use automated assembly to provide more uniform operator effort.

---

[*] M. M. Andreasen et al., ibid.

7.3   DESIGN FOR DISASSEMBLY (DFD)

7.3.1   INTRODUCTION

The growing concern for environmental protection, occupational health, and resource utilization has stimulated many new activities to cope with the problems created by the steadily increasing consumption of industrial and consumer products. One of the major problems is the disposal of used products. Even though recycling is steadily increasing, huge amounts of solid waste are still disposed in landfills creating serious pollution and health problems.

The disassembly of products—the separation by nondestructive or semidestructive means—appears to be one of the more serious problems impeding the reuse of products, since many products are not designed for easy disassembly. Integrated design (as opposed to modular design), certain fastening and assembly principles, and surface coatings, for example, can make it very difficult to disassemble a product and to separate materials into homogeneous groups.

The influence of design for disassembly (DFD) on recyclability and easy disassembly is that it makes it possible to reuse, remanufacture, and recycle materials in an efficient manner. Reuse and remanufacture can save many resources by prolonging the useful life of products. The automobile industry has used these principles for some time as demonstrated by the ubiquitous presence of automobile junk yards and the availability of a wide range of rebuilt automotive components. With regard to copiers, about 60% of copier toner cartridges are being returned to companies for refurbishing and resale.

When employing DFD, the life-cycle phases of development, production, distribution, usage, and disposal or recycling are considered simultaneously, from the conceptual product design stage through the detailed design stage. In this approach, policies must be established for environmental, occupational health, and resource issues, as well as for the disposal or recycling of the used products. The design for disassembly is therefore closely related to the design for the environment and to the design for maintainability, which are presented in Section 10.5 and Section 10.3, respectively.

Assembly and disassembly may also be relevant (1) for large construction projects where onsite assembly is required; (2) to make a product compatible with the modes of transportation with respect to size, weight, and packaging; and (3) for products that are assembled by the user.

7.3.2   DFD GUIDELINES AND THE EFFECTS ON THE DESIGN FOR ASSEMBLY

There are two strategies in planning the dismantling of a product[*]:

- Remove the most valuable parts first and stop dismantling when the marginal return on the operation becomes unfavorable.
- Maximize the yield of each dismantling operation by having the dismantling operation releasing many parts at once.

There are two hierarchies in the dismantling process: one for the destination of the components and the other for the quality of the dismantled components. The higher one is in the destination hierarchy the more the investment of raw materials, labor, and energy is conserved. This hierarchy is:

- Refurbish.
- Reuse.
- Remanufacture.
- Recycle to high-grade material.

---

[*] M. Simon, "Design for Dismantling," *Professional Engineer*, Vol. 17, No. 10, pp. 20–22, November 1991.

- Recycle to low-grade material.
- Incinerate for energy content.
- Dump in landfill site.

When designing for disassembly one should consider the following:

- **Identification:** Materials and parts must be identified. For example, in the United States polymers are classified into seven categories.
- **Recycling:** To recycle metals and alloys, their grade or levels of contaminants must be known. The general principle is to reduce variety.
- **Compatibility:** Ensure compatibility of materials. For example, swelling of a product component due to corrosion locks parts together and the corrosion (or wear) of the heads of fasteners can also make dismantling more difficult.
- **Overall form and structural strength:** Build in weak points for dismantling in such a way that service loads can still be carried.
- **Fasteners and adhesives:** Consider such means as fasteners that insert as a rivet, but can be removed as a screw, and adhesives that are strong in shear but are weak when peeled apart.

How these guidelines interact with, and affect, the design for assembly are summarized below.[*]

Positive Effects on Assembly
- Reduce the number of components.
- Reduce the number of separate fasteners.
- Provide open access and visibility for separation points.
- Avoid orientation changes during disassembly.
- Avoid nonrigid parts.
- Ensure that disassembly can be done with common tools and equipment.
- Design for ease of handling and cleaning of all components.

Negative Effects on Assembly
- Design two-way snap fits or break points on snap fits.
- Use joining elements that are detachable or easy to destroy.
- Design for ease of separation of components.
- Use water-soluble adhesives.

DFD Guidelines Having Relatively Little Effect on Assembly
- Design products for reuse.
- Eliminate need to separate parts.
- Reduce number of different materials.
- Enable simultaneous separation and disassembly.
- Place components in logical groups according to recycling and disassembly sequences.
- Identify separation points and materials.
- Facilitate the sorting of noncompatible materials.
- Use molded-in material identification in multiple locations.
- Provide a technique to safely dispose of hazardous waste.
- Select an efficient disassembly sequence.

---

[*] J. F. Scheuring, B. Bras, and K. M. Lee, "Effects of Design for Assembly on Integrated Disassembly and Assembly Processes," Proceedings of the Fourth International Conference on Computer Integrated Manufacturing and Automation Technology, Troy, NY, pp. 53–59, October 1994.

8 Material Selection

The important attributes of a wide range of engineering materials are summarized, and typical products that are made with these materials are given.

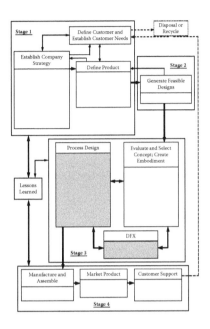

8.1 INTRODUCTION

8.1.1 IMPORTANCE OF MATERIALS IN PRODUCT DEVELOPMENT

The development of materials has been one of the major influences in the creation of modern products. New products greatly depend on the properties of the materials used to create them, so much so that entire eras in human development have been categorized by the dominant material in each era—from the Stone Age to the Bronze Age to the Iron Age and beyond. The development of new materials has accelerated so rapidly that of all the materials available to the product designer, most have been developed in the last 100 years. This includes all engineered polymers, most engineered ceramics, many new metallurgical alloys, and a host of composite materials.

Many properties of a material are a function of their atomic structure: chemistry, crystal structure, and atomic or molecular bonding. Examples of these atomic properties are the Young's modulus, coefficient of thermal expansion, and melting point. Other properties are primarily a function of the way in which the material has been processed, such as yield strength, ductility, fracture toughness, and corrosion resistance. Thus, in order to select the appropriate material for a specific task, it is necessary to specify its processing. The processes that are used to form the material affect its microstructure. Consequently, understanding these process-microstructure-property relationships is necessary in order to correctly select the appropriate material for a given application.

In this chapter, a number of broad categories of materials will be discussed,* and criteria will be presented that can be used in selecting the appropriate material for a wide range of applications. In addition, the ways in which these properties result from and are influenced by the material processing method and the material microstructure will also be addressed.

8.1.2 GUIDELINES FOR MATERIALS SELECTION

All products are composed of one or more materials. The range of properties that one has to consider when selecting a material is summarized in Table 8.1. For many products in the initial stages

---

* A list of properties of these materials is given in Appendix A, Table A.1.

TABLE 8.1
Examples of the Types of Materials Information Required during the Product Realization Process[a]

Material identification
  Class (metal, plastic, ceramic,
    composite)
  Subclass
  Industry designation
  Condition designation (heat
    treatment, etc.)
  Specification (grade)
  Alternative names
  Component designations
    (composite/assembly)
Material production history
  Manufacturability strengths and
    limitations
  Composition
  Condition (fabrication)
  Assembly technology
  Constitutive equations relating to
    properties
Temperature (cryogenic to elevated)
  Creep rates, rupture life at elevated
    temperatures
  Relaxation at elevated temperatures
Environmental stability
  Compatibility data
  General corrosion resistance
  Solvent resistance
  Chemical reactivity
  Stress corrosion cracking
    resistance
  Toxicity (at all stages of production
    and operation)
  Recyclable/disposal
Damage tolerance
  Fracture toughness
  Fatigue crack growth rates (define
    environment and load)
  Temperature effects
  Thermal shock

Material properties
  Density
  Specific heat
  Coefficient of thermal expansion
  Thermal conductivity
  Yield strength
  Elongation
  Reduction of area
  Moduli of elasticity (tension/
    compression)
  Stress strain curve or equations
  Hardness
  Fatigue strength (define test
    methods, load, and
    environment)
  Conductivity (electrical/thermal)
  Dielectric content
  Poisson's ratio
  Ultimate strength
  Damping coefficient
  Wear (galling, abrasion, erosion)
  Melting point
  Porosity
  Permeability
  Transparency
  Electric arc resistance
  Magnetic
Joining technology
  Fusion
  Adhesive bonding
  Fasteners
  Welding
  Soldering
  Brazing
Finishing technology
  Impregnation
  Painting
  Electroplating
  Oxidation

Processability information
  Finishing characteristics
  Weldability/joining technologies
  Suitability for forging, extrusion,
    and rolling
  Formability (finished product)
  Castability
  Repairability
  Flammability
  Machinability
  Hardenability
  Heat Treatability
  Shrinkage
Application history/experience
  Successful uses
  Unsuccessful uses
  Applications to be avoided
  Failure analysis reports
  Maximum life service
Availability
  Multiple sources
  Vendors
  Sizes
  Forms
Cost/cost factors
  Raw material
  Finished product or required added
    processing
  Special finishing/protection
  Special tooling/tooling costs
Quality control/assurance issues
  Inspectability
  Repair
  Repeatability
  Failure mechanism
  Defects (typical)
Material design properties
  Tension
  Compression
  Shear
  Bearing
  Controlled strain fatigue life

[a] Compiled from R. Bakerjian, Ed., *Tool and Manufacturing Engineers Handbook*, Vol. 6, *Design for Manufacturability*, SME, Dearborn, MI, 1993; "Computer-aided Materials Selection during Structural Design," National Materials Advisory Board, Washington, DC, 1995; E. H. Cornish, *Materials and the Designer*, Cambridge University Press, Cambridge, 1987; and G. Dieter, *Engineering Design: A Materials and Processing Approach*, 2nd ed., McGraw-Hill, New York, 1991.

of their development, materials are selected from a host of different options based upon four critical characteristics that are needed to satisfy the functional requirements and their constraints.

- Performance.
    Mechanical and physical properties.
    Electrical, thermal, and magnetic properties.
- Producibility.
- Reliability and environmental resistance.
- Cost.

8.1.2.1   Performance

This characteristic refers to those properties that are required for the product to satisfy its functional requirements. Materials typically perform one or more functions in a product such as carrying loads, providing heat conduction or thermal insulation, providing electrical conduction or insulation, or containing fluids. The selection of a material to meet the functional requirements of a product, or one of the product's subsystems, is an important part of the IP^2D^2 team's decision-making process.

From Table 8.1, it is seen that there are many different mechanical and physical properties of materials that can be used to assess the suitability of a material in meeting a wide range of functional requirements. However, the main mechanical concerns in preliminary design are those that deal with strength (e.g., yield strength), weight of the product (e.g., density), and deformation of the material under loading (e.g., Young's modulus). The main modes of mechanical failure are failure by yield, failure due to fracture, and creep failure due to elevated temperatures. Typical electrical, thermal, and magnetic properties are dielectric strength, electrical resistivity, thermal expansion, thermal conductivity, and magnetic permeability. Additional concerns are flammability and maximum/minimum operating temperatures, which can change the material structure (e.g., by melting), change its mechanical performance (e.g., creep), or change its electrical or magnetic performance (e.g., exceed the Curie temperature). Electrical resistivity and magnetic permeability frequently are selected as the main electrical and magnetic properties that need to be considered during the preliminary material selection process. A material's main thermal properties are described by its thermal distortion (change of shape with temperature) and its thermal insulation characteristics, which are based on thermal conductivity, specific heat, and the coefficient of thermal expansion.

8.1.2.2   Producibility

Material selection cannot be made independently of the selection of the manufacturing process, since the manufacturing process will affect the performance properties of the material. Furthermore, the selection of the manufacturing process will depend on certain properties of the materials. Material properties that can dictate the choice of a manufacturing process include ductility, toughness, formability, and castability. In addition, one must take into account the geometric attributes of the product (i.e., shape, size) and the quantity to be manufactured, since not all manufacturing processes are suitable for all product sizes and manufacturing volumes. Consequently, one must more or less simultaneous evaluate candidate materials and manufacturing processes in order to evaluate a product's producibility.

8.1.2.3   Reliability and Environmental Resistance

This characteristic relates to the durability of a material, which is its ability to resist deterioration in the environment in which it will be used. It includes such properties as fatigue resistance and resistance to radiation, chemical solvents, and corrosive agents.

8.1.2.4   Cost

This characteristic relates to the overall product cost of a specific material. A primary consideration is raw material cost.* Raw material cost metrics include cost per unit weight, cost per unit strength, or cost per unit elastic modulus, depending upon the application. Overall product cost often involves factors other than just raw material cost. These include the cost of processing the material into the desired shape and with the desired properties, the overall life cycle cost associated with the material in terms of its reliability and its need for replacement over the life of the product, and the salvage, recycling, and/or waste cost associated with its removal from service. These cost considerations are discussed in Chapter 3.

In selecting materials for many products, optimizing a single material property does not necessarily produce the best performance. Often, the desired performance criteria are combinations of two or more material attributes such as the strength-to-weight ratio (rather than just strength or weight) and the stiffness-to-weight ratio (rather than just stiffness or weight). Ashby[†] has developed a very straightforward way of presenting multiple material properties so as to permit evaluation of the material performance criteria for specific product applications. Example products include a spring, in which the greatest energy stored before fracture is desired, and a hinge, in which the greatest deflection before fracture is desired. In the case of the spring, the desired material performance criterion is the square of fracture toughness divided by the modulus. In the case of the hinge, it is the fracture toughness divided by the modulus.

The following materials selection parameters will satisfy many applications.

Mechanical and physical properties
- Failure mode: yield strength/density.
  - **or** fracture toughness/density.
  - **or** creep strength/density.
- Elastic modulus/density.
- High temperature strength/density.
- Density.

Electrical, magnetic, and thermal properties
- Electrical resistivity.
- Magnetic field strength.
- Thermal insulation.
- Thermal distortion.

Environmental and reliability factors
- Solvent resistance.

Cost factors
- Relative material cost.

The various materials discussed in this chapter have been summarized according to these categories in Table 8.2. The fifth major category, production methods, has been deliberately omitted at this point so that all materials that satisfy the above selection criteria can be considered independently of the manufacturing methods. The manufacturing processes are then coupled to the materials only after an equivalent set of manufacturing selection criteria, which are independent of the above material selection criteria, are considered. See Section 9.1 and Table 9.2.

The reason for dividing some of the mechanical and physical properties by the density is to find materials that can provide the same or higher levels of mechanical performance with lighter weight. Consider first the ratio yield strength/density. Strength is an extensible property; that is, the

---

* A list of the relative costs of these materials is given in Appendix A, Table A.2.
† M. F. Ashby, *Material Selection in Mechanical Design*, Pergamon Press, Oxford, 1992.

TABLE 8.2
Relationship of Various Materials to Several Material Selection Parameters[*]

Material	Yield Strength/ Density	Creep Strength/ Density	Fracture Toughness/ Density	Young's Modulus/ Density	Density	Magnetic Property	Resistivity	Thermal Distortion	Thermal Insulation	Solvent Resistance
Low carbon steel	M	M–H	H	H	H	M–H	L	L–M	L	M–H
Medium carbon steel	M–H	M–H	M	H	H	M–H	L	M	L	M–H
High carbon steel	M–H	M–H		H	H	M–H	L	M	L	M–H
Low alloy steel	M–H	M–H	L–H	H	H	M–H	L–M	L–H	L–M	M–H
Tool steel	M–H	M–H	L–H	H	H	M–H		M–H	N/A	M–H
Stainless steel	L–H	M–H	M–H	H	H	M–H	L	M–H	L–M	M–H
Gray iron	L	L	L	M–H	H	M–H		M		M–H
Malleable iron	L–M	M–H	L–M	H	H	M–H		M		M–H
Ductile iron	L–M	L–H	L–M	H	H	M–H	M	M	L	M–H
Alloy cast iron				H	H	M–H	M	L–H		M–H
Zinc alloys	L–M	L	L–M	M–H	H	L	L	L–M	L	M–H
Aluminum alloys	L–H	L–H	M–H	H	M	L	L	L–M	L	M–H
Magnesium alloys	L–H	L–H	M	H	L	L	L	L–H	L–M	M–H
Titanium alloys	M–H	M–H	L–H	H	M	L	M	L		M–H
Copper alloys	L–H		L–H	M–H	H	L	L	L–H	L–M	M–H
Nickel alloys	L–H	L–H	M–H	H	H	H	L–M	L–H	L–M	M–H
Tin alloys	L–M			L–H	H	L	L	M	L–M	M–H
Cobalt alloys	M			H	H	H	M	L		M–H
Molybdenum alloys	M			H	H	L	L	L	L	M–H
Tungsten alloys	L–M		L	H	H	L	L	L	L	M–H
Low expansion alloys	L–H			H	H	L	L–M	L–H	M	M–H
Permanent magnet alloys					M–H	H	L–M			M–H
Electric resistance alloys				H	H	L	L–M	M		M–H
ABS	L–M	L	L	L	L	L	M	H	L–H	L
Acetals	M	L		L	L	L	H	H	L–H	M

(continued)

TABLE 8.2 (CONTINUED)
Relationship of Various Materials to Several Material Selection Parameters*

Material	Yield Strength/ Density	Creep Strength/ Density	Fracture Toughness/ Density	Young's Modulus/ Density	Density	Magnetic Property	Resistivity	Thermal Distortion	Thermal Insulation	Solvent Resistance
Nylons	L–M	L–M	L	L	L	L	H	H	L–H	M
Fluorocarbons	L	L		L	L	L	H	H	L–H	M–H
Polycarbonates	M	L	L	L	L	L	H	H	L–H	L
Polyimides	L–M	L–M		L	L	L	H	L–H		M
Polystyrene	L–M	L	L	L	L	L	H	H	L–H	L
PVC	L–M	L	L	L	L	L	H	H	L–H	L
Polyurethane	L–M	L	L	L–M	L	L	H	H		L
Polythene	L–M	L	L	L	L	L	H	H	L–H	M
Polypropylene	L–M	L	L	L	L	L	H	H	L–H	M
Acrylics	M			L	L	L	H	H		L
Alkyds	L–M	L–M		L–H	L	L	H	M		
Epoxies	L–H	L–H	H	L–M	L	L	H	M–H	L–H	M–H
Phenolics	L–M	L–M		L–H	L	L	H	L–H		M
Silicones	L	L		L–M	L–M	L	H	H		L
Polyester	L–H	L–H	L	M–H	L	L	H	M–H		M–H
Rubbers					L	L				
Cellulosics	L–M	L		L	L	L	H	H	L–H	L
Units	KPa/kg/m³	MPa °C/ (kg/m³)	KPa m$^{1/2}$/ (kg/m³)	MPa/kg/m³	kg/m³		μ ohm cm	W/μm	KJ s$^{1/2}$/ (m² K)	
H	>100	>30	>10	>15	>5000		>10000	<2	<5	
M	30 to 100	10 to 30	5 to 10	5 to 15	2500 to 5000		25 to 10000	2 to 5	5 to 10	
L	<30	<10	<5	>5	<2500		<25	>5	>10	

* A. Kunchithapatham, "A Manufacturing Process and Materials Design Advisor," M.S. thesis, University of Maryland at College Park. May 1996.

load carrying capability of a component can be increased by increasing its size. This will result in increased volume, weight, and material costs. Materials that can provide high strength with reasonably sized components are said to have high structural efficiency. Structural efficiency thus favors materials that have a high strength and low weight. This is especially important in vehicles, such as automobiles and aircraft, where components must withstand high stresses without yielding, while possessing low weight for fuel economy and performance.

Next consider the ratio elastic modulus/density. For a given geometry, elastic modulus is a measure of the stiffness of a material, which is the ability to resist deflection, extension, or contraction when acted upon by a load. The stiffness of the material will determine the deflections it undergoes, the amount of energy it can reversibly absorb, and whether or not it will fail by buckling. In many applications, such as in the design of bridge beams and supports, failure may be defined by buckling or by the magnitude of deflections rather than by yielding or fracture. The materials metric of interest in such applications is of the form E^k/ρ, where E is the Young's modulus, k is a number that varies with the geometry of the section and the nature of the loading, and ρ is the density of the material. Thus, elastic modulus per density favors stiff materials with low weight.

Lastly, we consider the ratio high temperature strength/density. Many materials have sufficient strength to perform specific functions satisfactorily at room temperatures, but their yield strength decreases as the temperature increases. This is a result of many phenomena. For metals, this loss of yield strength is a result of the strengthening process itself. One way to increase the yield strength of a material is to introduce permanent deformation into the material during the manufacturing process, a process known as *work hardening*. When the material is subsequently heated to a temperature in excess of half its melting point in °K, the reverse process, known as *annealing*, occurs. During annealing, the material recovers and re-crystallizes, thereby removing the deformation that initially caused the strengthening. Another process that occurs in metals is *creep*. This is a flow-type phenomenon that also occurs at temperatures near half the melting point in °K. In this case, the material begins to deform with time under a constant stress. Polymers also lose their strength above their glass transition temperature, which is the temperature at which long-range molecular motion of the polymer chains is initiated. Glasses lose their strength above their softening temperature, where they begin to exhibit flow characteristics.

While these transition points are important for a fundamental understanding of material behavior, it is often more appropriate when selecting materials for specific applications to use other, more practical, metrics. These include the *heat deflection temperatures*, which are temperatures at which the material begins to deflect under specific load levels, and the *working temperature range*, a range over which there is a very small decrease in the yield strength. The product of the maximum working temperature (the temperature at which the yield strength drops to 80% of its value at room temperature) and the average yield strength in this working temperature range is a good measure of the high temperature strength capabilities. This parameter is useful in selecting cookware and chemical processing equipment, where a material with slightly lower yield strength and a larger working temperature range will be more useful than a material with higher yield strength and a narrow working temperature range. Dividing this product by the density gives the ratio strength/density, which is important, for example, in selecting materials for engine components.

In the following sections, the important attributes of metals, plastics, and ceramics are discussed. Typically, metals are usually strong, stiff, tough, and electrically and thermally conducting. Plastics are usually weak, compliant, durable, temperature sensitive, and electrically and thermally insulating. Ceramics are strong, brittle, durable, refractory (do not deform at high temperature), and electrically insulating. In addition, ceramics are often, but not always, thermally insulating.

The materials discussed in the following sections are organized as shown in Table 8.3. An example of how to use the information in this chapter is given at the end of Section 9.1, since material selection should not be performed without consideration of the manufacturing methods.

TABLE 8.3

Organization of the Materials Discussed in Chapter 8

Metals	Polymers	Other
Ferrous	**Thermoplastics—Partially Crystalline**	**Ceramics**
Plain carbon steel	Polyethylene	Structural
Alloy steels	Polypropylene	Electrically insulating
Low alloy steels	Acetals	Thermally conductive
Tool steels	Nylons	Magnetic
Stainless steels	Fluorocarbons	**Composites**
Cast irons	Polyimides	Metal matrix
Gray irons	Cellulosic materials	Fiber reinforced
Malleable irons	**Thermoplastics—Amorphous**	Carbon/carbon
Ductile irons	Polycarbonates	Cemented carbides
Alloy cast irons	Acrylonitrile butadiene styrene (ABS)	Functionally graded
Nonferrous Alloys	Polystyrene	Smart Materials
Light alloys	Polyvinyl chloride	Piezoelectric
Zinc alloys	Polyurethane	Magnetostrictive
Aluminum alloys	**Thermosets—Highly Crosslinked**	Shape Memory
Magnesium alloys	Epoxies	**Nanomaterials**
Titanium alloys	Phenolics	**Coating**
Heavy alloys	Polyesters	Wear and scratch resistant
Copper alloys	**Thermosets—Lightly Crosslinked**	Electrically conductive and insulating
Nickel alloys	Silicone resins	
Tin alloys	Acrylics	
Cobalt alloys	Rubbers	
Refractory metals	**Engineered**	
Molybdenum alloys		
Tungsten alloys		
Special purpose alloys		
Low-expansion alloys		
Permanent magnet materials		
Electrical resistance alloys		

8.2 FERROUS ALLOYS

8.2.1 Plain Carbon Steels

Steel refers to an iron-based alloy containing less than 2% carbon by weight. These alloys are formed by heating to a temperature between 723°C and 1148°C to fully dissolve the carbon in the interstitial spaces between the iron atoms, in a process known as *austenitization*, and then cooling the solid solution to room temperature. The hardness of the steel depends on the cooling rate. Fast cooling rates form a hard, brittle steel microstructure known as *martensite*, while slower cooling rates form a soft, ductile steel microstructure known as *pearlite*. The final product can be *cast*, that is, poured from the molten state to a near-net or final shape, or *wrought*, where the material is plastically deformed into the desired shape using such processes as rolling and forging. Steel is considered a plain carbon steel when it contains less than 2% total alloying elements and when no minimum content is either specified or required for additives such as chromium, cobalt, columbium (niobium), molybdenum, nickel, titanium, tungsten, vanadium, or zirconium. Variations in the carbon content have the greatest effect on the mechanical properties, with increasing carbon content leading to increased hardness and strength and reduced ductility. Plain carbon steels are generally subdivided into low-carbon steels, medium-carbon steels, and high-carbon steels according to their carbon content. Typical products made from plain carbon steels are given in Table 8.4.

- **Low-carbon steels** contain up to 0.3% carbon and are typically not heat treated to produce high strength microstructures. The largest application of this class of steel is flat-rolled products usually in the cold-rolled and the annealed condition. The carbon content for these high-formability steels is very low (i.e., less than 0.1% carbon).
- **Medium-carbon steels** are similar to low-carbon steels except that the carbon ranges from 0.3% to 0.6%. These steels are strengthened by quenching and tempering and are the most versatile.
- **High-carbon steels** contain from 0.6% to 1.0% carbon and are used to form spring materials. These steels are most costly to make, have poor formability and weldability, and have more carbon than is needed for maximum as-quenched hardness.

8.2.2 ALLOY STEELS

There are a wide range of reasons for placing alloying elements into steel. These include the following:

- **Increasing strength and hardness.** When steel is cooled, the cooling rate at the surface is faster than the cooling rate at the interior. This can make the surface hard while the interior remains soft and ductile. Alloying elements allow the harder phase to form at slower cooling rates, thereby increasing the depth to which a steel can be hardened, a property known as *hardenability*. Manganese, molybdenum, and chromium are the best alloying elements for increasing hardenability.
- **Increasing toughness.** The strength of steel increases and the toughness decreases with increasing carbon content. Since alloying elements increase strength and hardness, lower carbon content can be used to obtain the same tempered hardness, thereby providing more toughness.
- **Improving high temperature strength**.
- **Improving high temperature corrosion and oxidation resistance**.
- **Improving wear and abrasion resistance**.

Steels with alloying elements are placed in categories based on the type and amount of alloying elements used.

8.2.2.1 Low-Alloy Steels

Low-alloy steels constitute a category of ferrous materials that exhibit mechanical properties superior to plain carbon steel as a result of adding alloying elements such as nickel, chromium, manganese, and molybdenum. The total alloy content can range from 2.0% up to levels just below that of stainless steels, which contain a minimum of 12% chromium. The primary function of the alloying elements in most low-alloy steels is to increase hardenability in order to optimize mechanical properties and toughness after heat treatment.

The four major groups of alloy steels are listed. Typical products made from materials from each of these groups are given in Table 8.4.

1. **Low-carbon quenched and tempered steels**, which combine high yield strength and high tensile strength with good notch fracture toughness, ductility, corrosion resistance, and weldability.
2. **Medium-carbon ultra-high-strength steels**, which are structural steels with yield strengths that can exceed 1.4 GPa.

TABLE 8.4
Typical Products Made from Ferrous Alloys

Material	Products		Section
Steel			
Plain carbon			8.2.1
Low carbon	Automobile-body panels	Stampings	8.2.1
	Wire products	Forgings	
	Seamless tubes	Boiler plates	
Medium carbon	Automobile components		8.2.1
	Engines		
	Transmissions		
	Suspension systems		
	Shafts and couplings		
	Crankshafts		
	Axles		
	Gears		
High carbon	Suspension springs	High-strength wires	8.2.1
Alloy steels			8.2.2.1
Low alloy	Forgings	Ball and roller bearings	8.2.2.1
	Tubing in fossil fuel and nuclear power	Arbors	
	plants	Cams	
	Welding wire	Chucks and collets	
	Tubing		
High-strength	Oil and gas pipelines	Equipment for	8.2.2.1
low alloy	Ships	Railways	
	Offshore structures	Excavation	
	Automobiles	Chemical processing	
	Pressure vessels	Pulp and paper industry	
	Machine tools	Refineries	
	Steam turbines	Marine industries	
	Valves and fittings		
Tool	Cutlery	Embossing tools	8.2.2.2
	Woodworking	Punches	
	Cutting tools	Blanking and trimming dies	
	Drills, milling cutters, taps, and gear hobs	Forming and coining dies	
		Chisels	
Stainless steels			8.2.2.3
Wrought	Cooking utensils	Equipment in	8.2.2.3
	Fasteners	Chemical plants	
	Cutlery	Food-processing plants	
	Flatware	Textile plants	
	Decorative architectural hardware	Heat exchangers	
	Piping		
Ferritic	Decorative trim	Heaters	8.2.2.3
	Acid and fertilizer tanks	Mufflers	
	Transformer and capacitor cases	Restaurant equipment	

TABLE 8.4 (CONTINUED)
Typical Products Made from Ferrous Alloys

Material	Products		Section
Martensitic	Steam turbine blades	Rifle barrels	8.2.2.3
	Machine parts	Scissors	
	Bolts	Cutlery	
	Bushings	Surgical instruments	
	Shafts	Ball bearings and races	
	Hardware	Nozzles	
Austenitic	Decorative trim	Tanks	8.2.2.3
	Food-handling equipment	Chemical and gas handling equipment	
	Aircraft parts	High temperature furnace parts	
	Cookware	Heat exchangers	
	Building exteriors	Oven linings	
Cast	Metal treatment furnaces	Equipment for	8.2.2.3
	Gas turbines	Chemical processes	
	Aircraft engines	Power plants	
	Military equipment	Steel mills	
	Oil refinery furnaces	Glass manufacturing	
	Turbochargers	Rubber manufacturing	
	Petrochemical furnaces	Cement mills	
Cast irons			
Gray	Clutch plates	Ingot and pig molds	8.2.3.1
	Brake drums		
Malleable	Magnetic clutches and brakes	Parts to be pierced and cold formed	8.2.3.2
	Thin-section castings		
Ductile (Nodular)	Vehicular systems	High-temperature applications	8.2.3.3
	Gear boxes	Turbo housings	
	Crankshafts	Manifolds	
	Disk brake calipers		
	Engine connecting rods		
	Idler arms		
	Wheel hubs		
	Axles		
	Suspension system parts		
Alloy	Explosives and fertilizers industries	Handling highly corrosive acids	8.2.3.4
	Drain pipes	Pumps	
	Tubes	Valves	
	Towers	Mixing nozzles	
	Fittings	Tank outlets	
		Steam jets	

3. **Bearing steels**, which are used for ball and roller bearing applications, are low-carbon case-hardened steels, and high-carbon through-hardened steels. Case hardening is a process in which extra carbon is added at the surface of the steel part before heat treatment in order to produce a component with a hard and wear-resistant surface but a ductile and tough interior.

4. **Chromium-molybdenum heat-resistant steels**, which contain chromium and molybdenum to provide improved oxidation and corrosion resistance along with increased strength at elevated temperatures.

High-strength low-alloy (HSLA) steels and micro-alloyed steel are designed to provide better mechanical properties and/or greater resistance to atmospheric corrosion than conventional carbon steels. They are not considered alloy steels because they are designed to meet specific mechanical properties rather than a chemical composition. There are a wide variety of products manufactured from HSLA and micro-alloyed steels, and some of the more common ones are listed in Table 8.4.

8.2.2.2 Tool Steels

Tool steel is any steel used to make tools for cutting, forming, or otherwise shaping a material into a part or component. In many applications, tool steels must be able to withstand extremely high loads as well as to resist wear. Many tool steels are wrought products, but the powder metallurgy process (see Section 9.9.1) is also used in making tool steels. Wrought materials are those manufactured by applying large deformations to the original shape as in forging, extrusion, and rolling. See Sections 9.8.1, 9.8.3, and 9.8.2, respectively, for a description of these processes. Powder metallurgy provides a more uniform carbide size and distribution in large sections while allowing special compositions that cannot be obtained otherwise. Powder metallurgy tool steels perform much better than wrought tool steels in that they can operate at much higher cutting speeds, and they have a much longer life. Many cutting applications today use cemented carbides instead of, or together with, tool steels. Cemented carbides are composite ceramic materials made using powder processing that contain tungsten carbide, silicon carbide, and other ceramic carbide materials that are harder than steel. These materials are often applied as a hard, wear-resistant coating on to a tool steel base by using plasma and chemical vapor deposition processes. Diamond is also coated on tool steel in a similar manner. See also Section 8.7.4.

The different classes of tool steels that are used are listed below. Some of the more common products made from each of these tool steels are given in Table 8.4.

- **High-speed tool steels**: There are two classes of high-speed tool steels, the molybdenum high-speed steels and the tungsten high-speed steels.
- **Hot-work tool steels**: Hot-work steels have been developed to withstand the combinations of heat, pressure, and abrasion associated with those manufacturing operations that involve punching, shearing, or forming of metals at high temperatures.
- **Cold-work tool steels**: Cold work tool steels are the tool steels that are restricted to applications that do not involve prolonged or repeated heating above 205°C to 260°C. The inherent dimensional stability of these steels makes them suitable for gages and precision measuring tools. Their extreme abrasion resistance makes them suitable for brick and ceramic molds.
- **Shock-resisting steels**: Shock-resisting tool steels possess a combination of high strength, high toughness, and low-to-medium wear resistance. They are used primarily for applications requiring high toughness and resistance to shock loading.
- **Low-alloy special purpose steels**: These steels are used for machine parts and other special purpose applications requiring good strength and toughness.

- **Mold steels**: These steels have very low hardness and low resistance to work hardening in the annealed condition.
- **Quenched and tempered steels**: These tool steels have low resistance to softening at elevated temperatures.

8.2.2.3 Stainless Steels

Stainless steels are iron-based alloys containing at least 12% chromium. They achieve their stainless characteristics through the formation of a transparent, adherent chromium-rich oxide surface film. The selection of stainless steels may be based on corrosion resistance, fabrication characteristics, availability, mechanical properties in specific temperature ranges, and product cost. However, corrosion resistance and mechanical properties are usually the most important factors in selecting a grade for a given application. Other alloying elements are incorporated in addition to the chromium in order to provide additional advantages. These include the following:

- Nickel, which improves corrosion resistance in neutral (nonoxidizing) atmospheres.
- Molybdenum, which improves corrosion resistance in chloride containing atmospheres.
- Aluminum, which protects the thin surface oxide from flaking at high temperatures.

Stainless steels can be categorized by manufacturing process or by composition. When categorized by manufacturing process, stainless steels can be separated into wrought stainless steel and cast stainless steel. Typical products for each of these two classifications are listed in Table 8.4.

When categorized by composition, stainless steels can be separated into the following three groups, with examples of typical products that are made from each of these groups given in Table 8.4.

Ferritic—These stainless steels contain 11–30% chromium, no nickel, and less than 0.12% carbon. They are weldable, ductile, and corrosion resistant, but not heat treatable.

Martensitic—These stainless steels contain 12–17% chromium, no nickel, and up to 1% carbon. These steels are heat treatable, and they sacrifice some ductility and corrosion resistance for higher strength and hardness.

Austenitic—These stainless steels contain 16–25% chromium, 7–20% nickel, and less than 0.25% carbon. They comprise 65–70% of the total U.S. stainless steel production because of their high corrosion resistance and formability. The most common stainless steels are in this group as follows:
- Types 302 and 304 are the most widely used stainless steels. They are used primarily in decorative trim, food-handling equipment, aircraft parts, cookware, building exteriors, and tanks.
- Type 316 contains 2.5% molybdenum for higher corrosion resistance and high temperature creep strength. It is used widely in chemical and gas handling equipment.
- Types 309 and 310 contain 23–25% chromium for high temperature corrosion resistance. They are used in high temperature furnace parts, heat exchangers, and oven linings.

8.2.3 Cast Irons

Cast irons are alloys of iron, carbon, and silicon that contain more than 2% carbon by weight. They have a lower melting point than steels and are easily cast. However, the finished products tend to be brittle. The carbon is in the form of graphite, and the types of cast iron differ in the micro-structural form this graphite takes.

8.2.3.1 Gray Irons

Gray irons are cast irons in which the graphite appears in the form of rods or fibers. These irons are classified based on their tensile strength and exist in grades from 20 to 60 (tensile strengths from 20 ksi [135 MPa] to 60 ksi [405 MPa]). However, in many applications, strength is not the crucial factor. For parts such as clutch plates and brake drums, where resistance to heat checking (thermal cracking on the surface) is important, low-strength grades of iron are superior performers. Similarly, in heat shock applications such as ingot or pig molds, a class 25 gray iron shows better performance than a class 60 gray iron. In machine tools and other parts subject to vibration, the better damping capacity of the low-strength irons is often advantageous. In gray cast iron, the following properties increase with an increase from class 20 to class 60:

- All strengths, including strength at elevated temperatures.
- Ability to be machined to a fine finish.
- Modulus of elasticity.
- Wear resistance.

On the other hand, the following properties decrease with increase in tensile strength, so that low-strength irons perform better than the high-strength irons.

- Machinability.
- Resistance to thermal shock.
- Damping capacity.
- Ability to be cast with thin sections.

Gray iron is used for many different types of parts in a wide variety of machines and structures. However, gray iron is not recommended where high impact resistance is required. They have considerably lower impact strength than cast carbon steels, ductile iron, or malleable iron. However, they need minimal impact strength to avoid damage during shipping. The machinability of gray cast irons is superior to most other cast irons of equivalent hardness and to all steels. Typical products made from gray cast irons are given in Table 8.4.

8.2.3.2 Malleable Irons

Malleable iron is a type of cast iron that, like ductile iron, possesses considerable ductility and toughness. Consequently, malleable iron and ductile iron are suitable for some of the same applications requiring good ductility and toughness, with the choice between them being made on the basis of cost and availability, rather than on properties. Malleable and ductile iron also exhibit high resistance to corrosion, excellent machinability, good magnetic permeability, and low magnetic retention. The good fatigue strength and damping capacity of malleable iron are also useful for long service in highly stressed parts.

Malleable iron is preferred over ductile iron in the following applications:

- Thin-section castings.
- Parts that are to be pierced, coined, or cold formed.
- Parts that require maximum machinability.
- Parts that must retain good impact resistance at low temperatures.
- Parts requiring wear resistance (martensitic malleable iron only).

Ductile iron is more advantageous in cases where thick sections are required, and when low solidification shrinkage is needed.

8.2.3.3   Ductile (Nodular) Irons

Ductile cast iron, previously known as nodular or spheroidal-graphite cast iron because of the sphe-roidal shape of the graphite, has relatively high strength and toughness, which gives it an advantage over either gray iron or malleable iron in many structural applications. Ductile iron castings are used for many structural applications, particularly those requiring strength and toughness combined with good machinability and low cost. Ductile iron can be tempered to exhibit high tensile strength, high fatigue strength, high toughness, and excellent wear resistance. Due to the lower density of ductile iron, it weighs 10% less than steel for the same section size. The graphite content provides damping for the quiet running gears, and its low coefficient of friction produces more efficient gear boxes. Typical products made from ductile (nodular) irons are given in Table 8.4.

8.2.3.4   Alloy Cast Iron

Alloy cast irons are those casting alloys based on the iron-carbon-silicon system that contain one or more alloying elements intentionally added to enhance one or more useful properties. Small amounts of alloying elements such as chromium, nickel, or molybdenum are added to achieve high strength and hardness. Otherwise, alloying elements are used almost exclusively to enhance resistance to abrasive wear and chemical corrosion, or to extend the service life at elevated temperatures. Typical products made from alloy cast irons are given in Table 8.4. The main classes of alloy cast irons are abrasion resistant white irons, corrosion resistant irons, and heat resistant (gray or ductile) irons.

- **White cast irons** are so named because of their white fracture surface. White cast irons are usually very hard, which is responsible for their excellent resistance to abrasive wear.
- **Corrosion resistant irons** derive their resistance to chemical attack chiefly from their high alloy content. The corrosion resistance of cast irons is enhanced by the addition of apprecia-ble amounts of nickel, chromium, and copper, singly, or in combination, or silicon in excess of 3%. High silicon irons are the most common of the corrosion resistant alloy cast irons. They have poor mechanical properties and poor mechanical and thermal shock resistance.
- **Heat resistant cast irons** are alloys of iron, carbon, and silicon having markedly improved high-temperature properties.

8.3   NONFERROUS ALLOYS

In the following sections, the properties and applications of many nonferrous alloys are des-cribed. The properties of nonferrous metals are determined in part by their crystal structure. Materials with a face-centered cubic structure such as copper, nickel, and aluminum are ductile, malleable, and have high electrical and thermal conductivity. Materials with body-centered cubic or hexagonal structures such as chromium, titanium, and zinc exhibit significantly less ductility, especially at low temperatures, and have lower electrical and thermal conductivities. All nonfer-rous alloys except cobalt and molybdenum exhibit high corrosion resistance. Low weight structural applications take advantage of the high strength-to-density ratios of aluminum, magnesium, and titanium. Cobalt, nickel, and titanium are used in structures subject to high temperatures because of their high melting temperatures and high creep resistance. Nickel is also used in cryogenic applica-tions because of its low temperature ductility.

8.3.1   Light Alloys

8.3.1.1   Zinc Alloys

Zinc alloys have low melting points and low heats of fusion (heat input for melting), do not require fluxing or protective atmospheres to limit oxidation when molten, and are nonpolluting, the last being a particularly important advantage. For these reasons, zinc and its alloys are used extensively in both gravity and pressure die castings (see Section 9.2.1). Their high fluidity permits zinc alloys to be cast with much thinner walls than other die casting alloys, and they can be die cast to tighter dimensional tolerances. Zinc alloys also allow the use of very low draft angles. Zinc alloy castings can be joined by brazing, by certain welding techniques using zinc-base fillers, and by soldering by using new zinc base solders developed to replace the cadmium, tin, or lead base solders that are no longer suitable. Lead and cadmium are particularly harmful to zinc products, as impurity levels as low as 0.01% can promote intergranular corrosion and swelling. Adhesive bonding or mechanical fasteners are also excellent methods for joining castings.

In addition to their excellent physical and mechanical properties, zinc alloys offer the following advantages.

- Good corrosion resistance.
- Excellent vibration and sound damping properties, which increase exponentially with temperature.
- Excellent bearing and wear properties.
- Spark resistance, except when containing large amounts of aluminum.

When used as general casting alloys, zinc alloys can be cast using such processes as high-pressure die casting, low-pressure die casting, sand casting (see Section 9.3.1), permanent mold casting (using iron, graphite, or plastic molds), spin casting (using silicone rubber molds), investment casting (see Section 9.4.1), continuous or semicontinuous casting, and centrifugal casting (see Section 9.2.2).

In addition to being used in cast products, zinc, in either pure form or with small amounts of alloying additions, is used in three main types of wrought products: flat-rolled products, wire-drawn products, and extruded and forged products. Zinc is also widely used as a coating material to protect iron from rusting or corroding in mildly corrosive media. Zinc is applied to the iron by electroplating or in the liquid dip coating process known as *galvanizing*. Zinc provides corrosion protection to the iron electrochemically, leaving the iron unreacted as long as the zinc is present and electrically connected to the iron. Typical products made from zinc alloys are given in Table 8.5.

8.3.1.2   Aluminum Alloys

Properties of aluminum that make it and its alloys attractive for a wide variety of uses are its appearance, low density, ease of fabrication, mechanical properties such as strength and fracture toughness, and corrosion resistance. Aluminum surfaces can be highly reflective and have a number of decorative and functional uses. It has excellent electrical and thermal conductivity, it is nonferromagnetic, and it is nontoxic. It is extremely lightweight for its strength and some aluminum alloys exceed the structural efficiency of steel. A listing of the different applications for each of the aluminum wrought alloy series are given in Table 8.5.

Aluminum wrought alloys are classified as follows:

1. **1xxx Series**: Aluminum that has 99% or higher purity. These alloys have excellent corrosion resistance, high thermal and electrical conductivities, low strength, and excellent workability.
2. **2xxx Series**: Copper is the principal alloying element in this series, often with some amount of manganese added. When heat treated, these alloys have mechanical properties comparable to those of low-carbon steel. They do not have as good corrosion resistance as the other aluminum alloys, and are clad with high purity aluminum or with the 6xxx Series to obtain excellent corrosion resistance. However, they do have superior resistance to stress corrosion cracking and fair weldability. These alloys have superior machinability when compared to other aluminum alloys, especially when they contain lead.

TABLE 8.5
Typical Products Made from Nonferrous Light Alloys

Material	Products		Section
Zinc	Artifacts made by casting Investment Continuous Centrifugal	Products that are Flat-rolled Wire-drawn Extruded Forged	8.3.1.1
Aluminum: wrought	Heat exchangers Evaporators Electrically heated appliances Automotive cylinder heads Automotive radiators Containers for foods and beverages Aircraft frames Missile bodies	Satellite components Mirror frames Utensils Food trays Small appliances Furniture Foil Food packaging	8.3.1.2
1xxx Series	Chemical equipment Reflectors Heat exchangers Electrical conductors and capacitors	Packaging foil Architectural applications Decorative trim	8.3.1.2
2xxx Series	Truck and aircraft wheels Truck suspension parts Aircraft fuselage and wing skins	Structural parts requiring good strength to 50°C	8.3.1.2
3xxx Series	Beverage cans Cooking utensils Heat exchangers Storage tanks Awnings	Furniture Highway signs Roofing Architectural applications	8.3.1.2
4xxx Series	Welding wire Brazed components	Forged engine pistons	8.3.1.2
5xxx Series	Architectural, ornamental, and decorative trim Cans and can ends Household appliances Streetlight poles	Boats and ship components Cryogenic tanks Crane parts Automotive structures	8.3.1.2
6xxx Series	Architectural applications Bicycle frames Transportation equipment	Bridge railings Welded structures	8.3.1.2
7xxx Series	Aircraft structures Mobile equipment	Parts subjected to high stress	8.3.1.2
Aluminum: Cast 2xx.x Series	Engine cylinder blocks Automotive diesel engine pistons	Aircraft engine cylinder heads	8.3.1.2

(*continued*)

TABLE 8.5 (CONTINUED)
Typical Products Made from Nonferrous Light Alloys

Material	Products		Section

3xx.x Series	Automotive engines
	  Blocks
	  Pistons
	Jet engine compressor cases
		Engine cooling fans
		High-speed rotating parts such as impellers
			8.3.1.2

4xx.x Series	Castings
	  Sand
	  Permanent mold
	  Die
			8.3.1.2

5xx.x Series	Marine
	  Hatch covers
	  Ladders
	  Bulkheads
	  Lifesaving equipment
	  Canoes
	  Oars and paddles
	  Rowboats
		Architectural applications
		Ladders
			8.3.1.2

8xx.x Series	Bearings	Engine connecting rods	8.3.1.2
Magnesium			
Pressure die casting	Automobile wheels Some archery equipment	Crankcases of air cooled automotive engines Materials-handling equipment	8.3.1.3

Sand and
  permanent mold
  casting
	Automotive
	  Clutch and brake pedal support
	    brackets
	  Steering column lock housings
	  Manual transmission housings
	High-speed textile and printing
	  machines
		Hand-held tools
		Luggage
		Computer housings
		Ladders
			8.3.1.3

Titanium	Chemical and petrochemical
	  processing equipment
	    Vessels
	    Pumps
	    Fractional columns
	    Storage tanks
	  Marine
	    Propeller and rudder shafts
	    Pumps
	    Lifeboat parts
	    Deep sea pressure hulls
	    Submarine components
		Energy production and storage equipment
		  Plate-type heat exchangers
		  Condensers
		  Piping and tubing for sea water
		  Steam-turbine blades
		Flue gas desulfurization units
		  Canisters for low-level radioactive waste
		Biomedical applications
		  Implantable pumps and components for
		    artificial hearts
		  Hip and knee implants
			8.3.1.4

3. **3xxx Series**: Manganese is the major alloying element in this series. Alloys of this series are used in moderate-strength applications requiring good workability.

4. **4xxx Series**: Silicon, the major alloying element in this series, lowers the melting point of the alloy without increasing the brittleness. For this reason, these alloys are used in welding and brazing applications. Some alloys in this series have a low coefficient of thermal expansion and high wear resistance.

5. **5xxx Series**: Magnesium is the major alloying element in this series, which results in a moderate-to-high-strength work-hardenable alloy. Alloys of this series possess good forming and welding characteristics, and good resistance to corrosion in marine atmospheres.

6. **6xxx Series**: These alloys contain silicon and magnesium in the right proportion to form magnesium silicide. Although not as strong as the 2xxx or 7xxx Series, they have good formability, weldability, machinability, and corrosion resistance, with medium strength.

7. **7xxx Series**: Zinc is the major alloying element in this series. With small percentages of magnesium, heat treatable alloys of moderate to very high strength can be obtained. Some of these alloys develop the highest strength properties of any commercial aluminum-base alloys and are used in parts that are subjected to high stress.

In addition to these series there are aluminum-lithium alloys, which have been developed primarily to reduce the weight of aircraft and aerospace structures. More recently these alloys have been investigated for use in cryogenic applications because of their low density, high specific modulus, and excellent fatigue and cryogenic toughness properties. They have not directly replaced conventional aerospace aluminum alloys because of their reduced ductility and fracture toughness, their accelerated fatigue, and their high explosion potential with water. Furthermore, similar strength-to-weight ratios have been achieved through the use of polymer composite materials. These composite materials have been substituted in many cases for the conventional aerospace aluminum alloys because of their lower cost and improved fracture toughness.

Aluminum alloys can be cast as well as wrought. Casting qualities, such as fluidity, are important in developing and selecting cast aluminum alloys in addition to such properties as strength, ductility, and corrosion resistance. A listing of the different products made from each of the aluminum cast alloy series is given in Table 8.5.

Aluminum cast alloys are classified as follows:

1. **2xx.x Series:** Copper is the major alloying element in this series along with small amounts of magnesium. Though excellent high-strength and high-ductility castings have been obtained with these alloys, they are not very easily cast. They have high densities and they exhibit poor corrosion resistance. In many cases, they have been replaced by 3xx.x series alloys. The 2xx.x series finds use in applications that require high-temperature strength and good wear resistance.

2. **3xx.x Series:** These alloys contain both silicon and either copper or magnesium as alloying elements. These additions increase the strength of the alloy.

3. **4xx.x Series:** These alloys contain silicon as the major alloying element. The alloys in this series are the most important commercial casting alloys because of their superior casting characteristics, including high fluidity in the molten state, excellent feeding during solidification, and good filling of the mold cavity. They are used in applications requiring good castability and good corrosion resistance, especially in mildly acidic environments. The most widely used alloys in this series are 443 for sand and permanent mold casting, and 413 for die casting.

4. **5xx.x Series:** These alloys contain magnesium as the primary alloying element. They have moderate-to-high strength and toughness properties, and high corrosion resistance, especially to sea water and marine atmospheres, which is the primary advantage of these alloys.

5. **7xx.x Series:** Zinc is the major alloying element with small amounts of magnesium. They have moderate to good strength characteristics, and with annealing, they have good dimensional stability. They also have good machinability; however, their castability is poor.

6. **8xx.x Series:** Tin is the major alloying element in these alloys, and small amounts of copper and nickel are used for improving the strength. These alloys are used for cast bearings because of the excellent lubricity imparted by tin along with good fatigue strength and resistance to corrosion by the lubrication oils.

8.3.1.3 Magnesium Alloys

Magnesium is a lightweight, high-strength material with excellent machineability, but high cost. It is used in alloy form in similar applications as other nonferrous alloys. Typical products made from magnesium alloys are given in Table 8.5.

8.3.1.3.1 Pressure Die Casting Alloys

Magnesium is well suited to pressure die casting because of its low melting point and nonreactivity with iron and steel. There are three types of magnesium alloys used for high pressure die castings: magnesium-aluminum-zinc-manganese (AZ); magnesium-aluminum-manganese (AM); and magnesium-aluminum-silicon-manganese (AS). An AZ type alloy is the most commonly used die casting alloy. The AZ alloys exhibit good mechanical and physical properties in combination with excellent castability and saltwater corrosion resistance. The AM series of alloys is used in applications requiring greater ductility than can be provided by the AZ series. The tensile and yield strengths of the AM series are comparable to the AZ series. The AM series also exhibits excellent corrosion resistance. The AS series of alloys is used when excellent creep strength is required. It also has good elongation, yield strength, and tensile strength.

8.3.1.3.2 Sand and Permanent Mold Casting Alloys

The magnesium sand and permanent mold casting alloys that contain aluminum as the primary alloying ingredient exhibit good castability, good ductility, and moderately high yield strength at temperatures up to approximately 120°C. One alloy in particular, AZ92A, which has a high aluminum content, has better yield strength and less ductility. It is also a more generally used alloy because it has better castability with less microporosity.

Magnesium alloys that contain high levels of zinc develop the highest yield strengths of the casting alloys and can be cast into complicated shapes. The high zinc content magnesium-zinc-zirconium alloys exhibit castability characteristics that are associated with shrinkage, microporosity, and cracking. These characteristics, combined with their high cost, limit their applicability to situations where exceptionally good yield strengths are required. They are intended primarily for use at room temperature.

8.3.1.4 Titanium Alloys

Titanium and its alloys have become indispensable in a number of applications in the aircraft industry, and are still finding applications where corrosion resistance is important, such as in the marine and biomedical industries. Commercially pure titanium is more commonly used than titanium alloys for corrosion applications, especially when high strength is not a requirement. Typical products made from titanium alloys are given in Table 8.5.

Titanium alloys are classified as either alpha alloys, near alpha alloys, or alpha-beta alloys and beta alloys, depending on the crystal structure of the titanium in the alloy. Alpha alloys are less corrosion resistant but higher in strength than pure titanium. Near alpha alloys have good ductility and formability, higher corrosion resistance, and higher strength and fatigue properties than the other titanium alloys. The beta alloys have lower strengths than the alpha-beta alloys but have good ductility, superior creep properties, and higher fracture-toughness.

8.3.1.4.1   Manufacturing Methods

Titanium alloys are forged (see Section 9.8.1) into a variety of shapes. However, titanium alloys are considerably more difficult to forge than aluminum alloys and alloy steels, particularly with conventional forging techniques. Most titanium forgings are produced to less highly complex configurations than are aluminum alloy forgings. Titanium alloys are extruded (see Section 9.8.3) to obtain rod-like shapes and seamless pipe products. Titanium and titanium alloy sheets and plates are obtained by cold rolling (see Section 9.8.2), which increases tensile and yield strengths, and causes a slight drop in the ductility.

Since titanium is costly, it is preferred to use near-net-shape technologies to process titanium and its alloys. Precision casting is by far the most widely used near-net-shape technology to process titanium and its alloys. Titanium cast parts are generally comparable to wrought products in all respects, and often superior in properties associated with crack propagation and creep resistance. Investment casting (see Section 9.4.1) is the process that allowed titanium to become indispensable in the aircraft industry. Powder metallurgy (see Section 9.9.1) has also been used as a near-net-shape manufacturing process to produce a number of aerospace components.

8.3.2   Heavy Alloys

8.3.2.1   Copper Alloys

Copper and its alloys are widely used because of their excellent electrical and thermal conductivity, outstanding resistance to corrosion, ease of fabrication, and good strength and fatigue resistance. Copper's electrical conductivity is surpassed only by silver, and is used as the standard against which all other materials' electrical conductivity is measured. Copper alloys are generally nonmagnetic and nonsparking, and can be soldered, welded, and brazed. For decorative parts, different colors are available, and they can be polished and buffed to any desired texture and luster. They can also be plated to obtain a variety of finishes. Hardness is produced by cold working, and softening is produced by heating above the recrystallization temperature. Typical products made from copper and its alloys are given in Table 8.6.

Pure copper: Pure copper is alloyed with many other elements to produce minor changes in properties. Chromium, zirconium, cadmium, and tin are added to improve the strength. Silver is added in small quantities to obtain improved heat resistance.

Brass: Plain brasses are alloys of copper and zinc. There are different kinds of brasses depending upon the zinc content. *Red brass* contains about 15% zinc. *Yellow brass* contains around 30% zinc. Brass with around 40% zinc is called *muntz metal*. Brasses are produced as cast, forged, machined, stamped, drawn, or spun parts. The comparatively high price of brass is offset by the unique properties that it offers, including high ductility; rapid drawing speeds in presses and dies; lower maintenance cost on dies; ease of plating or use for plating; ease of machining, forming, and casting; high value of scrap; and ability to be alloyed with many metals to give desirable characteristics.

Bronze: Initially, the term *bronze* referred to copper-tin alloys. However, bronzes now can contain silicon or aluminum as the alloying element instead of tin. Bronzes with tin are called *phosphor bronzes* because they contain a noticeable amount of phosphorus from the refining operation. Their strength and hardness increase as the amount of tin increases. They have a low coefficient of friction with most material surfaces.

Silicon bronze is an important alloy for marine applications and high-strength fasteners. Silicon-bronzes can be easily cast, forged, welded, stamped, rolled, and spun. They find principal use where resistance to corrosion and high strength are required.

Aluminum bronzes contain aluminum and no tin. Iron is added to aluminum bronzes to increase the strength and hardness. These alloys have excellent corrosion resistance, and their high-strength allows them to be used as structural materials.

TABLE 8.6
Typical Products Made from Nonferrous Heavy Alloys

Material	Products		Section
Copper			
Pure copper	Cables and wires	Home heating systems	8.3.2.1
	Electrical contacts	Panels for absorbing solar energy	
	Automobile radiators	Applications requiring rapid	
	Heat exchangers	conduction of heat	
Brass	Plumbing products	Memorial markers	8.3.2.1
	Flanges	Plaques	
	Feed pumps	Statuary	
	Meter casings and parts	Low-pressure valves	
	Hydraulic and steam valves	Air and gas fittings	
	Valve disks and seats	Hardware,	
	Impellers	Ornamental castings	
	Injectors		
Bronze	Springs	Bearing plates	
Silicon bronze	Applications requiring corrosion		8.3.2.1
	resistance and high strength		
Aluminum bronze	Structural applications	Agitators	8.3.2.1
	Valve nuts	Crane gears	
	Cam bearings	Connecting rods	
	Impellers		
Manganese bronze	Marine propellers and fittings	Gear shift forks	8.3.2.1
	Pinions	Architectural application	
	Ball bearing races	Bridge trunnions	
	Worm wheels	Gears and bearings	
Beryllium bronze	Corrosion-resistant spring	Molds for plastics	8.3.2.1
	materials		
	X-ray window	Diaphragms	
Cupronickels	Applications requiring corrosion		8.3.2.1
	resistance		
Nickel silvers	Nameplates	Silver-plated cutlery and	8.3.2.1
	Bezels	silverware	
Nickel alloys	Applications requiring corrosion	Chemical and petrochemical	8.3.2.2
	resistance and/or heat resistance	industries	
	Aircraft gas turbines	Bolts	
	Combustion chambers	Fans	
	Bolts	Valves	
	Casings	Reaction vessels	
	Shafts	Piping	
	Exhaust systems	Pumps	
	Blades	Pollution control equipment	
	Afterburners	Scrubbers	
	Thrust reversers		

TABLE 8.6 (CONTINUED)
Typical Products Made from Nonferrous Heavy Alloys

Material	Products		Section
	Steam turbine power plants	Flue gas desulfurization equipment	
	Bolts	Liners	
	Turbine blades	Fans	
	Stack gas reheaters	Stack gas reheaters	
	Reciprocating engines	Ducting	
	Turbochargers	Metal processing plants	
	Exhaust valves	Ovens	
	Hot plugs	Afterburners	
	Valve seat inserts	Exhaust fans	
	Hot-work tools and dies	Coal gasification and liquefaction systems	
	Prosthetic devices		
	Dental tools	Tubing	
	Aerodynamically heated skins	Bleaching circuit equipment	
	Rocket engine parts	Scrubbers in pulp and paper plants	
	Furnace mufflers in heat-treating equipment	Nuclear power system ducting	
Tin alloys	Fusible alloys in automatic safety devices	Pewter products	8.3.2.3
		Costume jewelry	
	Fire sprinklers	Decorative artifacts	
	Boiler plugs	Dishes	
	Furnace controls	Cups	
	Solders	Bowls	
	Potable water plumbing	Decanters	
Cobalt alloys	Steam turbine erosion shields	Chain saw guide bars	8.3.2.4
	Rudder bearings	Carpet knives	
	Steel industry bar mill guide rolls	Rotary drill bearings in oil exploration machinery	
	Pump seal rings		

Manganese bronzes are similar to aluminum bronzes, but their properties are slightly inferior to them.

Beryllium bronzes are used where a nonferrous, a nonsparking, or a good electrical conductor with high strength is required. They comprise some of the best corrosion-resistant and/or high electrical conductivity spring materials.

Cupronickels: Cupronickels contain nickel as the main alloying element. They are ductile and are primarily hardened and strengthened by cold work. They are used in applications requiring corrosion resistance.

Nickel silvers: Nickel silvers are alloys of copper, nickel, and zinc. They can be ductile and soft, or less ductile and harder depending on their composition. They have moderate strength in the cold worked condition, and thus find application as springs and mechanical components. With the right combination of nickel and zinc, alloys that have the appearance of silver can be obtained; thus, the name "nickel silvers."

8.3.2.2 Nickel Alloys

Many applications of nickel alloys involve the unique physical properties of special purpose nickel-base or high-nickel alloys. Some typical products made from nickel alloys are given in Table 8.6. Nickel-base alloys offer excellent corrosion resistance to a wide range of corrosive media. Nickel-copper alloys, the most common of which contain about two-thirds nickel and one-third copper and are known as Monels, are work-hardenable but not heat treatable alloys. They have high strength, weldability, excellent corrosion resistance, and toughness over a wide temperature range. They perform very well in sea water where they are resistant to the effects of erosion and cavitation. Furthermore, they are extremely corrosion resistant to most acids and nearly all alkalis, and thermally stable to over 500°C in an oxidizing environment. Nickel-chromium alloys are used to provide resistance to oxidation and carburization at temperatures exceeding 760°C. The addition of iron to these alloys can help prevent internal oxidation that occurs in atmospheres that are oxidizing to chromium but reducing to nickel. An example is Incoloy, which was introduced as a sheathing material for electric stove elements.

One special category of nickel alloys are the nickel-based super alloys. These are nickel-chromium-cobalt alloys into which up to 4% aluminum and up to 4% titanium are added. These are alloys designed to retain high strengths at very high temperatures, while providing good corrosion and oxidation resistance, and superior resistance to creep and rupture at elevated temperatures. These alloys are widely used in aircraft and industrial gas turbines.

8.3.2.3 Tin Alloys

Tin alloys can be divided into three categories that are discussed below: fusible alloys, solders, and pewters. Typical products made from these tin alloys are given in Table 8.6.

Fusible alloys are any of the more than 100 white metal alloys that melt at relatively low temperatures. Most commercial fusible alloys contain bismuth, lead, tin, cadmium, indium, and antimony. These alloys find important uses in automatic safety devices such as fire sprinklers, boiler plugs, and furnace controls. Under ambient temperature, these alloys have sufficient strength to hold parts together, but at a specific elevated temperature the fusible alloy link will melt, thus disconnecting the parts.

Solders: Tin is an important constituent in solders because it wets and adheres to many common base metals at temperatures considerably below their melting points. Commercially pure tin is used for soldering side seams of cans for special food products and aerosol sprays. The electronics and electrical industries employ solders containing 40–97% tin that provide strong and reliable joints under a variety of environmental conditions. High tin solders are used for joining parts in electrical systems because their electrical conductivity is higher and their wettability greater than that of high-lead solders. Some solders are used to fill crevices at seams and welds in automotive bodies, thereby providing smooth joints and contours. Tin-zinc solders are used to join aluminum. Tin-antimony and tin-silver solders are employed in applications requiring joints with high creep resistance and in applications requiring a lead-free solder composition, such as potable-water plumbing. Recent environmental legislation has outlawed the use of lead-containing solders in electronic applications as well. Tin-bismuth, tin-zinc, and tin-silver-copper solders have been substituted for the tin-lead solders, with the most widely accepted replacement being a version of a tin-silver solder, known as SAC, containing 3 to 4% silver and less than 1% copper.

Pewter: Pewter is a tin-base white metal containing antimony and copper. Pewter is malleable and ductile, and it is easily spun or formed into intricate designs and shapes. Pewter parts do not require annealing during fabrication and have good solderability. Pewter can be formed either by rolling, hammering, spinning, or drawing. Much of the costume jewelry produced today is made of pewter alloys centrifugally cast in rubber or silicone molds. Pewter tarnishes in soft water, but not in hard water.

8.3.2.4   Cobalt Alloys

Cobalt is a tough silver-gray magnetic metal that resembles iron and nickel in appearance and in some properties. Cobalt is useful in applications that utilize its magnetic properties, its corrosion resistance, its wear resistance, and/or its strength at elevated temperatures. Some cobalt-base alloys are also biocompatible, which has prompted their use as orthopedic implants. Wear-resistant alloys form the single largest application area of cobalt-base alloys. Cobalt-base wear-resistant alloys have moderately high yield strength and hardness. Typical products made from cobalt alloys are given in Table 8.6.

8.3.3   Refractory Metals

8.3.3.1   Molybdenum Alloys

Molybdenum is used as an alloying element in cast irons, steels, heat-resistant alloys, and corrosion-resistant alloys to improve the following properties: hardenability, toughness, abrasion resistance, corrosion resistance, and strength and creep resistance at elevated temperatures. Molybdenum and its alloys have high thermal conductivity and low specific heat.

Molybdenum is used as a filler metal for brazing tungsten and for resistance heating elements in electric furnaces that operate at temperatures up to 2200°C. In addition, molybdenum alloys are also particularly well-suited for use in airframes because of their high stiffness, retention of mechanical properties after thermal cycling, and good creep strength. Pure molybdenum has good resistance to hydrochloric acid and is used for acid service in the chemical process industries.

Molybdenum and its alloys can be forged using either open or closed dies, can be fabricated in sheet form by conventional rolling and cross-rolling processes, and are readily extruded into a number of shapes including tubes, round to round bars, round to square bars, and round to rectangular bars. Typical products made from molybdenum alloys are given in Table 8.7.

8.3.3.2   Tungsten Alloys

Tungsten is combined with cobalt as a binder in cemented carbide composites used in cutting and wear applications. The high melting point of tungsten makes it an obvious choice for structural

TABLE 8.7
Typical Products Made from Refractory Metals

Material	Products		Section
Molybdenum alloys	Nozzles	Boring bars	8.3.3.1
	Leading edges of aerodynamic control surfaces	Tool shanks	
		Resistance welding electrodes	
	Reentry cones	Cladding	
	Pumps	Truing grinding wheels	
	Turbine wheels	Molds	
	Hot work tools	Thermocouples	
	Piercer points	Heat-radiation shields	
	Extrusion die	Heat sinks	
	Isothermal forging die		
Tungsten alloys	Counterweights	In wire form	8.3.3.2
	Flywheels	Lighting	
	Governors	Electronic devices	
	Radiation shields	Thermocouples	

applications exposed to very high temperatures. Tungsten is used at lower temperatures for applications that can take advantage of its high elastic modulus, density, or shielding characteristics. At room temperatures, tungsten is resistant to most chemicals. Wrought tungsten has high strength, strongly directional mechanical properties, and some room-temperature toughness.

Tungsten and tungsten alloys can be pressed and sintered into bars and subsequently fabricated into wrought bar, sheet, or wire. Tungsten alloys called *tungsten heavy metals* (tungsten-nickel-copper and tungsten-nickel-iron alloys) have been developed that are relatively ductile and with lower melting points, good machinability, and good mechanical properties. Tungsten and tungsten alloys are used extensively in applications for which a high density material is required. Typical products made from tungsten alloys are given in Table 8.7.

8.4 SPECIAL PURPOSE ALLOYS

8.4.1 Low Expansion Alloys

Low-expansion alloys include various iron-nickel alloys and several alloys of iron combined with nickel-chromium, nickel-cobalt, or cobalt-chromium. Some of the trade names of the low-expansion alloys used and their compositions are

- **Invar**, which is a 64% iron–36% nickel alloy, has the lowest thermal coefficient of iron-nickel alloys.
- **Kovar**, which is a 54% iron–29% nickel–17% cobalt alloy, has coefficients of expansion closely matching those of hard glass.
- **Elinvar**, which is a 52% iron–36% nickel–12% chromium alloy, has a constant modulus of elasticity over a wide temperature range.
- **Super Invar**, which is a 63% iron–32% nickel–5% cobalt alloy, has an expansion coefficient smaller than Invar, but over a narrower temperature range.

High-strength, controlled expansion alloys: There is a family of iron-nickel-cobalt alloys strengthened by the addition of niobium and titanium that show a combination of exceptional strength and low coefficient of expansion that make this family useful for applications requiring close operating tolerances over a range of temperatures. Several components for gas turbine engines are produced from these alloys. Some of these alloys are sold under the trade name Incoloy, which was discussed under nickel alloys in Section 8.3.2.2.

Low expansion alloys are used as rods and tapes for geodetic surveying; compensating pendulums and balance wheels for clocks and watches; moving parts that require control of expansion such as pistons for some internal combustion engines; bimetal strips; glass-to-metal seals; thermoelastic strips, vessel and piping for storage and transportation of liquefied natural gas; superconducting systems in power transmissions; integrated-circuit lead frames; components for radios and other electronic devices; and structural components in optical and laser measuring systems. They are also used along with high-expansion alloys to produce movements in thermo-switches and other temperature-regulating devices.

8.4.2 Permanent Magnet Materials

Permanent magnet is the term used to describe solid materials that have sufficiently high resistance to demagnetizing fields and sufficiently high magnetic flux output to provide useful and stable magnetic fields. Permanent magnets are normally used in a single magnetic state. This implies insensitivity to temperature effects, mechanical shock, and demagnetizing fields. Permanentic. magnets are typically made from metal alloys of cobalt or iron, both of which are ferromagnetic. Ferromagnetism is the ability of microstructural domains of a material to orient the magnetic spins of their constituent atoms in the direction of an applied magnetic field, and then to keep those spins oriented in that direction after the applied magnetic field is withdrawn. However, permanent magnet materials include a variety of alloys, intermetallics, and ceramics. Each type of magnet material possesses unique magnetic and mechanical properties, corrosion resistance, temperature sensitivity, fabrication limitations, and cost. These factors provide a wide range of options in designing magnetic parts. Permanent magnet materials are developed for their chief magnetic characteristics: high induction, high resistance to demagnetization, and maximum energy content. Maximum energy content is most important because permanent magnets are used primarily to produce a magnetic flux field. The various magnetic materials are the following.

- **Magnet alloys**: This alloy is represented by Cunife, which contains approximately 20% iron, 20% nickel, and 60% copper. The material is extremely anisotropic, with the superior magnetic properties in the direction of rolling, a factor that must be considered in magnet design. The mechanical softness of this alloy permits easy cold reduction and working, thus leading to many applications in the form of wire or tape.
- **Alnico alloys**: Alnico alloys are one of the major classes of permanent magnet materials. The Alnico alloys vary widely in composition and in preparation, to give a broad spectrum of properties, costs, and workability. Generally, Alnico is superior to other permanent magnet materials in resisting temperature effects on magnetic performance.
- **Platinum-cobalt alloys**: Although platinum-cobalt magnets are expensive, they are used in certain applications. Platinum-cobalt is isotropic, ductile, easily machined, resistant to corrosion at high temperatures, and has magnetic properties superior to all except the rare-earth-cobalt alloys. Best magnetic properties are obtained at an atomic ratio of 50% platinum and 50% cobalt.
- **Cobalt and rare-earth alloys**: Permanent magnet materials based on combinations of cobalt and the lighter rare-earth lanthanide metals, such as samarium-cobalt, are the materials of choice for most small, high-performance devices operating between 175 and 350°C. These materials are manufactured by powder metallurgy methods, and have low-temperature coefficients.
- **Hard ferrites**: Known also as *ceramic permanent magnet materials*, they have high electrical resistivities, are poor conductors of heat, and are not affected by high temperatures or by atmospheric corrosion; however, their magnetic properties are more temperature dependent than other permanent magnet alloys. See also Section 8.6.4.2.
- **Neodymium-iron-boron alloys**: These alloys have become the material of choice for a wide range of permanent magnet applications. They are processed by powder metallurgy or by the consolidation of rapidly solidified materials.

8.4.3 Electrical Resistance Alloys

Electrical resistance alloys are used to convert the heat generated into mechanical energy. They are classified as resistance alloys, thermostat metals, and heating alloys.

8.4.3.1 Resistance Alloys

Resistance alloys have uniform resistivity, stable resistance (no time-dependent aging effects), reproducible temperature coefficients of resistance, and low thermoelectric potential when compared to copper. The types of resistance alloys that are used commercially are as follows.

- **Copper-nickel resistance alloys** are generally referred to as radio alloys. These alloys have very low resistivities and moderate temperature coefficient of resistance. They are chiefly used for resistors that carry relatively high currents and are restricted to applications involving low operating temperatures.

- **Copper-manganese-nickel resistance alloys** are generally referred to as manganins. These alloys have been adopted almost universally for precision resistors, slide wires, and other resistive components with values of 1 kΩ or less, and are also used for components with values up to 100 kΩ. The resistance values of manganins change no more than 1 ppm per year when the material is properly heat treated and protected.

- **Constantan** has, like manganin, become a generic term for a series of alloys that have moderate resistivities and low-temperature coefficients of resistance. Constantan is considerably more resistant to corrosion than the manganins. Use of constantan as electrical resistance alloys is restricted largely to ac circuits because the thermoelectric potential when compared to copper is quite high (about 40 μV per degree centigrade at room temperature). Constantan is also widely used in thermocouples to generate a voltage difference that is dependent on a temperature difference.

- **Nickel-chromium-aluminum resistance alloys** contain small amounts of metals such as copper, manganese, or iron. These alloys have resistivities about two and one-half to three and one-half times that of manganin. Nickel-chromium-aluminum resistance alloys have been adopted almost universally for the construction of wire-wound precision resistors having resistance values around 100 kΩ, and are also used for resistors with values as low as about 100 Ω.

8.4.3.2 Thermostat Metals

A thermostat metal is a composite material that is usually in the form of a strip or a sheet and consists of two or more materials bonded together, one of which may be a nonmetal. Because the materials bonded together to form the composite differ in thermal expansion, the curvature of the composite is altered by changes in temperature; this is the fundamental characteristic of any thermostat material. A thermostat material is, therefore, a system that transforms heat into mechanical energy for control, indicating, or monitoring purposes. In applications such as circuit breakers, thermal relays, motor overload protectors, and flashers, the change in temperature necessary for the operation of the element is produced by the passage of current through the element itself. This means that the material should have a high electrical resistivity. They are available in strips or sheets in thickness ranging from 0.13 to 3.2 mm.

8.4.3.3 Heating Alloys

Resistance heating alloys are used in many varied applications, from small household appliances to large heating systems and furnaces. The primary requirements of materials used in heating elements are high melting point, high electrical resistivity, reproducible temperature coefficient of resistance, good oxidation resistance, absence of volatile components, and resistance to contamination. Other desirable properties are good elevated-temperature creep strength, high emissivity, low thermal expansion, and low modulus (both of which help minimize thermal fatigue), good resistance to thermal shock, and good strength and ductility at fabrication temperatures. There are four main groups of resistance heating materials.

- **Nickel-chromium and nickel-chromium-iron alloys** serve the greatest number of applications. The ductile wrought alloys in this group have properties that enable them to be used at both low and high temperatures in a wide variety of environments.

- **Iron-chromium-aluminum alloys**, which are also ductile alloys, play an important role in heaters for the higher temperature ranges, and are constructed to provide more effective mechanical support for the element. They have excellent resistance to oxidation at elevated temperatures, and they have low tensile strengths.

- The **pure metals**, which consist of molybdenum, platinum, tantalum, and tungsten, have much higher melting points, low resistivities, and very high temperature coefficients of resistance. Molybdenum and tantalum have high tensile strengths, even at high temperatures. All of them, except platinum, are readily oxidized and are restricted to use in non-oxidizing environments. They are valuable for a limited range of application, primarily for service above 1370°C. The cost of platinum prohibits its use except in small, special furnaces.

- The **nonmetallic heating-element materials** are used at still higher temperatures, and include silicon carbide, molybdenum disilicide, and graphite. Graphite and silicon carbide have high resistivity and negative temperature coefficients of resistance. Molybdenum disilicide elements have low resistivity and very high positive temperature coefficients of resistance. Silicon carbide can be used in oxidizing atmospheres at temperatures up to 1650°C. Graphites have poor oxidation resistance and should not be used at temperatures above 400°C. Molybdenum disilicides are effective to temperatures of 1700 to 1900°C. Molybdenum disilicide heating elements are gaining increased acceptance for use in industrial and laboratory furnaces. Among their desirable properties are excellent oxidation resistance, long life, constant electrical resistance, self-healing ability, and resistance to thermal shock. Nonmetallic heating elements are considerably more fragile than the metal heating element alloys. All of them have low tensile strengths. However, because of their high resistivity, silicon carbide elements can be made with large cross sections to reduce resistance and, as a result, can withstand high mechanical loads.

8.5 POLYMERS

8.5.1 INTRODUCTION

Organic in nature, polymers are made from large chain-like molecules built up from atoms of carbon, hydrogen, oxygen, and nitrogen. Properties of polymers depend greatly on the molecular weight and the arrangement of atoms within the molecule. Other types of polymers (the amino resin types: melamine and urea) have large three-dimensional molecular structures. Consequently, a very wide range of polymers with differing properties is available and these polymers find many uses. Typical properties are low density and low cost, moderate strength, corrosion resistance, heat and chemical resistance, good impact strength, insulating properties, and formability.

Polymers may be grouped into three categories:

- **Thermoplastics**. Solid at room temperature, these materials soften with heating and eventually become fluid. Its fluidity is utilized in processing finished parts. Like molten metal they take the shape of the mold that they are placed in and then remain in that shape after cooling. They can be remelted and reshaped multiple times.
- **Thermosetting resins**. These materials consist of a soft and sticky resin of little use on its own and a hardening agent. When the resin and hardening agent are mixed, heated, and cured, the final product becomes sufficiently stiff and hard to be a useful engineering material. In the heating and curing process, the molecular structure is altered by a process known as *crosslinking*. These materials cannot then be reheated and reformed. Thermosets remain stable and the process is irreversible.
- **Elastomers**. These materials withstand repeated elongation (up to 200%), yet may show complete strain recovery upon release of the applied stress.

X-ray diffraction studies show that many of the polymers, specifically the thermoplastics, indicate sharp features associated with regions of three-dimensional order (crystallinity) and diffuse features characteristic of disordered (amorphous) regions. Most thermoplastics are only either partially crystalline with adjacent amorphous regions or they are completely amorphous.

Polymer molecules may also be crosslinked. Crosslinking is the formation of strong primary bonds between polymer molecules that converts them into large three-dimensional networks that pull the chains closer, leaving less free volume and restricting their mobility. Crosslinking increases the plastic's modulus, hardness, and wrinkle resistance. Crosslinked polymers are less soluble in solvents, are thermally stable, and have lower coefficient of thermal expansion.

Heavy crosslinking joins the polymer together at so many points that even the motion of small segments is restricted. Such polymers have high modulus and high creep resistance.

Light crosslinking joins the molecules together sufficiently to prevent them from flowing completely past each other, but not enough to restrict the motion of large segments within the polymer molecule. Lightly crosslinked polymers have excellent strength, toughness, elasticity, and resistance to tearing and abrasion. However, long-term stress will cause a gradual disentanglement of the molecules. This is the case for elastomers.

Plastics have several manufacturing advantages. They can be used to form very complex parts in one operation. Both rigid and flexible elements can be incorporated in one part; for example, integral hinges and snap fit elements. Color can be incorporated in the material and many plastics have natural lubricity at low loads and velocities.

On the other hand, plastics are not as strong as metals. They have very high coefficients of thermal expansion, and they cannot usually be used at elevated temperatures where they will soften or decompose. They have poor resistance to creep and, therefore, should not be highly stressed at elevated temperatures. For those plastics that are recycled, there is some degradation of their properties. Some of the important characteristics for several of the thermoplastics and thermosets discussed in the following sections are as follows: ABS (acrylonitrile butadiene styrene) has high impact strength and scratch resistance, good weatherability, and is readily plated. Acrylics, polypropylene, and polystyrene resist acids, alkalis, and solvents, and have good colorability. Acrylics also have low moisture absorption. Epoxies, fluorocarbons, phenolics, polyesters, and silicones have good electrical properties and good corrosion resistance. Fluorocarbons, phenolics, and silicones also have high temperature capability, as do nylons, polycarbonates, and polypropylene.

8.5.2 THERMOPLASTICS—PARTIALLY CRYSTALLINE

8.5.2.1 Polyethylene

Polyethylene has low heat conductivity and ranges from very flexible to stiff depending on its molecular weight and crystallinity. High density polyethylene is highly crystalline, opaque, and stiff. Low density polyethylene is mostly amorphous, transparent, and flexible. There is a wide plasticity range, and the material becomes molten and moldable over a wide range of conditions. Polyethylene has no moisture absorption and does not require any storage precautions or drying operations. It can, however, develop static electrical charges and attract dust particles and other atmospheric contaminants. The shrinkage of polyethylene is generally high, but is greater in the direction of flow than the direction across flow. Polyethylene is used for a wide range of products, some of which are listed in Table 8.8.

8.5.2.2 Polypropylene

Polypropylene has a low density, and its products are characterized by rigidity, good surface hardness, and reasonable environmental stress resistance, which gives a good tear strength. Polypropylene cools much faster than polyethylene, which is an important factor that must be considered when designing parts with this material. Stresses may develop during cooling, accompanied by distortion of the part. Polypropylene is a notch-sensitive material and sharp corners should be avoided; instead, small radii should be provided. Due to the inertness of polypropylene, it is very difficult to cement. Typical products made from polypropylene are listed in Table 8.8.

TABLE 8.8
Typical Products Made from Polymers

Material	Products		Section
Thermoplastics: partially crystalline			
Polyethylene	Consumer products 　Buckets 　Bowls 　Baby baths 　Watering cans 　Funnels 　Mixing bowls	Industrial products 　Kick plates in buses 　Sealing plugs on valves 　Pipe fittings 　Packing material	8.5.2.1
Polypropylene	Bottles Bowls	Tubing Valves	8.5.2.2
Acetals	Gears Bearings	Hardware components Business machine 　assemblies	8.5.2.3
Nylons	Hosiery and undergarments Prepared food packaging Gears	Rollers Cams Door latch components	8.5.2.4
Fluorocarbons	Low friction and stick-free coatings Waterproof fabrics Wire and cable insulation	Printed wiring boards Cooling liquids	8.5.2.5
Polyimides	Self-lubricating polyimides 　High speed, high load bearings 　Bearing cages for gyroscopes	Jet engines High-temperature electrical 　connectors Air compressor piston rings	8.5.2.6
Cellulosic materials	Transparent film for 　Cinema 　Overhead projection 　Still photography		8.5.2.7
Thermoplastics: amorphous			
Polycarbonates	Vandal-proof public 　Lighting fittings 　Windows 　Doors Automotive 　Interior panels 　External bumper extensions	Appliance and business 　machine housings Crash helmets Photographic equipment Binocular bodies Water pumps	8.5.3.1
ABS (acrylonitrile butadiene styrene)			8.5.3.2
Polystyrene	Molding Food packaging Toilet cleaners	Flame retardant grade 　Housing for motors 　Television cabinets 　Appliance parts	8.5.3.3
PVC (polyvinyl chloride)	Imitation leather Decorative laminates Upholstery	Ducts Tanks Fume hoods	8.5.3.4

(*continued*)

TABLE 8.8 (CONTINUED)
Typical Products Made from Polymers

Material	Products		Section
	Wall coverings	Pump parts	
	Flexible tubing	Handles	
	Corrosion-resistant coatings on metals	Plumbing pipes and elbows	
	Insulation on electrical tool handles and wires	Building siding	
		Window frames	
	Rigid piping for cold water and chemicals	Gutters	
	Guards	Interior molding and trim	
	Automobile tops	Shower curtains	
		Refrigerator gaskets	
		Appliance components	
Polyurethane	Polyurethane foam		8.5.3.5
	Cushioning		
	Thermal insulation		
Thermosets: highly crosslinked			
Epoxies	Adhesives for	Laminating resins for	8.5.4.1
	Aircraft honeycomb structures	Airframe and missile applications	
	Paintbrush bristles		
	Concrete topping compounds	Filament-wound structures	
	Caulking and sealant compounds		
	Casting compounds for	Tooling fixtures	
	Prototype molds	Fiber fabric reinforced composites	
	Patterns		
	Tooling	Epoxy-based solutions for	
	Electrical and electronic equipment	Product finishes	
	Potting and encapsulation compounds	Marine finishes	
	Impregnating resins	Masonry finishes	
	Varnishes	Structural steel	
		Aircraft finishes	
		Automotive primers	
		Can and drum linings	
Phenolics	Laminates	Adhesives	8.5.4.2
	Molding	Molded parts	
	Surface coatings	Printed wiring boards	
Polyesters	Composite with fiber glass		8.5.4.3
	Boat hulls		
Alkyd	Circuit breakers	Distributor caps	8.5.4.3
	Transformer housings	Rotors	
Polybutylene terephthalate	Automotive	Motor brush holders	8.5.4.3
	Body components	Fuse holders	
	Ignition components	Terminal blocks	
	Window and door hardware	Food processor blades	
	Transmission components	Fans	
	Switches	Gears	
	Relays	Frame and bracket parts	

TABLE 8.8 (CONTINUED)
Typical Products Made from Polymers

Material	Products		Section
Polyethylene terephthalate (PET)	Containers for Beverages Pharmacy items		8.5.4.3
Thermosets: lightly crosslinked			
Silicone Resins	Mold-release agents Seals Gaskets Tubing	Hoses Wire insulation Coated fabrics Contact lenses	8.5.5.1
Acrylics	Impact resistant windows		8.5.5.2
Rubbers			
Natural rubber	Vibration isolators Shock absorbers Tires Shoe heels	Hoses Gaskets Electrical insulators	8.5.5.3
Synthetic rubbers			
Styrene-butadiene rubber	Automobile tires		8.5.5.3
Acrylonitrile-butadiene rubber	Gaskets Seals Gasoline hoses	Oil-pump seals, Carburetor and fuel-pump diaphragms	8.5.5.3
Neoprene	Garden hoses Insulation for wire and cable Gasoline-pump hoses	Packing rings Motor mountings Oil seals	8.5.5.3
Butyl rubber	Inner tubes Tubeless tires	Dairy hose Gas masks	8.5.5.3

8.5.2.3 Acetals

Acetals are a relatively new thermoplastic, first becoming available in 1960. In appearance, they are similar to nylon, with which they compete for similar applications. Typical products made from acetyls are listed in Table 8.8.

Acetals have excellent mechanical properties, which enable them to be used in place of metals in many applications. They have a high tensile strength, which is retained over at least several years in air/water exposures at temperatures as high as 95°C. They have high impact resistance, which is only slightly affected by subzero temperatures. Their stiffness is high, and creep and fatigue resistance exceed that of any other thermoplastic. The coefficient of friction is low and abrasion resistance is comparable to metals. Acetals resist solvents and most alkalis, but are attacked by acids. Colorability is good and discoloration by industrial oils and grease is not a problem. Moisture absorption is low, weatherability is good, but short-term exposure to ultraviolet radiation causes chalking of the surface, while prolonged exposure affects strength.

8.5.2.4 Nylons

Polyamide is the chemical term used to describe linear polymers in which the structural units are connected by amide groups. The first polyamide produced in continuous filament form was used for hosiery and undergarments as a replacement for natural silks. The trade name Nylon was adopted, but today the term *nylon* is used generally to describe any polyamide capable of forming polymers. Typical products made from nylons are listed in Table 8.8.

General characteristics of nylons are as follows.

- **Transparency**: The natural state of nylon molding and extrusion resins is translucent, beige, or off-white. Extruded films find applications in prepared food packaging.
- **Antidrag and antifriction properties**: Extruded nylons provide abrasion and cut-through resistance, and they can be pulled through complicated conduit paths because of inherent antidrag properties. Low friction, a property common to all nylons, makes them suitable for moving and bearing parts.
- **Heat resistance**: Most grades are self-extinguishing and impart an odor of burning wool. Continuously exposed to dry heat, nylons embrittle at about 120°C.
- **Moisture absorption**: All nylons absorb moisture to a degree depending on formulation, and reach an equilibrium between 0.20 and 0.25%. Type selection and design tolerances thus become critical when moving parts are required to operate in a humid environment.
- **Chemical resistance**: Nylons are generally resistant to gasoline, liquid ammonia, acetone, benzene, and organic acids. They are attacked, lose strength, and swell when exposed to chlorine and peroxide bleaches, nitrobenzene, and hot phenol. Nylons are not recommended for extended exposure to ultraviolet light, hot water, and alcohols. They are resistant to moth larvae, fungus, and mildew.
- **Melting point**: Unlike most thermoplastics, nylons are highly crystalline with sharply defined melting points at which they become extremely fluid and free flowing.

8.5.2.5 Fluorocarbons

Fluorocarbons are compounds of carbon in which fluorine instead of hydrogen is attached to the carbon chain. The resulting compounds are very stable to heat and resistant to chemical attack. Fluorocarbon plastics include polytetrafluoroethylene (PTFE or Teflon®), fluorinated ethylene propylene (FEP), and polyvinylidene fluoride (PVF). Typical products made from fluorocarbons are listed in Table 8.8.

8.5.2.6 Polyimides

Polyimides are a family of some of the most heat- and fire-resistant polymers known. Polyimides have good wear resistance and low coefficients of friction. Both of these properties are improved by using PTFE fillers. Self-lubricating parts containing graphite powders have flexural strengths above 70 MPa, which is considerably higher than most thermoplastic gearing materials. Electrical properties are outstanding over a range of temperature and humidity conditions. Parts are unaffected by exposure to dilute acids, hydrocarbons, esters, ethers, and alcohols. They are attacked by dilute alkalis and concentrated organic acids.

Molded glass reinforced polyimides are used in jet engines and in high-temperature electrical connectors. High-speed, high-load bearings for business machines and computer printout devices use self-lubricating polyimides. Other products made from polyimides are given in Table 8.8.

8.5.2.7 Cellulosic Materials

These are a wide range of compounds that have cellulose as the base material, the most popular being cellulose acetate. Cellulose acetate is characterized by high resilience, high toughness, high surface finish, good impact strength, low heat resistance, low moisture resistance, and poor dimensional stability, even though it shrinks uniformly in both directions. It can be either transparent or opaque.

8.5.3   Thermoplastics—Amorphous

8.5.3.1   Polycarbonates

Polycarbonates possess very high impact strength, glass clear transparency, heat resistance, high dimensional stability, and chemical resistance. These properties make polycarbonates ideal for vandal-proof public lighting fixtures, windows, and doors. Risk of breakage can be virtually eliminated by glazing with Lexan®, a polycarbonate sheet several hundred times stronger than glass. It resists hammers, bricks, sparks, sprayed liquids, and heat.

Polycarbonates are also available in a variety of colors produced with pigments. They have good creep resistance and are excellent electrical insulators. Because of their high heat resistance (up to 145°C) polycarbonates are one of the few thermoplastics suitable for use with high output lighting where temperatures can exceed 100°C. They are also nonflammable and self-extinguishing. Being resilient, the material resists denting. Structural foam polycarbonate has a strength-to-weight ratio between two and five times that of conventional metals. Typical products made from polycarbonates are listed in Table 8.8.

8.5.3.2   Acrylonitrile Butadiene Styrene (ABS)

Acrylonitrile butadiene styrene (ABS) is a copolymer in which three different polymers are combined to produce a family of opaque thermoplastics offering a balance of properties, the most outstanding being impact resistance, tensile strength, and scratch resistance.

The ABS plastics are ductile and are characterized by high impact strength, moderate tensile and compressive strengths, marked dimensional stability, and extremely good corrosion and chemical resistance. They are very rigid and retain their properties over a wide range of temperatures, frequently as low as −40°C. ABS is easily coated with metal using electroless plating techniques. Its flow characteristics give good moldability, and unlimited colors are available with high gloss, stain resistant finishes. Weatherability is outstanding, with most environments having little or no effect on ABS, although prolonged strong sunshine may cause embrittlement. Resistance to creep and cold flow is moderate to good. Acoustic damping characteristics are excellent. Flammability is classified as slow burning, but ABS may be made flame resistant by blending with polyvinyl chloride.

8.5.3.3   Polystyrene

Polystyrene is one of the most popular and widely used thermoplastics. It is cheap in its basic form, and can be supplied in any color ranging from crystal clear to opaque black. Moldings made from polystyrene have bright colors, hard surfaces, excellent dimensional stability, freedom from distortion, and a metallic noise when dropped on a solid surface. Polystyrene is free from taste and smell, nontoxic when in contact with food, noncontaminating when in contact with toilet preparations, and does not support the growth of mold or fungus. It can be reground and reused. Polystyrene is resistant to most inorganic materials, but is attacked by organic solvents, which render it useless. Filled grades of polystyrene use mica and glass fiber as fillers to enhance the electrical properties and impact strength. Typical products made from polystyrene are listed in Table 8.8.

8.5.3.4   Polyvinyl Chloride

Polyvinyl chloride (PVC) is one of the most widely used plastics. It is used to produce products by extrusion (see Section 9.8.3), compression molding (see Section 9.2.3), blow molding (see Section 9.7.1), injection molding (see Section 9.2.4), powder coating, and liquid processing. Plasticized PVC is more commonly used as vinyl; in this form, it has low strength. Typical products made from PVC and vinyl (plasticized PVC) are listed in Table 8.8.

PVC is preferred over other plastics for electrical insulation applications because it is not flammable, does not sustain combustion, and has good dielectric properties. Rigid PVC has sufficient strength and stiffness to almost qualify as an engineering plastic. It is a thermoplastic that has good abrasion resistance, is easily molded, and is easily fabricated. In the extruded form, rigid PVC is widely used for low-cost piping for cold water and chemicals. Chlorinated PVC (CPVC) is used for hot water piping (temperatures up to 82°C). PVC is also used for components that require corrosion resistance. PVC is very resistant to many acids and bases, but it has poor resistance to organic solvents and organic chemicals.

8.5.3.5 Polyurethane

Polyurethane is best known as polyurethane-foam. Polyurethane is more expensive than nylon, has lower heat resistance, is not as tough, is more flexible, has better chemical properties, has better chemical resistance, has much lower moisture absorption, and has better dimensional stability than nylon. The range of applications of polyurethane is wide, with special emphasis on jobs that require resilience and resistance to impact. Typical applications of polyurethane are listed in Table 8.8.

8.5.4 THERMOSETS—HIGHLY CROSSLINKED

8.5.4.1 Epoxies

Epoxies are thermoset molding materials that are made by combining an epoxy resin with a hardener. The term is used to indicate the resins in both the thermoplastic (uncured) and thermoset (cured) state. Cured epoxies are a major structural type in plastics technology, widely used as glues, structural materials, and insulating encapsulates for microelectronic devices. A list of some of the application of epoxies is given in Table 8.8.

8.5.4.1.1 Liquid Resins

These are low-viscosity liquids that readily convert to the thermoset phase upon mixture with a curing agent. There are other liquid resins—acrylics, phenolics, polyesters, etc.—that cure in a similar fashion, but epoxy resins possess a unique combination of the following properties:

- **Low viscosity**. The liquid resins and their curing agents form low-viscosity, easy-to-process systems.
- **Easy cure**. Epoxy resins cure quickly and easily at practically any temperature from 5 to 150°C, depending on the selection of the curing agent.
- **Low shrinkage**. One of the most important and advantageous properties of epoxy resins is their low shrinkage during cure.
- **High adhesive strengths**. Epoxy resins are excellent adhesives, and provide the best adhesive strengths in contemporary plastics technology without the need for long curing times or high pressures.
- **Good mechanical properties**. The strength of properly formulated epoxy resins usually surpasses that of other casting resins. This is in part due to their low shrinkage, which minimizes curing stresses.
- **High electrical insulation**.
- **Good chemical resistance**. The chemical resistance of cured epoxy resin depends considerably on the curing agent used. Outstanding chemical resistance can be obtained by proper specification of the material.
- **Versatility**. The basic properties may be modified in many ways: by the blending of resin types; by the selection of curing agents; and by the use of modifiers and fillers.

8.5.4.1.2 Solid Resins

The chief use of solid epoxy resins is in solution coatings. The high molecular weight materials are cooked with conventional drying oils or reacted with other resins resulting in toughness, scuff resistance, and chemical resistance. Room temperature curing films have been produced to provide properties equal to or exceeding those of many baked-type finishes.

The excellent adhesion of epoxy resins, ease of cure, mechanical strength, and high chemical resistance are advantages of the solid and liquid resins.

8.5.4.2 Phenolics

Phenolic resins are among the oldest of plastics and were the first to be commercially exploited. They are used principally in reinforced thermoset molding materials. Combined with organic and inorganic fibers and fillers, the phenolics provide dimensionally stable compounds with excellent moldability.

Phenolics are formed by compression and injection molding and extrusion (see Sections 9.2.3, 9.2.4, and 9.8.3, respectively). Injection molding usually provides the fastest cycle time, but it may not produce the best properties. For example, in long fiber-filled compounds, compression molding gives the greatest strengths. Phenolic resins are used as bonding and impregnating materials. Phenolics are used in laminates, molding, surface coatings, and adhesives. Laminates are usually reinforced with paper, fiber, or cloth.

Phenolic molding compounds are characterized by low cost, superior heat resistance, high heat distortion temperature, good flame resistance, excellent dimensional stability, good water and chemical resistance, and excellent moldability. Phenolic compounds are classified as general purpose, nonbleeding, heat resistant, impact, electrical, and special purpose. Most phenolic compounds are black. Typical products made from phenolics are listed in Table 8.8.

8.5.4.3 Polyesters

Polyesters have good strength, toughness, resistance to chemical attack, low water absorption, and the ability to cure at low pressures and temperatures. These resins have wide applications in low-pressure laminates. One widespread application is the manufacture of boat hulls, using glass fibers as the laminating material and polyester as the resin. Additional products made from polyesters are listed in Table 8.8. Three common polyesters are given below.

- **Alkyds** are thermoset molding materials that are made by combining a monomer with unsaturated polyester resins and additives or fillers. Alkyds are useful in applications requiring high-temperature electrical properties, arc resistance, and dimensional stability. Like most thermosets, alkyds are hard and stiff and retain their mechanical and electrical properties at elevated temperatures. Only small changes in dielectric loss factor are caused by temperature. Their properties depend greatly on the fillers and manufacturing processes used. Typically, tensile strength is low, whereas compressive strength is higher by a factor of 4 to 5. Their outstanding properties are their electric arc resistance, low moisture absorption, and retention of electrical properties when wet. Specific products made from alkyds are listed in Table 8.8.
- **Polybutylene terephthalate** (PBT) is a polyester-type resin that provides high heat resistance, good mechanical strength and toughness, general chemical resistance, good lubricity and wear resistance, low moisture absorption, and aesthetic appeal. It is mainly fabricated using injection molding (see Section 9.2.4). Several products made from PBT are listed in Table 8.8.
- **Polyethylene terephthalate** (PET) is a typical polyester that is used in blow molding operations (see Section 9.7.1). The key properties of PET include excellent gloss and clarity, high toughness and impact strength, good chemical resistance, low permeability to carbon dioxide, good processability, and good dimensional stability. The significance of PET in the bottle or packaging market was achieved through plastic blow molding technology. Several products made from PET are listed in Table 8.8.

8.5.5 Thermosets—Lightly Crosslinked

8.5.5.1 Silicone Resins

Silicone resins have excellent resistance to heat, water, and certain chemicals. The resins are usually filled or used in laminated parts. Good physical properties are maintained to 230 to 260°C. Silicones are classified as fluids, elastomers, and resins. Thus, silicone polymers may be fluid, gel, elastomeric, or rigid in their final form. The properties that make silicones attractive include low surface tension, high lubricity with rubber and plastic surfaces, excellent water repellence, good electrical properties, thermal stability, chemical inertness, and resistance to weather. Silicone fluids are used mainly in the plastics and rubber industry as mold-release agents. When added to organic plastics in molding, they provide water repellence, abrasion resistance, lubricity, and flexibility. Typical products made from silicone resins are listed in Table 8.8.

8.5.5.2 Acrylics

Acrylics are thermoplastics that offer excellent optical characteristics, resistance to environmental conditions, and ease of forming. They are known by trade names such as Lucite®, Perspex®, and Plexiglas®.

Acrylics transmit about 92% of light and have a high refractive index. They offer very good dimensional stability and are serviceable to 110°C; however, they show a large thermal expansion, about seven times that of steel. Resistance to weathering is excellent, but high intensity ultraviolet light exposure produces crazing, which may be alleviated by annealing. (Crazing is the opening of small voids throughout the volume of a polymer that can develop at stress levels below the breaking stress.) High tensile strength and good impact resistance enhance its usefulness. Acrylics are unaffected by alkalis, industrial oils, and inorganic solvents, but acid and alcohol result in deterioration. Moisture absorption is low. One application of acrylics is listed in Table 8.8.

8.5.5.3 Rubbers

Rubber is the term that is applied to substances, either natural or synthetic, that are characterized by exceptional elastic deformity. Properly prepared rubbers can deform more than 200% of their original length when stretched and can return to their original length when the stress is removed. Thus, in addition to high elasticity, rubbers possess high resilience.

8.5.5.3.1 Natural Rubber

Natural rubber is not used in the form in which it is extracted from the rubber-producing plants. In this state it is temperature sensitive, soft in summer and hard in winter. Natural rubber is, therefore, vulcanized, which involves the addition of sulfur to cross-link the various molecules, thus reinforcing the solid properties of natural rubber. Ingredients are also added to natural rubber to minimize attack by oxygen, ozone, and sunlight. Vulcanized rubber products are used in structures that must withstand severe and repeated flexure. They are also used in applications that require good wear resistance. This is due to the unique property of rubber to yield to an abrasive particle, rather than get cut. This is especially useful in the manufacture of tires. Rubbers possess high coefficients of friction when in contact with dry surfaces. This has increased their use in the tire and shoe industry, as well as in many mechanical applications. Rubbers can be molded under heat and pressure into a variety of intricate shapes during the vulcanization or curing operations. They can be extruded into hoses and gaskets, and they can be made to adhere to metals, glass, plastics, and fabrics. They are poor conductors of electrical current and can be used as insulators. Rubbers are attacked by oxygen, sunlight, and ozone and need to be protected by the addition of antioxidizing agents. Fluctuations in temperature can cause rubber to lose its elasticity and resilience. Rubber is easily degraded by mineral oils, gasoline, benzene, and other organic solvents and lubricants. They are usually resistant to most inorganic acids and bases and salts. Some products made from natural rubbers are given in Table 8.8.

Synthetic rubbers have been developed to overcome some of the shortcomings of natural rubber. They equal natural rubber in its elasticity and resilience. Some of the commonly used synthetic rubbers are listed below and typical products made from them are given in Table 8.8.

8.5.5.3.2 *Styrene-Butadiene Rubber*

Styrene-butadiene rubber, frequently referred to as Buna-S, is one of the earlier synthetic rubbers. It is about 70% as resilient as natural rubber, has a much lower tensile strength, and has a poorer tear resistance. However, it has good wear and abrasion resistance along with excellent water resistance. It is less expensive than natural rubber. The most common product made from styrene-butadiene rubber is listed in Table 8.8.

8.5.5.3.3 *Acrylonitrile-Butadiene Rubber*

Acrylonitrile-butadiene rubber has low tensile strength when compared to natural rubber and is two-thirds as resilient as natural rubber, but can be compounded to give tensile strengths approaching natural rubber. It loses its resilience as temperature decreases, but its resilience increases as temperature increases. For this reason, it is used instead of natural rubber in installations where temperatures are much higher than normal room temperature. Another important property is its resistance to deterioration when exposed to gasoline and oils. Several products made from acrylonitrile-butadiene rubber are listed in Table 8.8.

8.5.5.3.4 *Neoprene*

Neoprene has the same resilience and tensile strength as natural rubber. However, it stiffens considerably at lower temperatures and is costlier than natural rubber, and does not soften at elevated temperatures as do natural rubbers. Neoprene is more resistant to attack by oxygen, ozone, and sunlight, it is impermeable to air and other gases, and it is flame resistant and will not support combustion. It has good resistance to corrosion of chemicals and water. Several products made from neoprene are listed in Table 8.8.

8.5.5.3.5 *Butyl Rubber*

Butyl rubber has tensile strength, resistance to tear, and resilience inferior to natural rubber. It is not resistant to gasoline and oils, and stiffens considerably at low temperatures. However, it is highly resistant to ozone and oxidation and is used when resistance to aging is required. It has excellent dielectric properties, and is resistant to sunlight and all forms of weathering. It is also practically impermeable to air and many gases. Several products made from butyl rubber are listed in Table 8.8.

8.5.5.3.6 *Silicone Rubber*

The mechanical properties of silicone rubber, including tensile strength, tear and abrasion resistance, resistance to set, and resilience, are inferior to that of natural rubber. Its important property is resistance to deterioration at elevated temperatures. Intermittent temperatures as high as 288°C and continuous temperatures of 232°C produce negligible change in flexibility and surface hardness of silicone rubber, whereas natural rubber and other rubber-like materials would quickly harden or deteriorate. Silicone rubber also has excellent resistance to oxidation, ozone, and aging. It is fairly resistant to oils, and does not deteriorate in gasoline. Because of its high cost (about 20 times that of natural and most synthetic rubbers), it has limited applications.

8.5.6 Engineered Plastics

Engineered plastics refer to commercially available polymer-based composite materials whose mechanical, chemical, and physical properties have been significantly enhanced or modified by the intentional addition of a wide variety of fillers. These additives serve a number of different roles, the most common being the following.[*]

8.5.6.1   Mechanical Property Enhancement

The most common fillers in polymers are added to improve tensile and compressive strength, stiffness, toughness, and dimensional stability. These fillers can be introduced as particulates of varying sizes or as fibers. Typical fillers include silica (SiO_2), alumina (Al_2O_3), silicon carbide (SiC), and carbon (C) particles, glass beads, and glass and carbon fibers. The higher strength, stiffness, and dimensional stability of fibers relative to the matrix polymer allow them to improve these properties in the polymer along the longitudinal direction of the fiber. Particulates, on the other hand, strengthen the polymer by dispersion hardening. As with other dispersion hardening systems, the property modification is a function of the filler size, with smaller, well-distributed particles providing better properties than fewer, larger particles. The property modifications are also a function of the percentage of filler, which can range from 5% by weight to 90% by weight. A typical percentage of filler in thermoplastics for mechanical property enhancement is 30% by weight, while some thermosetting encapsulants used in microelectronic devices can be as high as 82% silica.

8.5.6.2   Conductivity Enhancement

High thermal and electrical conductivity filler particles can be added to polymers to improve these properties as well. Silver particles are often added to thermoset polymers, such as epoxy, to produce thermally and electrically conductive glue, while carbon fibers, nickel-coated carbon fibers, or stainless steel fibers are often added to thermoplastic polymers to provide just enough electrical conductivity for the polymer to resist accumulating a static charge or to provide EMI/RFI shielding. Particles that enhance thermal conductivity without altering the electrically insulating properties of the polymer include boron nitride and aluminum nitride. Again, the properties are a function of the percentage of filler used. However, in this case, conductivity increases slightly until a percolation threshold is reached at which point the conductive particles begin to touch each other and create a continuous network within the polymer. The conductivity jumps at the percolation threshold and then continues to increase slowly above it. Adding pyrolytic graphite fibers to the polymer can create exceptional increases in the thermal conductivity of the polymer along the longitudinal direction of the fiber with little or no change in the orthogonal directions.

8.5.6.3   Wear Resistance

Wear-resistant compounds offer protection from surface marring and scratching, reduction in noise between mating parts, and elimination of the slip/stick phenomenon that is prevalent in sliding parts. These compounds are used in such applications as bearings and gears. Typical additives that improve wear resistance include perfluoropolyether oil, PTFE (Teflon), silicone, molybdenum disulfide, and graphite, along with aramid, carbon, and glass fibers.

8.5.6.4   Color

Colorants can be added to impart specific colors to the polymer and include carbon particles, which are used to make the polymer black and opaque.

8.5.6.5   Flame Retardant Increase

Fillers can be added to interfere with the combustion process by limiting the availability of oxygen, by initiating endothermic chemical reactions that cause cooling, or by catalyzing reactions that form moisture that can douse the flame. These additives can be in the form of particulates added to the polymer and are often composed of metal oxides, metal hydroxides, or phosphorus. This property

---

[*] For access to a very large amount of useful technical information about engineered plastics and the very wide range of products that are made from these materials, see http://www.rtpcompany.com/.

is especially important in the electronics industry, where sparks and electrical discharges cannot be allowed to cause the electronic systems to catch fire.

8.5.6.6 Plasticizers

Another category of polymer additives is plasticizers. These are low-molecular-weight organic compounds that are added to polymers to improve flexibility, durability, and toughness by occupying positions that separate the long polymer chains. Common plasticizers are liquids with low vapor pressure based on esters and phthalates. These additives change hard and stiff polymers such as polyvinyl chloride (PVC) into tough and flexible polymer composites, such as vinyl, by permitting long-chain molecular motion.

8.6 CERAMICS

8.6.1 STRUCTURAL CERAMICS

Ceramics are inorganic, nonmetallic, solid elements or compounds. Structural ceramics are formed by combining two or more of these ceramic materials in powder form, and either mechanically or metallurgically binding them together. There are several classes of ceramics.[*]

1. **Technical ceramics** are mixtures of different ratios of silicon dioxide and other light metal oxides of elements such as aluminum and magnesium. They are used to form products with good stiffness, hardness, wear resistance, and chemical inertness. Some of them also provide good electrical insulation.
2. **Engineering ceramics** consist of sintered combinations of oxides of aluminum, magnesium, zirconium, silicon, and several other materials. The products made with these powders have much better high temperature properties and greater hardness than the technical ceramics, although not necessarily greater wear resistance.
3. **Advanced ceramics** are mainly based on oxides, carbides, and nitrides of silicon, lithium, zirconium, and aluminum. The main advantage of these mixtures is their toughness (the ability to resist fracture). They are very resistant to thermal shock, and some combinations are almost as tough as white cast iron.

Ceramics are produced by forming a powder mixture of the desired composition into the desired shape, and then firing it at a high enough temperature and for a long enough time to sinter the powder. During sintering, the powder shrinks, coalesces, and recrystallizes to form a solid. It does this without melting to form a liquid. The sintering makes the part dense, with shrinkage sometimes of the order of 25%. The large amounts of shrinkage can make it difficult to form components to precise dimensions. Also, ceramics are brittle materials and, as such, cannot withstand tensile loads. Ceramics are used in applications in which wear and high temperature are primary concerns. Some products made from ceramics are listed in Table 8.9.

8.6.2 ELECTRICALLY INSULATING CERAMICS

One important property of ceramic materials is the ability to withstand electric fields with very little conduction of electricity. Ceramics are therefore widely used in electronics as insulating materials, where they serve as the dielectric in capacitors, as an insulating substrate for patterning circuits in a multichip module, and as the case material for electronic devices.

---

[*] The term *ceramics* also covers materials such as furnace linings, pottery, tiles, glass, etc., but these applications are infrequently encountered by product design engineers.

TABLE 8.9
Typical Products Made from Ceramics

Material	Products		Section
Ceramics			
Structural	Coatings	Cements	8.6.1
	Insulators	Water pump seals	
	Dielectrics	Automotive rocker arm wear pads	
	Abrasives	Molds and nozzles for molten metals	
	Cutting tools	Substrates for microcircuits	
Electrically insulating	Dielectric in capacitors		8.6.2
	Insulating substrate in a multichip module		
Thermally conductive	Improve heat transfer in		8.6.3
	Component to epoxy molding compounds in plastic encapsulated microcircuits		
	Sheets of polymer film placed between uneven surfaces in heat dissipating devices		
Magnetic ceramics			
Soft ferrites	Cores of switched-mode power supplies		8.6.4.1
	RF transformers		
	Inductors where the magnetic field switches frequently		
Hard ferrites	Radios	Principal component in	8.6.4.2
	Component to reduce high frequency electrical noise (RFI) in computer cables	Transformers	
		Inductors	
		Electromagnets	
	Recording tape coatings	Switched-mode power supplies	

The most common materials for electrical insulation in microelectronic devices are silicon dioxide, silicon nitride, aluminum oxide, and aluminum nitride. These materials have high volume resistivity ($>10^{10}$ Ω-cm at room temperature) and high breakdown voltages, allowing them to withstand large electric fields with low leakage current. In addition, they have low dielectric constants, so they do not accumulate charge that could distort the fields. Their properties are stable to high temperatures, and they have low chemical reactivity and high corrosion resistance. Silicon dioxide is used as the insulator in microelectronic devices on the chip. Aluminum oxide, aluminum nitride, and silicon nitride are widely used as substrate materials for electronic packaging.

The most widely used electronically insulating ceramic is aluminum oxide because of its low cost. In addition to a high volume resistivity and a large breakdown voltage, aluminum oxide has a low coefficient of thermal expansion, fair fracture toughness, and fair thermal conductivity. Aluminum nitride is preferred in applications where significant amounts of heat must be transferred through the ceramic because it has a higher thermal conductivity than alumina and a lower coefficient of thermal expansion, the latter providing better dimensional stability with temperature. These characteristics offset its significantly higher cost. Silicon nitride is preferred in applications requiring resistance to thermal or mechanical shock, because it has higher fracture toughness than the other two.

8.6.2.1 Ferroelectrics

Ferroelectrics undergo a change in their crystal structure in response to an electric field that acts to reduce the field, thereby permitting more charge to be stored across them before breakdown occurs. In addition to their high dielectric constants, these materials have low dielectric losses. However, the process by which these materials alter their crystal structure becomes more random at higher temperatures, so that at temperatures above a certain level, known as the Curie temperature, the materials no longer show this effect. Examples of this type of material are barium titanate, which is widely used in capacitors, and lead zirconate titanate, which is widely used in capacitors and piezoelectric devices.

8.6.3 THERMALLY CONDUCTIVE CERAMICS

Aluminum nitride, silicon nitride, and boron nitride are good thermal conductors. Aluminum nitride's thermal conductivity is more than half of that of copper. Therefore, aluminum nitride, silicon nitride, and boron nitride are often added as fillers to improve the thermal conductivity of polymer-based composite materials. They are added to epoxy molding compounds to improve heat transfer in plastic encapsulated microcircuits and to sheets of polymer film to improve heat transfer between uneven surfaces when heat dissipating devices are attached to heat sinks.

8.6.4 MAGNETIC CERAMICS

Ceramic materials known as ferrites have begun to replace metallic alloy permanent magnets in a number of applications. Ferrites are nonconductive, ferrimagnetic compounds known as spinels. They are derived from iron oxides, such as hematite and magnetite, as well as oxides of other metals. They are used in applications ranging from inductors and transformers to microelectronics, some specific examples of which are listed in Table 8.9. They share common mechanical properties with other ceramics, in that they are hard and brittle. Ferrites are known by the two metals that are present in the compound. There are two main groupings of ferrites: soft, which refers to its low coercivity, and hard, which refers to its high coercivity. Coercivity is a measure of a material's coercive force when the material has been magnetized to saturation.

Ferrites are produced by powder processing in which a mixture of powdered precursors are heated and pressed in a mold. The precursors are usually carbonates of the metals desired, which are calcined to form the oxides during the powder processing. Once cooled, the mixed and sintered cake is milled into new composite particles, which are then pressed into shape, dried, and refired to create the final ferrite. The shaping is often done in the presence of an external magnetic field to achieve a preferred orientation of particles.

8.6.4.1 Soft Ferrites

These ferrites contain nickel, zinc, or manganese and are used in transformer or electromagnetic cores. The high electrical resistance of the ferrites leads to very low eddy current losses and the low coercivity reduces losses at high frequencies. Products utilizing these characteristics are given in Table 8.9.

8.6.4.2 Hard Ferrites

Permanent ferrite magnets have a high remanence (remaining magnetic induction after the magnetizing influence has been removed) and are formed from oxides of iron, barium, or strontium. In the magnetically saturated state, they conduct magnetic flux well and have a high magnetic permeability. This permits them to store stronger magnetic fields than pure iron. They are commonly used in radios. Other products made from hard ferrites are listed in Table 8.9.

8.7   COMPOSITES

A composite material is one that is composed of two or more physically distinct materials that are combined in such a way as to produce properties that are not available in the individual materials. Composite materials are found in many distinctly different combinations, examples of which include: (1) metal matrix composites, which include mixtures of ceramic particles and fiber reinforcement in a continuum and metals such as aluminum and magnesium and (2) polymer matrix composites, which include both laminates of thermosetting polymers with fiber reinforcing sheets or webs, and thermoplastics with fillers.

Composites have found application in a wide range of products; some of the more common ones are listed in Table 8.10.

8.7.1   METAL MATRIX COMPOSITES

Metal matrix composites are mixtures of a metal or alloy and ceramic fillers that are used to modify the properties of the metal. The ceramic fillers can increase the strength or stiffness of the metal, decrease its coefficient of thermal expansion, or increase its thermal conductivity. One of the most common metal matrix composites is aluminum-silicon carbide, which consists of silicon carbide particles or fibers that are placed in molten aluminum, which then solidifies around the fibers. One application of a metal matrix composite is as a "heat spreader" in which a heat generating device is placed on the spreader to dissipate the heat. The silicon carbide fibers decrease the coefficient of thermal expansion of the aluminum, making it a better match for what is mounted on it, while only slightly decreasing the thermal conductivity. The particles also strengthen and stiffen the aluminum without reducing its fracture toughness.

8.7.2   FIBER-REINFORCED COMPOSITES

In fiber-reinforced composites, the various roles of the matrix portion of the composite are (1) to provide the bulk form of the part; (2) to hold the embedded material in place; (3) to transfer the load to the reinforcing material; (4) to contribute to the electrical and chemical properties; (5) to carry

TABLE 8.10
Typical Products Made from Composites

Material	Products		Section
Composites	Aircraft components	Sports equipment	8.7
	Wings	Diving boards	
	Blades	Tennis rackets	
	Fuselages	Skis	
	Safety equipment	Surf boards	
	Military helmets	Bobsleds	
	Pressure vessels	Sailboats	
	Rocket casings		
	Storage tanks		
Metal Matrix Composites	Heat spreader		8.7.1
Carbon/Carbon	Sports equipment	Airframes	8.7.3
	Tennis racquets	Aircraft brakes	
	Golf club shafts		
Cemented Carbides	Cutting tools		8.7.4

interlaminar shear; and (6) to govern shrinkage and the coefficient of thermal expansion in directions orthogonal to the fiber.

Commonly used thermoset plastic matrices are polyester and epoxy. The polyesters are usually used in small-scale low technology applications such as boat building. The epoxies are widely used for advanced composite applications. Some of the advantages of epoxy resin composites are the following: no by-products; low shrinkage; resistance to solvents and chemicals; resistance to creep and fatigue; and good electrical insulating properties. Some of their disadvantages are high coefficient of thermal expansion, high degree of smoke liberation during fire, slow curing, and, in some cases, degradation with exposure to ultraviolet light.

The reinforcing fibers can be polymers, glass, or metals, which provide the strength, ductility, and toughness. The fiber diameters are on the order of 0.01 mm. The most common fibers are glass, graphite, and Kevlar® (an aramid fiber). These reinforcing fibers are usually oriented in a direction that will permit the object to carry the load most effectively. To make the application of the fiber to a specified orientation easier, these fibers are woven into cloth with specific patterns.

Increases in the filler content increase the strength of the material through dispersion strengthening, increase the modulus of the material by sharing some of the elastic load, and increase the density of the material. Long thin fibers provide higher levels of strengthening and higher modulus than spherical powders or short stubby fibers. Woven fiber sheets provide strengthening and stiffening in the plane of the sheet but no changes to the properties in the transverse direction.

8.7.3 CARBON/CARBON COMPOSITES

Carbon fiber reinforced epoxy composites have some unique properties and are a subset of fiber-reinforced composites. The high strength and modulus of the carbon fibers combines with the toughness of the epoxy to produce a composite with high strength, toughness, and stiffness. Carbon/carbon composites are also used for their good heat conductive properties. Pyrolytic graphite fibers have a very high thermal conductivity in the longitudinal direction and a very small thermal conductivity in the transverse dimensions. Placing these fibers in a highly oriented manner in an epoxy matrix creates composites with high thermal conductivity superior to silver and copper in the longitudinal fiber direction. These carbon/carbon composites are used in a variety of applications including in aircraft brakes where it is necessary to have a material that combines high strength, toughness, and heat resistance with wear resistance and high thermal conductivity. Several products made from carbon/carbon composites are listed in Table 8.10.

8.7.4 CEMENTED CARBIDES

One type of metal matrix composite is the cemented carbide. These composites consist of particles of tungsten carbide, silicon carbide (sapphire), or other hard ceramic embedded in a matrix of cobalt metal. These materials are used for cutting tools, and they combine the wear and abrasion resistance of the carbides with the strength and ductility of the metal matrix. The carbide cutting particles are placed in a powder of the metal and the mixture is processed using powder metallurgy techniques (see Section 9.9.1). The entire powder compact is fired and sintered to produce the final composite material. Key properties of the resulting metal matrix composite are high hardness, excellent impact resistance, and good toughness.

8.7.5 FUNCTIONALLY GRADED MATERIALS

Functionally graded materials are materials whose microstructure and/or composition are engineered to continuously vary with location within a material in order to provide different mechanical and physical properties to meet specific functional requirements at different locations in the material. These modern materials are created using new manufacturing techniques to produce the appropriate spatial distributions of material phases, porosity, filler concentration, and/or composition. Functionally graded materials have gotten their inspiration, in part, from nature. Two examples of functionally graded materials in nature are (1) bone, where the interior structure of bone changes, depending on the principal stress directions and the magnitude of shear stress that it carries; and (2) bamboo, where between 20% and 30% of the cross sectional area of the stem is made of longitudinal fibers that are distributed nonuniformly through the wall thickness, the concentration being most dense near the exterior.[*] An example of a common functionally graded metal is steel that is produced by carburization, a process where additional carbon is diffused into the steel before heat treatment. After heat treatment, the resultant structure is hard and brittle at the outermost layer where the carbon content is the greatest and, as the center of the piece is approached, the carbon concentration decreases and consequently the hardness decreases and the material's toughness increases. The addition of coatings to carbide cutting tools to greatly improve their wear resistance is another example.[†] Military armor is sometimes thought of as a functionally graded material, even though the constitutive components do not vary continuously through its depth. Each layer comprising the material has a different function to perform: the outer layer for durability and heat resistance, a layer to satisfy ballistic requirements, a structural layer, and finally a layer for flame protection.

The process of creating functionally graded engineered materials is still relatively new, but several products have been successfully introduced. They include (1) glass plates and glass fibers with functionally graded refractive index, which are used as miniature focusing lenses and for wide band optical signal transmission, respectively; and (2) polymer skin foams, which have a dense surface layer and increasing porosity toward the interior. The resulting material has high impact strength and finds use in instrument panels. Functionally graded composites are still a commercial rarity, but a few promising ones may result in higher efficiency thermoelectric coolers, solid oxide fuel cells, and temperature stable capacitors.[‡]

8.8   SMART MATERIALS

Smart materials are materials that provide a mechanical response to an electromagnetic or thermal input. This allows the material to change its shape in response to an external stimulus, thereby permitting it to be used as a sensor or as an actuator. Smart materials also experience the converse phenomenon. They put out an electromagnetic or thermal signal in response to forced changes in their shape. Three main types of smart materials are now discussed.

8.8.1   PIEZOELECTRIC MATERIALS

Piezoelectric materials are materials that change their shape in response to an electrical signal or conversely output an electrical signal in response to a change in shape. The phenomenon is due to the complex crystal structure of the material in which mechanical stress rearranges the atoms in such a way as to affect charge distribution. This change in charge distribution results in the creation of a current that can be sensed and read by output circuitry. Changing shape in response to a signal allows these materials to be used in loudspeakers where they vibrate in response to the action of an AC input signal. Generating an electrical signal in response to a mechanical stress allows these materials to be used in such devices as microphones and sonar sensors, where they convert the mechanical vibratory forces into electrical signals. The most widely used piezoelectric materials are the ferroelectrics, and of them the most common is lead zirconate titanate (PZT). Several products made from PZT are listed in Table 8.11.

---

[*] Silva, E.C.N., Walters, M.C., and Paulino, G. H., "Modeling bamboo as a functionally graded material: Lessons for the analysis of affordable materials," *Journal of Materials Science*, Vol.41, No.21, pp. 6991–7004, 2006.

[†] http://www.sme.org/cgi-bin/find-articles.pl?&02ocm002&ME&20021001&&SME&#article.

[‡] Mortensen, A., *Concise Encyclopedia of Composite Materials*, Elsevier, Oxford, 2007.

TABLE 8.11
Typical Products Made from Smart Materials and Nanomaterials

Material	Products		Section
Piezoelectric	Microphones	Hydrophones	8.8.1
Shape Memory Alloys	Eyeglass frames	Dental braces	8.8.3
	Pipe couplings	Robotic devices	
	Peripheral vascular stents	Aircraft noise reduction devices	
Nanomaterials	Electrical connections	Materials employing carbon nanotubes	8.9
	Magnetic materials	Thermal pastes	
		Greases	
		Pads	

8.8.2 MAGNETOSTRICTIVE MATERIALS

Magnetostrictive materials are the magnetic analog of piezoelectric materials. They change their shape in the presence of a magnetic field. This effect is the result of a distortion of the crystal structure when the material is placed in a magnetic field. Conversely, they create a magnetic field when they are deformed: extended, bent, or compressed. These effects allow these materials to be used in sensing and actuating applications. Many metals exhibit a magnetostrictive effect, including nickel and iron. However, the most common magnetostrictive material is a nickel-titanium alloy known as NiTiNOL. This material has the ductility and formability of a metal alloy, thereby permitting it to be drawn into wires or rolled into sheets, and it has one of the highest magnetostrictive effects measured. The material is used in sonar sensors and other sensing and actuating applications. A new material, GalFeNOL, which is a gallium-iron alloy, has the potential of even larger magnetostrictive effects. However, it does not yet have the same level of ductility and formability as NiTiNOL.

8.8.3 SHAPE MEMORY MATERIALS

Shape memory materials are materials that, after being deformed at room temperature, regain their original shape by the application of heat. Shape memory materials are a lightweight, self-contained alternative to conventional hydraulic or pneumatic actuators. There are three alloys that show this behavior: copper-aluminum-nickel, copper-zinc-aluminum-nickel, and the most widely used, nickel-titanium. In addition, there are a number of polymers that were developed in the 1990s that have this behavior. The nickel titanium alloys are initially supplied in the austenitic phase. When plastically deformed, instead of deforming by dislocation glide, they deform by shear transformation to the martensitic phase. When heated to 500°C, the martensite is changed back to austenite and the sample regains its original predeformation shape. Repeated use of the shape memory effect may cause the temperature of the austenite to martensite transformation to change in a process known as functional fatigue. Both one-way and two-way shape memory effects have been observed. In the one-way effect, a small amount of room temperature deformation is reversed by heating. The sample then stays in the undeformed state when returned to room temperature. In the two-way effect, a large amount of deformation is performed at room temperature. When the sample is heated, it returns to its original shape, but when cooled, it returns again to its deformed shape. One of the most useful properties of shape memory alloys is "pseudo-elasticity." This is the ability of the material to undergo repeated applications of stress and never lose its original shape. This property is made use of, for example, in eyeglass frames, where the frame will not remain bent regardless of the amount of twisting and bending it encounters.

Shape memory alloys are typically made by casting, using vacuum arc melting, or induction melting. The high temperature shape is established by training the metal; that is, by deforming it at a temperature between 400 and 500°C. The properties of shape memory alloys are similar to other metal alloys. The yield strength is lower than steel but higher than aluminum, sometimes reaching 500 MPa, but the true advantage of this material is the high level of recoverable strain. Products made shape memory alloys are listed in Table 8.11.

8.9   NANOMATERIALS

Nanomaterials are materials whose properties are determined by their structure at the nanometer scale. Some of these materials contain nanoparticles, while others have nanometer-sized grains or magnetic domains. Many of these materials are made by incorporating or sintering nanoparticles, which are particles that have one or more dimensions in the nanometer range. These particles can be used as fillers in composites or sintered into a final product. Several products made from nanometals are listed in Table 8.11.

8.9.1   Sintered Nanoparticle Solids

Because of their very small size, nanoparticles have a large surface energy, which drives the particles to combine into larger agglomerates. This has been used to permit sintering of these particles into a porous solid structure at lower temperatures and pressures than would be necessary for processing conventional micron-sized powders. For example, this nanoparticle sintering process has been used to make permanent electrical connections between contacting surfaces using silver and gold nanoparticles, at pressures 10 times lower than for standard particles.

8.9.1.1   Nanocrystalline Magnetic Materials

Nanocrystalline magnetic materials are magnetic materials whose crystal grains are in the nanometer range. They have excellent properties at frequencies from 10 to 100 kHz. These materials possess saturation induction values of 1.6 to 2.1 T, which are three to five times those of ferrites and twice those of amorphous magnets. Furthermore, nanocrystalline soft magnetic alloys have been developed over the past decade with high permeabilities, high Curie temperatures, and without the magnetic hardening associated with the crystallization of amorphous alloy, which limits core losses. Core losses can reduce efficiency and increase the temperature of surrounding components leading to failure well below the Curie temperature.

8.9.1.2   Carbon Nanotubes

A widely investigated nanomaterial is the carbon nanotube. These structures are a phase of carbon in which the carbon atoms are arranged in a tube with a length in the micrometer range and a diameter in the nanometer range. These thin walled tubes have a very high modulus and strength, as high as the theoretical limit in the longitudinal direction. In addition, carbon nanotubes have very high thermal conductivity in the longitudinal direction. The high strength and thermal conductivity have resulted in the incorporation of carbon nanotubes into a wide range of composite materials, some of which are listed in Table 8.11.

8.10   COATINGS

Coatings are placed on a material to provide the surface with properties that are different from the bulk properties. Bulk materials are often chosen for strength and toughness, but may not have optimal corrosion resistance, wear resistance, hardness, or electrical properties. Surface coatings are used to impart these characteristics to the product. Coatings can be applied in a number of ways including spin coating, dip coating, electroplating, electroless plating, chemical vapor deposition, and plasma vapor deposition.

8.10.1 WEAR AND SCRATCH RESISTANCE

Hard, abrasion-resistant coatings are often added to a bulk metal to provide wear and scratch resistance. Coating materials include thin film diamond and carbide coatings, which are deposited by chemical or plasma vapor deposition.

8.10.2 ELECTRICALLY CONDUCTIVE/INSULATING

Gold is often electroplated onto copper-alloy connectors to provide a soft, electrically conductive interface for mating. The gold used is called *hard gold*, because it has cobalt additions to increase its wear and scratch resistance, permitting multiple matings. Soft gold, which is pure gold, has higher conductivity and better thermal stability, but poor wear and galling resistance. Other metals that can be used for electrical coatings include platinum, palladium, silver, and tin. These metals are often plated onto copper connectors on printed wiring boards. These coatings are applied by using an electroplating process in which the copper connector is made the cathode and the metal is plated onto it from solution. Coatings can also be applied to make a material's surface electrically insulating, such as anodizing aluminum, to make a dark, corrosion-resistant, insulating coating for surface appearance.

BIBLIOGRAPHY

M. F. Ashby, *Material Selection in Mechanical Design*, Pergamon Press, Oxford, 1992.

R. M. Brick, A. W. Pense, and R. B. Gordon, *Structure and Properties of Engineering Materials*, 4th ed., McGraw Hill, New York, 1977.

K. G. Budinski, *Engineering Materials, Properties and Selection*, 4th ed., Prentice Hall, Engelwood Cliffs, NJ, 1992.

J. A. Charles and F. A. A. Crane, *Selection and Use of Engineering Materials*, 2nd ed., Butterworths, London, 1989.

E. H. Cornish, *Materials and the Designer*, Cambridge University Press, Cambridge, 1987.

R. J. Crawford, *Plastics Engineering*, 2nd ed., Pergamon Press, Oxford, 1987.

R. D. Deanin, *Polymer Structure, Properties and Applications*, Cahners Books, Boston, MA, 1972.

L. Edwards and M. Endean, Eds., *Manufacturing with Materials*, Butterworths, London, 1990.

M. M. Farag, *Selection of Materials and Manufacturing Processes for Engineering Design,* Prentice Hall, Englewood Cliffs, NJ, 1989.

D. P. Hanley, *Selection of Engineering Materials*, Van Nostrand Reinhold Co., New York, 1980.

C. A. Harper, *Handbook of Plastics, Elastomers, and Composites*, 2nd ed., McGraw-Hill, New York, 1992.

N. C. Lee, Ed., *Plastic Blow Molding Handbook*, Van Nostrand Reinhold, New York, 1990.

Metals Handbook, 10th ed., prepared under the direction of the ASM International Handbook Committee, ASM International, Materials Park, OH, 1990.

B. W. Neibel and A. B. Draper, *Product Design and Process Engineering*, McGraw-Hill, New York, 1974.

B. W. Niebel, A. B. Draper, and R. A. Wysk, *Modern Manufacturing Process Engineering*, McGraw-Hill, New York, 1989.

W. F. Smith, *Structure and Properties of Engineering Alloys*, McGraw Hill, New York, 1981.

J. S. Walker and E. R. Martin, *Injection Molding of Plastics*, Iliffe Books Ltd., London, 1966.

9 Manufacturing Processes and Design

The most important attributes, limitations, and material compatibility of 19 manufacturing processes and six layered manufacturing technologies are summarized and typical products made by each are given.

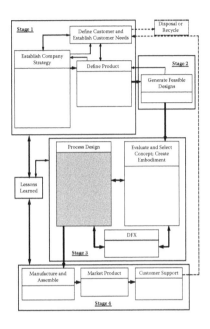

9.1 INTRODUCTION

9.1.1 COMMON DESIGN ATTRIBUTES

Manufacturing processes are selected with certain product attributes in mind: size, form (shape complexity), surface finish, strength, dimensional tolerances, cost, and the number of pieces to be made. Surface condition affects appearance, the ability to assemble the part to other components, and, in some cases, protection against corrosion. Dimensional accuracy can affect performance and interchangeability of components. Shape complexity and size often determine whether a manufacturing process is the most suitable one to manufacture a part. Production run (or production rate) will govern whether a process is the most economical one to manufacture the desired quantity of parts. Cost is affected by all of the preceding attributes.

One or more of the attributes mentioned above can be used to determine candidate manufacturing processes for the product's manufacture. Secondary operations such as heat treatment, machining, grinding, painting, etc., are not considered at this stage. As a general rule, these secondary operations should be avoided or minimized to reduce cost and handling and to increase the production rate. For many manufacturing processes, the satisfaction of these criteria often involves compromises, and not all of the attributes may be satisfied to the degree desired. In order to minimize the severity of these compromises, an attempt has been made to identify the capabilities of the most common manufacturing processes in such a way that the most important attributes can be satisfied to the highest degree. The results are summarized in Table 9.1, where each of the criteria is divided into three ranges: low (*L*), medium (*M*), and high (*H*). The corresponding numerical values represented by each range are also given, along with their engineering units.

The last link, which is between the candidate manufacturing processes as determined from Table 9.1 and the candidate materials as determined from Table 8.2, is given in Table 9.2. The data in this table relate the degree of suitability of the candidate materials to each of the candidate manufacturing processes. Four degrees of suitability have been used: *E* means that the material has excellent compatibility with the manufacturing process and is very well-suited for this manufacturing method; *G* means that the material has good compatibility with the manufacturing process, and has been used in this manufacturing process; *S* means that this material is seldom used in this process; and "-" indicates that the material is neither used nor recommended for use with that manufacturing process.

TABLE 9.1
Manufacturing Processes and Their Attributes[a]

				Attribute			
Process	Surface Roughness	Dimensional Accuracy	Complexity	Production Rate	Production Run	Relative Cost	Size (Projected Area)
Pressure die casting	L	H	H	H/M	H	H	M/L
Centrifugal casting	M	M	M	L	M/L	H/M	H/M/L
Compression molding	L	H	M	H/M	H/M	H/M	H/M/L
Injection molding	L	H	H	H/M	H/M	H/M/L	M/L
Sand casting	H	M	M	L	H/M/L	H/M/L	H/M/L
Shell mold casting	L	H	H	H/M	H/M	H/M	M/L
Investment casting	L	H	H	L	H/M/L	H/M	M/L
Turning/facing	L	H	M	H/M/L	H/M/L	H/M/L	H/M/L
Milling	L	H	H	M/L	H/M/L	H/M/L	H/M/L
Grinding	L	H	M	L	M/L	H/M	M/L
EDM	L	H	H	L	L	H	M/L
Blow molding	M	M	M	H/M	H/M	H/M/L	M/L
Sheet metal working	L	H	H	H/M	H/M	H/M/L	L
Forging	M	M	M	H/M	H/M	H/M	H/M/L
Rolling	L	M	H	H	H	H/M	H/M
Extrusion	L	H	H	H/M	H/M	H/M	M/L
Powder metallurgy	L	H	H	H/M	H	H/M	L
Units	μm	mm		parts/hr	parts		m²
H	>6.4	<0.13	High	>100	>5000	High	>0.5
M	>1.6	>0.13	Medium	>10	>100	Medium	>0.02
	<6.4	<1.3		<100	<5000		<0.5
L	<1.6	>1.3	Low	<10	<100	Low	<0.02

[a] A. Kunchithapatham, A Manufacturing Process and Materials Design Advisor, M.S. thesis, University of Maryland, College Park, May 1996.

9.1.2 GENERAL GUIDELINES FOR REDUCED MANUFACTURING COSTS

The following principles and guidelines are applicable to virtually all manufacturing processes and will usually result in components and products that can be manufactured at reduced cost.[*][†]

1. **Simplicity**. The simplest design will usually be the most reliable, the easiest to service, have the fewest parts, have fewer intricate shapes, require the fewest adjustments, and most likely be the least costly to produce.
2. **Standard materials and components**. The use of standard components simplifies inventory management, purchasing, reduces tooling and equipment investments, and has the advantage that much is known about them: reliability, environmental suitability, etc.

---

[*] J. G. Bralla, Ed., *Handbook of Product Design for Manufacturing*, McGraw-Hill, New York, 1986.
[†] J. G. Bralla, Ed., *Design for Manufacturability Handbook*, 2nd ed., McGraw-Hill, New York, 1999.

TABLE 9.2
Suitability of Materials and Manufacturing Processes[a]

Material	Pressure Die Casting	Centrifugal Casting	Compression Molding	Injection Molding	Sand Casting	Shell Mold Casting	Investment Casting	Single Point Cutting	Milling	Grinding	EDM	Blow Molding	Sheet Metal Working	Forging	Rolling	Extrusion	Powder Metallurgy
Low carbon steel	-	E	-	-	E	E	E	G	G	E	E	-	G	G	G	G	E
Medium carbon steel	-	E	-	-	E	E	E	G	G	E	E	-	G	G	G	G	E
High carbon steel	-	E	-	-	E	E	E	G	G	E	E	-	G	G	G	G	E
Low alloy steel	-	E	-	-	E	E	E	-	G	E	E	-	G	G	G	G	E
Tool steel	-	-	-	-	G	E	E	-	-	-	E	-	-	-	-	-	E
Stainless steel	-	G	-	-	E	E	E	G	-	-	E	-	G	G	G	G	E
Gray iron	-	E	-	-	E	E	E	G	G	E	E	-	-	S	S	S	E
Malleable iron	-	E	-	-	E	E	E	G	G	E	E	-	G	S	S	S	E
Ductile iron	-	E	-	-	E	E	E	G	G	E	E	-	G	S	S	S	E
Alloy cast iron	-	E	-	-	E	G	S	G	-	S	E	-	-	S	S	S	E
Zinc alloys	E	-	-	-	G	G	E	E	E	G	E	-	G	S	S	G	E
Aluminum alloys	E	E	-	-	E	G	E	E	E	S	E	-	E	E	G	E	E
Magnesium alloys	E	-	-	-	E	G	E	G	-	S	E	-	G	S	S	E	E
Titanium alloys	-	-	-	-	-	G	S	-	-	G	E	-	-	G	E	E	E
Copper alloys	G	E	-	-	E	G	G	E	E	S	E	-	E	E	G	S	E
Nickel alloys	-	E	-	-	E	G	G	-	-	S	E	-	G	S	S	G	E
Tin alloys	E	-	-	-	G	G	S	-	-	S	E	-	S	-	-	S	E
Cobalt alloys	-	-	-	-	-	G	G	-	-	S	E	-	-	-	-	-	E
Molybdenum alloys	-	-	-	-	-	G	G	-	-	S	E	-	-	-	-	-	E
Tungsten alloys	-	-	-	-	E	G	G	-	-	S	E	-	-	S	-	-	E
Low expansion alloys	-	E	-	-	E	G	G	-	-	S	E	-	G	S	G	G	E
Permanent magnet alloys	-	E	-	-	E	G	G	-	-	S	E	-	-	S	G	G	E
Electric resist. alloys	-	E	-	-	E	G	G	-	-	S	E	-	G	S	G	G	E

(continued)

TABLE 9.2
Suitability of Materials and Manufacturing Processes[a]

Material	Pressure Die Casting	Centrifugal Casting	Compression Molding	Injection Molding	Sand Casting	Shell Mold Casting	Investment Casting	Single Point Cutting	Milling	Grinding	EDM	Blow Molding	Sheet Metal Working	Forging	Rolling	Extrusion	Powder Metallurgy
ABS	-	-	-	-	-	-	-	G	G	G	-	G	-	-	-	E	-
Acetals	-	-	-	-	-	-	-	G	G	G	-	G	-	-	-	G	-
Nylons	-	-	-	E	-	-	-	G	G	G	-	G	-	-	-	G	-
Fluorocarbons	-	-	-	-	-	-	-	G	G	G	-	G	-	-	-	S	-
Polycarbonates	-	-	-	-	-	-	-	G	G	G	-	G	-	-	-	G	-
Polyimides	-	-	-	-	-	-	-	G	G	G	-	G	-	-	-	S	-
Polystyrene	-	-	-	E	-	-	-	G	G	G	-	G	-	-	-	E	-
PVC	-	-	-	-	-	-	-	G	G	G	-	G	-	-	-	E	-
Polyurethane	-	-	E	-	-	-	-	G	G	G	-	G	-	-	-	G	-
Polyethylene	-	-	-	E	-	-	-	G	G	G	-	E	-	-	-	E	-
Polypropylene	-	-	-	-	-	-	-	G	G	G	-	E	-	-	-	E	-
Acrylics	-	-	-	-	-	-	-	G	G	G	-	-	-	-	-	S	-
Alkyds	-	-	E	-	-	-	-	G	G	G	-	-	-	-	-	S	-
Epoxies	-	-	E	E	-	-	-	G	G	G	-	-	-	-	-	S	-
Phenolics	-	-	E	-	-	-	-	G	G	G	-	-	-	-	-	G	-
Silicones	-	-	E	-	-	-	-	-	-	-	-	-	-	-	-	S	-
Polyester	-	-	E	-	-	-	-	G	G	G	-	-	-	-	-	S	-
Rubbers	-	-	E	E	-	-	-	-	-	-	-	-	-	-	-	S	-

Manufacturing Process

[a] A. Kunchithapatham, A Manufacturing Process and Materials Design Advisor, M. S. thesis, University of Maryland, College Park, MD, May 1996.

E = Excellent: The material is one of the most suitable for the process.
G = Good: The material is a good candidate for the process.
S = Seldom used: The material is seldom used in the process.
"-" Unsuitable: The material is neither used nor is suitable for use in the process.

3. **Standardize the components of a product line**. For a family of similar products, use as many of the same materials, parts, and subassemblies as possible.
4. **Use liberal tolerances**. Higher costs can result from the indiscriminate use of tight tolerances. They should be specified only when the product's performance requires it.
5. **Use the materials that are most compatible with the manufacturing process selected**.
6. **Minimize the number of individual parts in a product**.
7. **Avoid secondary operations**. Eliminate secondary operations such as inspection, plating, painting, heat treating, unnecessary machining, and material handling.
8. **Make the design commensurate with the level of production**. The part should be suitable for a production method that is economical for the quantity forecast.
9. **Utilize special process characteristics**. Use injection-molded plastics to create colored and surface texture directly in the mold, use appropriate plastics to provide hinges as an integral part of the component, use powder metallurgy to create porous parts that allow the retention of lubrication, etc.

9.1.3 RELATIONSHIP TO PART SHAPE

Associated with each manufacturing process is a set of design guidelines that indicate specific geometric limitations, which if adhered to result in a product that is easy to make and of lower cost. These guidelines have evolved through experiences with the various manufacturing processes. To illustrate the significance of these guidelines, consider the idealized (conceptualized) part shown in Figure 9.1a. If the part were to be made by the six processes indicated in Figures 9.1b to 9.1g, then the modifications to the conceptual part would be those shown in the respective figures.

For almost all manufacturing processes, there exist geometrically-based manufacturing guidelines. These guidelines indicate at the detailed level of the design that certain geometric attributes of a part are easier to obtain with one manufacturing process than with another one. If too many geometric details run counter to good practice, as indicated by the guidelines for that process, then

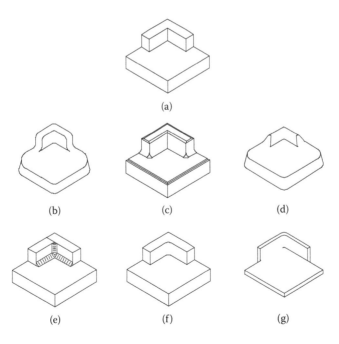

FIGURE 9.1 Effects of the manufacturing processes on geometry: (a) original concept, (b) casting, (c) powder metallurgy, (d) forging, (e) welding, (f) milling, and (g) sheet metal bending. Process-dependent features are exaggerated.

either another manufacturing process should be selected or portions of the part geometry should be altered so as to better accommodate these geometric limitations.

9.1.4   EXAMPLE—STEEL FRAME JOINING TOOL

To illustrate how the information in Chapters 8 and 9 can be used, we continue with the steel frame joining tool, and present the reasoning that goes into the selection of the materials and manufacturing processes for the following components described at the end of Section 6.4: (1) tool shell, (2) impact piston, and (3) compressor piston.

9.1.4.1   Tool Shell

9.1.4.1.1   *Material Requirements*

The power tool shell provides support and mechanical and electrical protection for the internal components of the tool. It must survive impact if dropped and withstand exposure to sunlight, moisture, and temperature extremes. To satisfy its ergonomic requirements, it must be capable of being formed into the complex shapes that human hands like to grasp. To be aesthetically pleasing, it must be formed into an attractive shape and have a pleasing color.

9.1.4.1.2   *Manufacturing Requirements*

The shell's surface condition should be smooth, but its actual finish does not have to satisfy any critical criteria. The dimensional accuracy of the shell is critical in that its two halves have to fit "seamlessly" together. Also, its shape will be complex.

9.1.4.1.3   *Candidate Material and Manufacturing Process*

A good candidate material is polycarbonate, which provides high strength, especially in impact situations, high electrical resistance, and high resistance to heat. This material can be molded into complex shapes using injection molding, and can be made in a variety of colors using the appropriate pigments.

9.1.4.2   Impact Piston

9.1.4.2.1   *Material Requirements*

The impact piston acts as the link between the potential energy in the compressed air and the linear force applied to the fastener. It must be able to withstand high impact loads without failure, and must also have resistance to fatigue. This means that the material must have a high toughness and it must resist wear on the surface that impacts the fastener. Elevated operating temperatures are not a factor. However, close tolerances are required to provide adequate sealing between the impact piston and its housing. Because of these requirements a metal will most likely be required; therefore, good machinability may also be required.

9.1.4.2.2   *Manufacturing Requirements*

The piston's surface condition requires a very smooth surface and a high dimensional accuracy to maintain the pneumatic seal. However, the part is cylindrical and its complexity is low.

9.1.4.2.3   *Candidate Material and Manufacturing Process*

The resistance to impact loads is the major criterion to be satisfied. Therefore, medium carbon steel, which is routinely used in shafts, axles, and gears, is a good candidate. However, market considerations may also dictate the use of other materials, such as case-hardened steel. This may be an effective advertising attribute, since many hand tools are presented in terms of their toughness and reliability. The use of such phrases as "case-hardened components" could be used to differentiate this product from its competition. Another possibility is a high-strength, low-alloy steel. The use of this type of steel may mean higher manufacturing costs, but could be justified if it results in an appreciable increase in sales. The high dimensional accuracy and good surface finish requirements dictate that a turning and/or grinding process be used. Ideally, this part would be sized to take advantage of readily available bar stock. If a high-strength, low-alloy steel is selected, then the hardness of the material may decrease tool life and extend machining and grinding times. Material waste may also be a concern.

9.1.4.3   Compression Piston Chamber

9.1.4.3.1   Material Requirements

The piston used to generate the compressed air will be subjected to moderately high compressive loads, and must have good fatigue properties and high internal dimensional accuracy to maintain the pneumatic seal. Its interior has a cylindrical geometry.

9.1.4.3.2   Manufacturing Requirements

The piston's surface condition requires a very smooth interior surface and a high dimensional accuracy to maintain the pneumatic seal. However, the part complexity is medium, and is primarily cylindrical. The outside of the chamber has no critical dimensional requirements.

9.1.4.3.3   Candidate Material and Manufacturing Process

A candidate material is a 3xx.x series aluminum alloy. A die casting process, followed by suitable machining of the interior cavities, are good candidate processes.

9.2   CASTING—PERMANENT MOLD

9.2.1   Pressure Die Casting[*]

Process description: Molten metal is injected under pressure into a permanent metal mold, which has water-cooling channels as shown in Figure 9.2. When the metal solidifies, the die halves open, and the casting is ejected. (The process is similar to plastic injection molding.) In the cold-chamber process, molten metal is poured into the injection cylinder, but the shot chamber is not heated. In general, the hot chamber method is suited primarily for the casting of alloys having melting temperatures in the neighborhood of 700°C or less and which do not have an affinity for iron. The cold-chamber method is used for handling alloys having higher melting points approaching 1060°C, for alloys that have an affinity for iron such as aluminum, and for the casting of parts that require the highest possible density.

Material utilization: Pressure die casting is a near-net-shape process. There is some scrap in the form of spurs, runners, and flash, but these can be easily recycled. Therefore, the process has excellent material utilization.

Flexibility: The tooling is dedicated and, therefore, the flexibility of the process is limited by the machine setup time.

Cycle time: Production rate is less than 200 pieces/hr, and less than 40 pieces/hr if pore-free die casting is required. The solidification time is typically less than 1 s. Therefore, the cycle time is basically controlled by the time taken to fill the mold and remove the casting.

Operating costs: The die and equipment cost is high, and the labor cost varies from low to medium. Pressure die casting is most advantageous at high production rates, because the substantial equipment/tooling costs can be amortized at high production levels. The required number of products to justify the use of the process is in excess of 10,000, and can go as high as 100,000. The lower the temperature of the molten metal, the longer is the die life.

---

[*] Additional information can be found at the North American Die Casting Association Web site: http://www.diecasting.org.

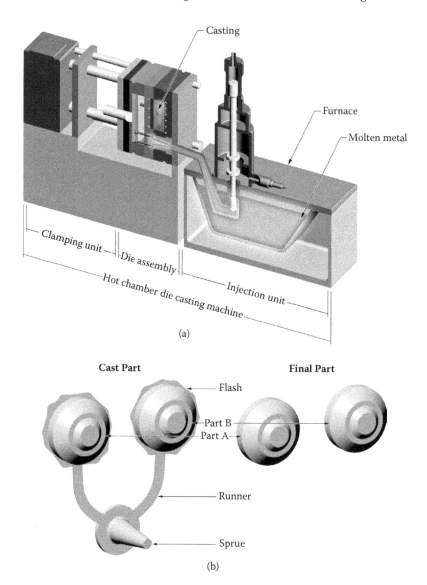

(a)

(b)

FIGURE 9.2 (a) Pressure die casting system; (b) representative part before and after flash removal. (Copyright © 2008 CustomPartNet. Courtesy of CustomPartNet, Inc., www.custompart.net.)

Dimensional accuracy: Pressure die casting can produce good surface texture, but turbulence while filling the mold produces a high degree of internal porosity. Typical surface roughness values are in the range 0.4–1.6 μm.

Shapes: Pressure die casting does not allow the use of complex cores, but it can be used for contoured surfaces. Intricate shapes and details can be incorporated, and it can be used to process a flat surface with a deviation from flatness of 1.1 mm/m. The maximum area that can be manufactured is 0.7 m^2. Minimum section thickness for die casting is between 25 and 50 mm. The normal dimensional accuracy is ± 4 mm/m. If undercuts are required, then the dies are considerably more costly. Pressure die casting can be used for round and nonround holes. The maximum depth to width ratio is 1:1 for blind holes and 4:1 for through holes. The minimum width of the work piece that can be manufactured

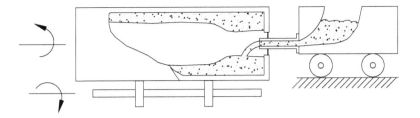

FIGURE 9.3 Centrifugal casting.

using pressure die casting is 1.5 mm for ferrous materials and 2.2 mm for nonferrous materials. Pressure die casting can also be used to manufacture hollow shapes. The maximum economical weight for an aluminum work piece is 45 kg, and somewhat smaller for other materials.

Products: Pressure die casting is used to manufacture intricately shaped parts for automobiles, appliances, outboard motors, hand tools, hardware, business machines, optical equipment, and toys.

Materials: Die casting allows the use of a wide range of materials, including high melting point metals. Although lead and tin alloys cast well, they are seldom used as they find no applications in machines and appliances. Materials that are either seldom or never used are iron, carbon steel, nickel alloys, titanium, and precious metals. Light alloys are the preferred materials because of their high fluidity and low melting temperature. The hot chamber method is restricted to very low melting temperature alloys such as magnesium and zinc. Aluminum alloys and magnesium alloys and copper alloys (in that order) have higher melting points than zinc alloys and are, therefore, more expensive to cast.

Advantages: High production rates are achievable, with castings having smooth surfaces and excellent dimensional accuracy.

Disadvantages: Pressure die castings require a large investment for equipment and tooling. The process is normally suited to metals with lower melting temperature. The size of the casting is limited.

9.2.2 CENTRIFUGAL CASTING*

Process description: Molten metal is introduced into a sand- or a copper-lined cylindrical steel mold, which is rotated about its long axis as shown in Figure 9.3. The centrifugal forces press the molten metal against the mold walls and cavities.

Material utilization: Since runners and risers are not used in the process, there is almost 100% material utilization.

Flexibility: The time to set up the equipment is relatively short.

Cycle time: Production rate is less than 50 pieces/hr. The cycle time is governed by the rate of introduction of the molten metal into the mold, as well as the rate of solidification of the metal. In the cases where sand lined molds are used, the solidification time is much longer and, therefore, the production rate is decreased.

Operating costs: The die costs are moderate, the equipment is expensive, and the labor cost varies from low to moderate. With sand molds the process is competitive in small quantities. Metal or graphite molds require a moderate production quantity to be cost effective.

---

* Additional information can be found at the North American Die Casting Association Web site: http://www.diecasting .org.

Dimensional accuracy: If the inner surface finish is not important, centrifugal casting can be used. The surface finish is poor because porosity and nonmetallic inclusions migrate toward the inner surface due to their lower density. However the outer surface has a good finish and the product has good dimensional tolerances. Typical exterior surface roughness values range from 6.3 μm upwards when sand molds are used, and are below 6.3 μm when permanent molds are used.

Shapes: This process is useful for casting cylindrical parts, especially if they are long, hollow, and need no cores. Except for small parts, the number of intricate shapes that can be cast is limited. Cylinders from 13 mm to 3 m in diameter with lengths to 16 m have been cast. The wall thickness of the cast parts ranges from 5 to 125 mm. Tolerances are comparable to those required for sand mold or permanent mold castings, depending on the centrifugal mold used.

Products: Typical parts made by centrifugal casting are large rolls, gas and water pipes, engine-cylinder liners, wheels, nozzles, bushings, bearing rings, and gears.

Materials: Frequently used materials are iron, carbon steel, alloy steel, aluminum alloys, copper alloys, and nickel alloys. Those that are either seldom or never used are stainless steel, tool steel, magnesium alloys, zinc alloys, tin alloys, lead, titanium, and precious metals. Basically, all metals, excluding refractory and reactive metals, can be used.

Advantage: Useful for casting cylindrical parts, especially if they are long. Process is applicable to all sand-cast metals.

Disadvantage: Except for small intricate parts, shapes that can be cast are limited. Spinning equipment is required and can be expensive.

9.2.3 COMPRESSION MOLDING

Process description: A molding material, usually preformed, is manually placed between mold halves and closed under pressure as shown in Figure 9.4. Either the material is heated (softened) or the molds are heated. The material then flows to fill the mold cavity, polymerizes from the heat, and hardens. The mold opens and the part is ejected or removed. No sprue, gates, or runners are required. The material used by the process must be measured

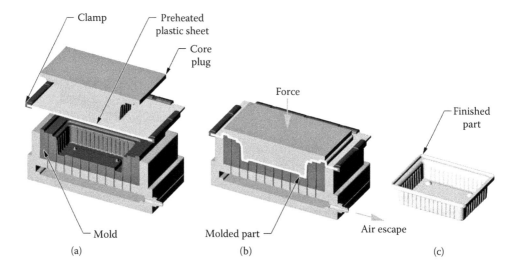

(a)   (b)   (c)

FIGURE 9.4 Compression molding: (a) placement of sheet material in die, (b) die closed under pressure, and (c) polymerized and hardened part ejected. (Copyright © 2008 CustomPartNet. Courtesy of CustomPartNet, Inc., www.custompart.net.)

accurately to avoid excess flash and to maintain uniform size within the mold. Metallic inserts may be molded into the product. Presses typically utilize hydraulic rams to produce sufficient force during molding. The factor limiting the maximum size of the product is the press size. The main parts of the presses are the plunger and the mold. The different types of molds that can be used are (a) flask type molds, which produce horizontal flash and require accurate charge of the plastic material; (b) straight type plunger molds, which produce vertical flash but allow inaccuracy in the charge of the plastic material; and (c) landed plunger type, which produces no flash and must have an accurate charge of the plastic material.

Material utilization: The process has a high material utilization rating. The scrap consists of flash only. However, the flash is not recyclable in the case of thermosets and, therefore, there is a small amount of waste.

Flexibility: The process uses dedicated tooling. The time taken to change the molds is rather fast and the changeover time is dictated by the time it takes to preheat the mold.

Cycle time: The process cycle time is limited by the rate of heat transfer, and in thermosets, by the rate of the reaction of the polymer. Multiple cavity molds can be used to increase the production rate and to reduce the cycle time. The solidification time is determined by the time taken for the polymer to react, and decreases as the temperature of the material increases. Production runs for compression molding vary from 1,000 to 1,000,000 pieces. Material handling is more time consuming because each cavity is usually loaded individually.

Operating costs: When thermosets are used, mold costs are high since the molds must withstand the molding pressures and temperatures. Molding cycles are longer than with injection molding, but the cost of the finished parts is low. When rubber is used, the equipment is not highly complex and the cost of parts is reasonable, especially for moderate quantities. However, machine and tooling costs are low when compared to most other permanent mold casting processes.

Dimensional accuracy: The surface finish of the product is very good and its dimensional accuracy is good. The tolerances vary with material, nominal part dimensions, and processing conditions. Problems that can occur include premature chemical reaction of the thermosets, air entrapment, and inadequate filling of the cavity. Typical tolerances are ±3.4 mm/m.

Shapes: Intricate shapes with side draws and undercuts are difficult to produce by this process because of the short flow lengths in the molds. Reentrant angles are, however, possible in materials that are flexible at temperatures of the mold when the part is released. Thermoset parts can be intricate and have undercuts, although these are more costly. Sizes range from miniature electronic components to large appliance housings. Rubber parts as small as miniature pads for instrument components and as large as tires for large mining trucks can be produced. Parts can be irregular, but normally they are not highly intricate. Tolerances vary with the size, material, and processing conditions. Typical tolerances are ±0.5 mm. It is possible to manufacture parts with controlled wall thickness, molded-in holes, and threads.

Products: The most common products made by compression molding with thermosets are electrical and electronic components, dishes, washing machine agitators, utensil handles, container caps, knobs, and buttons used in higher-temperature environments. Typical rubber products are cushioned and antiskid mounting pads, gaskets and seals, tires, and diaphragms.

Materials: Typical thermosetting plastics are alkyds (DAP and DAIP), aminos (urea and melamine), epoxides, phenolics, polyesters, polyamides, polyurethanes, and silicones. Rubber is also commonly used. More recently, the process is being used to manufacture polymer-matrix composites.

Advantages: Tooling is relatively inexpensive; the process produces fewer flow lines and lower internal stresses. Since there are no runners, gating, and sprues, these potential waste by-products are eliminated, which is advantageous when thermosets are used since they cannot be recycled. The shrinkage rate is lower and more uniform, which allows molding of thin walled parts with a minimum amount of warpage and dimensional deviation.

Disadvantages: Intricate shapes with side draws and undercuts are difficult to produce. Phenolic, a common material for this process, has limited molded-in color choice. Larger parts take longer to cure than in injection molding. Material handling is more time consuming as each cavity is usually loaded individually. Also, close tolerances are difficult to achieve.

9.2.4   PLASTIC INJECTION MOLDING[*]

Process description: Plastic injection molding is a near net shape process in which thermoplastic material is fed by gravity through a hopper into a heated barrel, where it melts and mixes as shown in Figure 9.5. The material is then forced into a mold cavity through a gate and runner system, where it cools and hardens to the configuration of the cavity. Since the mold is kept cold, the plastic solidifies almost as soon as the mold is filled. Molds can be designed either for a single part or for multiple parts. Once the product's material is chosen and the corresponding mold manufactured, the material cannot be changed because the shrinkage rates of different materials differ. Heavy clamping forces are necessary to keep the mold sections together to prevent the plastic from escaping under the high injection pressures. The cavity can produce either a solid or open-ended shape. The injection molding machine uses either a ram or screw-type plunger to force the molten plastic material into the mold cavity. In many products, the injection molding process produces a parting line, sprue, gate marks, and ejector pin marks. After release from the mold, the finished parts are trimmed and are ready for immediate use. Machines are rated according to the weight of material displaced in one stroke of the plunger. Almost all machines are operated on an automatic cycle. Injection molding is used to produce more thermoplastic products than any other process. In some cases, thermosets can also be used.

Material utilization: The material utilization of the process is high. However, there is some scrap in the form of runners and sprues. Thermoplastic scrap can be recycled with a small amount of degradation in the properties of the molded materials.

Flexibility: The process flexibility is restricted by mold changeover and the machine setup time. Depending on the new product's material and any color changes in the product, changeover time can be very long and costly.

Cycle time: The production rate of injection molding is fast, with typical product runs varying from 10,000 to 10,000,000 parts. The cycle time is limited by the solidification time and the demolding time. These are affected by the cooling channels provided around the mold and by the size of the part.

Operating costs: At high production rates the process can be very economical. In addition, no secondary machining operations are needed. However, equipment and tooling costs are high, and mold costs can be very high.

---

[*] There is a U.S. company that has developed a method called the Protomold Rapid Injection Molding process to provide CNC machined advanced aluminum alloy molds directly from a CAD drawing of a part. The parts typically have low to medium complexity and are produced in quantities of a few prototypes to up to 10,000. The larger quantities are typically for pilot production runs or for market testing. In some cases, the part will not be needed in quantities larger than this upper limit and this can be an economical way to produce them. The upper limit is due to the life expectancy of the aluminum mold. Additional information can be found at http://www.protomold.com/ProtomoldProcess.aspx.

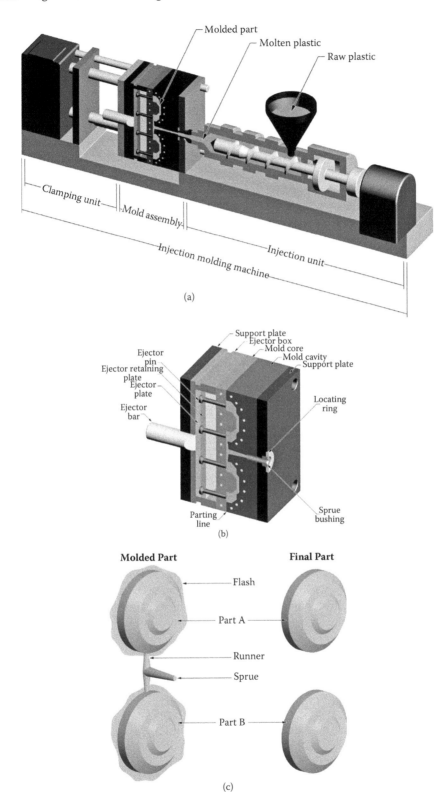

FIGURE 9.5 (a) Injection molding machine, (b) closed mold, and (c) representative parts before and after flash removal. (Copyright © 2008 CustomPartNet. Courtesy of CustomPartNet, Inc., www.custompart.net.)

Dimensional accuracy: The quality of the products that can be manufactured using injection molding can be very good, but it is sometimes sacrificed in order to attain high production rates. Pebbled-like and rough surfaces can be made with injection molding. These surfaces are frequently used to mask scratches. Typical tolerances are ±0.2 mm for thermoplastics and ±0.05 mm for thermosets.

Shapes: Complex shapes are possible. Part sizes vary between 0.05 and 0.5 m in length or width and 0.05 and 0.4 m in depth. Small reentrant angles are possible if the material is flexible.

Products: Typical products made by injection molding are containers of all types, covers, knobs, tool handles, plumbing, fittings, and lenses.

Materials: Typical materials used by injection molding are epoxy, nylon, polyethylene, rubber, and polystyrene. Recently, composites are being used increasingly in this process.

Advantages: The major advantages of injection molding are rapid production rates, intricate parts, molded-in color, and finished parts. Millions of parts can be manufactured with virtually no mold wear.

Disadvantages: The major disadvantages of injection molding are that tooling is costly, large parts are not possible, and small quantities are not economical. Large clamping forces are needed to prevent the mold from producing flash.

Design guidelines:
- Maintain uniform wall thickness and provide for gradual changes in wall thickness.
- Use a stepped diameter for a deep blind hole.
- Maintain uniform wall thickness in thermoset parts.
- Use a bead at the parting line to facilitate removal of mold flash.
- Use decorative designs on surfaces to hide shrinkage.
- Avoid undercuts.
- Observe minimum spacing between holes and sidewalls, between holes, and between holes and the edge of the part.

For additional guidelines see Bralla[*] and Niebel, Draper, and Wysk.[†]

9.2.5 METAL INJECTION MOLDING[‡]

Process description: Metal injection molding is a near net shape process that is very similar to plastic injection molding. The differences are as follows. Heated micron-sized metal powders are mixed with a multiple component thermoplastic binder system. When the mixture is cooled to room temperature, the mixture consists of granulated pea-sized particles that are subsequently fed into an injection molding machine. The thermoplastic binder system ensures that when the granules are reheated as part of the injection molding process, they have the proper flow characteristics to fill the mold cavity. When the parts are released from the mold they are placed in a controlled solvent system to remove some of the thermoplastic binder. Just enough of the binder is removed from the part so that the part still retains its shape. The last part of the operation is sintering where the part is heated to below the melt point of the metal powders in a controlled atmosphere to remove the residual binder and to cause the whole part to shrink uniformly until it reaches 93 to 99% of its full density. The parts shrink from 15 to 30% of their original size.

---

[*] Bralla, 1986, ibid.

[†] B. W. Niebel, A. B. Draper, and R. A. Wysk, *Modern Manufacturing Process Engineering*, McGraw-Hill, New York, 1989.

[‡] For more detailed descriptions of the process, see http://www.formphysics.com/process.htm and http://www.gknsinter-metals.com/technology/mim.htm.

Dimensional accuracy: The process is best suited to small complex parts of quantities greater that 10,000 and with tolerances greater than ±0.05 mm.

Products: Small complex parts for a wide variety of applications such as orthodontic appliances, controlled expansion hermetically sealed electronic packaging, tremolos for guitars, components in high caliber pistols, components in automotive engines and transmissions, and component connectors in hydraulic systems.

Materials: Part materials include ferrous alloys, copper, stainless steels, iron-nickel, and iron-cobalt composition. The properties of the process are comparable to those of wrought products because of their high density and controlled microstructure.

Advantages: Metal-injected molded parts can have properties equal or superior to those made by powder metallurgy, die casting, and investment casting.

9.2.6 IN-MOLD ASSEMBLY

Many plastic products are created by injection molding individual parts and then manually assembling the parts together to form products. This approach frequently requires fasteners and adhesives, which in turn contribute to increased cycle time and labor costs. In addition, manual assembly usually requires that the parts have some minimum size so that they can be handled by the assemblers. A new alternative called *in-mold assembly* is emerging to overcome some of the challenges associated with assembly operations. In-mold assembly processes combine the molding and assembly steps so that the need for postmolding assembly operations is eliminated. Hence, the size- and shape-related constraints imposed by manual assembly operations are eliminated. The in-mold assembly takes place by having the material flow into the final location so that it is either geometrically constrained by the surrounding components or chemically bonded to another component. This eliminates the need for fasteners and/or adhesives.

Process description: There are three fundamentally different types of in-mold assembly processes: multicomponent injection molding, multishot injection molding, and overmolding. Multicomponent injection molding is the simplest and most common form of in-mold assembly. As shown in Figure 9.6, it involves either simultaneous or sequential injection of two different materials through either the same or different gate locations in a single mold. As shown in Figure 9.7, multishot injection molding is the most complex and versatile of the in-mold assembly processes. It involves injecting the different materials into the mold in a specified sequence, where the mold cavity geometry changes between sequences. The sections of the mold that are not to be filled during a molding stage are temporarily blocked. After the first injected material sets, one or more blocked portions of the mold are opened and the next material is injected. This process continues

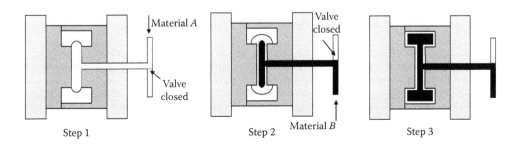

FIGURE 9.6 Multicomponent molding.

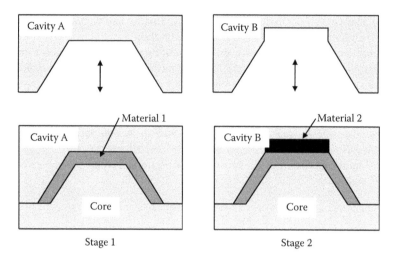

FIGURE 9.7 Multishot molding. (Material feed not shown.)

until the required multimaterial part is created. Over-molding involves placing a resin around a previously-made injection-molded plastic part. Each of the three classes of processes is considerably different, and each process requires its own set of specialized equipment.

Products: Many industries, such as the automotive, toy, electronic, power tool, and appliance industries, that currently make use of traditional single-material injection molding processes are beginning to use multimaterial molding processes. Some common applications include multicolor objects, skin-core arrangements, in-mold assembled objects, soft-touch components (with rigid substrate parts), and selective compliance objects.

Advantages: In-mold assembly processes can be used to create multimaterial structures when the two different materials adhere to each other because their molecular weights are compatible.[*] On the other hand, the process can create rigid body joints by using materials with incompatible molecular weights, which ensures that the two materials do not adhere to each other.[†]

Disadvantages: There are some disadvantages of in-mold assembly processes that may not make it always the best choice. Usually, mold costs are much higher for in-mold assembly operations and there are fewer molding shops that offer in-mold assembly capabilities. Also, the presence of a plastic part inside the mold fundamentally changes the nature of the injection molding process and, therefore, additional considerations must be taken into account when designing products involving the in-mold assembly process.[‡]

---

[*] For a detailed discussion of how to use in-mold assembly for creating compliant joints, see R.M. Gouker, S.K. Gupta, H.A. Bruck, and T. Holzschuh, "Manufacturing of multi-material compliant mechanisms using multi-material molding," *International Journal of Advanced Manufacturing Technology*, 30(11-12): 1049–1075, 2006.

[†] For a detailed discussion of how to use in-mold assembly for creating common rigid body joints, see A.K. Priyadarshi, S.K. Gupta, R. Gouker, F. Krebs, M. Shroeder, and S. Warth, "Manufacturing multi-material articulated plastic products using in-mold assembly," *International Journal of Advanced Manufacturing Technology*, 32(3-4): 350–365, March 2007.

[‡] For a detailed discussion of design for manufacturing considerations associated with the in-mold assembly processes, see A. Banerjee, X. Li, G. Fowler, and S.K. Gupta, "Incorporating manufacturability considerations during design of injection molded multi-material objects," *Research in Engineering Design*, 17(4): 207–231, March 2007.

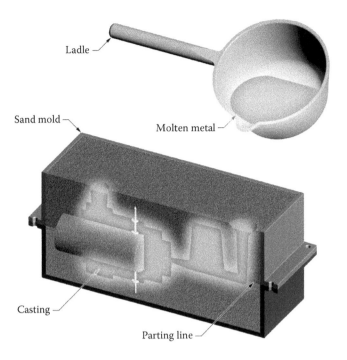

FIGURE 9.8 Sand casting. (Copyright © 2008 CustomPartNet. Courtesy of CustomPartNet, Inc., www .custompart.net.)

9.3 CASTING—PERMANENT PATTERN

9.3.1 SAND CASTING*

Process description: A mixture of sand, clay binder, and other materials is packed around a wood or metal pattern with cores to form a mold. The pattern is removed and the cavity left by it is filled with molten metal as shown in Figure 9.8. Normally the mold is in two pieces, which are fitted together before pouring. Cored sand casting is a process that involves the assembly of various mold sections and cores (made of sand or composites) to form a casting mold. With cored sand casting, shapes not obtainable through other casting processes can be produced. The proper design of the running and gating systems reduces problems related to turbulence while pouring.

Material utilization: Up to 50% of the molten metal used goes to form the sprues, runners, and risers. Both the mold and scrap metal can be remelted and used again. However, due to the amount of scrap produced per cycle, the material utilization rating for this process is poor. Recycling sand molds helps reduce the costs for large production volumes; however, this can be difficult since the removal of binders and hardening agents is costly.

Flexibility: Patterns are cheap and easy to make, and the process is very flexible.

Cycle time: The cycle time is limited by the rate of heat transfer through the casting and the mold. The use of multiple molds can increase production rates. Typical production rates are less than 20 pieces/hr. The number of parts for a production run ranges from 10 to 100,000.

---

* Additional information can be found at the American Foundry Society Web site: http://www.moderncasting.com/.

Operating costs: The cost of dies and equipment is low, and the labor cost varies from low to medium. Pricing is, in part, weight-based.

Dimensional accuracy: The surface texture of sand cast parts is poor. Porosity cannot be avoided, and it is difficult to eliminate nonmetallic inclusions from the work piece. Surfaces are irregular and grainy with average surface roughness varying from 12 to 25 μm.

Shapes: Complex pieces such as engine blocks can be manufactured using sand casting. Cores can have points built into them to aid in their alignment. Undercuts are possible with mold cores. Parts can weigh as little as 0.03 kg or as large as 180 metric tons, with sizes ranging to 3 m by 3 m. Deviation from flatness is typically 4 mm/m. Intricate shapes with undercuts, reentrant angles, and complex contours are practicable to cast, as are contoured surfaces and hollow shapes. Typical dimensional accuracy is ±16 mm/m. Normal dimensional accuracy is ±1.5 mm for holes 75 mm in diameter. The minimum practical cast hole diameter varies from 13 to 25 mm and requires a draft angle. The maximum depth to width ratio is 1.5:1 and also requires a draft angle. Minimum wall thickness should be greater than 6 mm.

Products: Typical products made by casting are engine blocks, machine frames, compressor and pump housings, valves, pipe fittings, and brake drums.

Materials: Most frequently used sand casting materials are iron, carbon steel, alloy steel, stainless steel, aluminum alloys, brass, copper alloys, magnesium alloys, and nickel alloys. Those that are sometimes used are tool steels, zinc alloys, tin alloys, and lead.

Advantages: Intricate shapes, wide range of sizes, and most metals can be used. Process can obtain shapes not attainable through other casting processes, and can be used for low production run products. Unit cost is low.

Disadvantages: Surface finish and dimensional accuracy are poor, and secondary operations are needed in many cases.

Design guidelines:
- Avoid sharp corners by providing generous fillets and radii.
- Keep walls uniform and thin.
- Keep cross-sections uniform.
- Keep the intersection of two walls at right angles.
- Reduce the number of ribs that intersect at a common point.
- Provide tapers or draft angles from the parting line for easy removal of mold and keep parting lines straight rather than stepped.

For additional guidelines see Bralla[*] and Pahl et al.[†]

9.3.2 SHELL MOLD CASTING

Process description: Molten metal is poured into a heat-cured, nonreusable shell mold made from silica sand on a resin binder and held until solidification of the molten metal occurs as shown in Figure 9.9. Sand mixed with a thermosetting-plastic resin is poured onto a heated metal pattern. The metal patterns have high tolerances, have a smooth surface finish, and are in two halves with sprues, gates, and runners. They may not have undercuts, but, otherwise, can have very complex shapes. Upon the application of heat, the sand mixture adjacent to the pattern fuses together forming a shell that is removed from the pattern. Metal is poured into the cavity formed when two shell halves are fastened together. The shell is

---

[*] Bralla, 1986, ibid.
[†] G. Pahl, W. Beitz, J. Feldhausen, and K.-H. Grote, ibid., Sections 7.5.8 and 7.5.9.

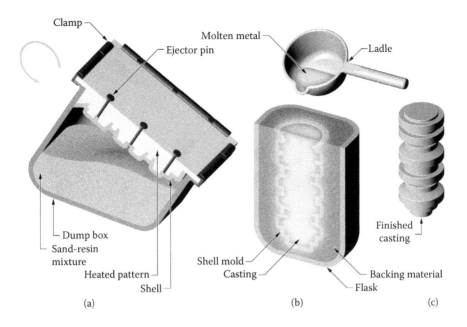

FIGURE 9.9 Shell mold casting: (a) making of shell, (b) casting, and (c) cast part. (Copyright © 2008 CustomPartNet. Courtesy of CustomPartNet, Inc., www.custompart.net.)

broken from the finished casting. Shell halves are either clamped or bonded together, and the mold is placed in a flask and surrounded with shot, sand, or gravel. This backing material reinforces the mold during the casting process.

Material utilization: The process requires nonreusable molds having runners, gates, and sprues; therefore, the process has poor material utilization.

Flexibility: Since a shell mold making machine is used, the process is not very flexible.

Cycle time: Production rates are typically less than 50 pieces/hr. Production volumes vary from 1,000 to 100,000 parts.

Operating costs: Since tooling costs are high, this process is used for large production volumes. The least economical quantity that can be produced is around 500. Sand cost is higher than for other processes due to the use of a resin. While tooling cost is high, the die cost is low to moderate. The labor cost is also low to moderate.

Dimensional accuracy: Fine-grained, strong castings with good surface finish and dimensional accuracy are obtained. This process is superior to other sand casting processes, and provides an accurate duplication of intricate shapes with high dimensional accuracy. Surface finishes on the order of 1.6 to 12.5 μm are feasible.

Shapes: Parts can be as heavy as 100 kg, but weights are normally less than 9 kg. Shapes can be complex with bosses, undercuts, holes, and inserts. Cores are used to produce holes and cavities. A section thickness as small as 2.5 to 6 mm is feasible.

Products: Typical products made by this process are connecting rods, lever arms, gear housing, cylinder heads, and support frames.

Materials: Many metals, metal alloys, and non-ferrous alloys can be cast. The most common metals are iron and steel.

Advantages: Best for high production runs. Process provides good surface finish and dimensional accuracy, and cleaning is considerably reduced.

Disadvantages: Casting size is limited. Equipment and tooling require a large investment, and resin also adds to the cost. Not all metals can be cast and the molds cannot be reused.

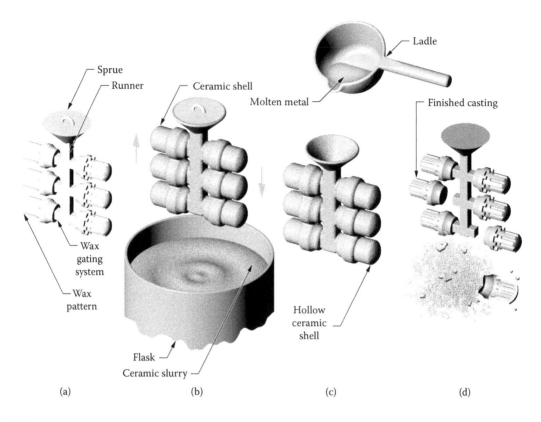

FIGURE 9.10 Investment casting: (a) pattern tree, (b) shell creation, (c) casting operation, and (d) cast parts. (Copyright © 2008 CustomPartNet. Courtesy of CustomPartNet, Inc., www.custompart.net.)

9.4 CASTING—EXPENDABLE PATTERN

9.4.1 INVESTMENT CASTING

Process description: A metal mold is made of the part. Wax is injected into the metal mold. As shown in Figure 9.10, the resulting wax parts are then gated to a sprue to form a cluster of wax molds, which are used to increase the production rate. The cluster is dipped into a ceramic slurry and dried. The dipping process is repeated until the required ceramic shell thickness is achieved. After the slurry sets, the mold is baked at high temperatures to remove the wax pattern. The resulting ceramic mold is now ready for the pouring of the casting material. Molten metal is poured into the ceramic mold, which has been preheated, thereby producing a casting. Pressure, vacuum, or centrifugal force is used to fill the cavity completely. The thin ceramic mold material is removed by vibration, chipping, sand blasting, or water jet. The gating is removed from the casting by cutting and grinding. The casting itself can have intricate internal and external features with little or no draft angles. Investment casting is a process used for metals that melt at temperatures too high for metal molds.[*]

Material utilization: The process is a near net shape process with little material lost in the feeding systems. The wax can be recycled. Therefore, the process has high material utilization.

Flexibility: The flexibility of the process is good because of the ease with which the patterns can be produced.

---

[*] A video demonstrating the steps in the process can be found at http://www.investmentcasting.org/.

Cycle time: The cycle time is limited by the rate of heat transfer from the casting. Production rates are low because of the complexity of the process. The production rate can be increased and cycle time lowered by using cluster molds and patterns. Production rates are typically less than 1,000 pieces per hour. The total number of parts per run range from 100 to 100,000, with 50 parts being the smallest number for the run to be economical.

Operating costs: The equipment costs can be high if reactive alloys are used. The labor costs are high due to the many stages in the process. The main cost savings come from the elimination of secondary operations such as machining. Otherwise, this process is costlier than other casting processes. It is most economical for low and medium production quantities.

Dimensional accuracy: There are no parting lines, and the product is a single piece that can have a complex shape. Excellent dimensional tolerances and surface finishes are obtained. Higher mold temperatures decrease the porosity of the part, but may produce coarse microstructures. Care must be taken when making the patterns to account for shrinkage during cooling and solidification. Surface finishes range from 1.6 to 3.2 μm.

Shapes: Part size is normally small, but can range in weight from 0.001 to 35 kg. Minimum wall thickness can vary from 0.75 to 1.8 mm. Tolerances vary from ±0.8 mm to ±1.5 mm. Shapes can be very intricate with such intricacies as contours, undercuts, bosses, and recesses.

Products: Typical parts made by this process are precision parts for sewing machines, firearms, surgical and dental devices, wrench sockets, pawls, cams, gears, turbine blades, valve bodies, and radar wave guides.

Materials: This process is suitable for almost all metals. Reactive metals can be cast under a vacuum. The most commonly used materials are carbon steel, alloy steel, stainless steel, tool steel, and aluminum-copper-nickel alloys. Aluminum, cast iron, and magnesium are very suitable because of their high fluidity. Cast steels and brass can also be used. Sometimes magnesium alloys and precious metals are used, but iron, zinc and tin alloys, lead, and titanium are seldom used.

Advantages: Almost all metals can be cast with high dimensional accuracy, smooth finishes, and intricate shapes.

Disadvantages: It is most suitable for small parts, and is justifiable only when secondary operations can be eliminated. Labor costs are high and patterns may be expensive.

9.5 CUTTING—MECHANICAL MACHINING

9.5.1 SINGLE POINT CUTTING: TURNING AND FACING

Process description—turning and facing: Turning is a material removal process that uses a single point cutting tool whose major motion is parallel to the axis of rotation of the work piece. Facing is a special case of turning in which the major motion is perpendicular to the axis of rotation. Turning produces cylindrical external surfaces that are straight cylinders, tapered cylinders, or any combination of these. Facing produces flat surfaces perpendicular to the axis of rotation. The single point tool produces fine helical marks on the cylindrical surfaces that are a function of the tool's feed rate. The work piece is held in a chuck and the tool is mounted on a tool post. Chips result from the cutting and chip removal can be a problem. The dimensional accuracy and surface finish are affected by tool geometry; cutting speed and feed rate; rigidity of the tool, work piece, and machine; alignment of machine components and fixtures; and cutting fluids. Cutting fluids are used to lubricate and cool simultaneously, which reduces tool wear, allows high cutting speeds and feed rates, flushes chips, prevents rust, and prevents a built-up edge on the cutting tool. Turning creates residual surface stresses, may create micro-cracks, and may cause surface or work hardening in unhardened materials.

Material utilization: Material utilization is extremely poor in turning. The scrap that is produced during machining is expensive to recycle due to the contamination of the lubricant and the changes in the microstructure as a result of the chip creation process.

Flexibility: The flexibility of this process is extremely high. It is ideal for the production of individual pieces and small batches.

Cycle time: The cycle time is controlled by the relative hardness of the work piece and the tool. The cycle time can be reduced somewhat by using automation. A typical material removal rate is 21 cm³/min.

Operating costs: Tooling is not dedicated and the machine costs depend on the degree of flexibility and automation selected. The operating costs can range from low to very high depending on the type of machine used and the complexity of the part and its tolerances.

Dimensional accuracy: Turning is often used to improve the surface texture, and is only limited by the time and the effort expended. Tolerances in both length and diameter are ±0.025 mm for most applications and for precision applications ±0.005 mm for diameter and ±0.0125 mm for length. Typical surface finishes vary over the range 0.4 to 3.2 μm.

Shape: Geometric possibilities within the cylindrical geometry are planes and faces, tapers, contours, radii, fillets, and chamfers.

Products: Typical products made by turning are rollers, pistons, pins, shafts, rivets, valves, tubing, and pipe fittings.

Materials: Aluminum, brass, and plastics have excellent machinability ratings; cast iron and mild steel have good machinability ratings. Stainless steel has a poor to fair machinability (due to work hardening).

Design guidelines:
- Provide for adequate clamping of the part.
- Avoid sharp corners and let radius dimensions be determined by the cutting tool's tip radius.
- Provide for adequate tool runout.
- Keep parts short and stocky to minimize deflection.

For additional guidelines see Bralla[*] and Pahl et al.[†]

9.5.2 MILLING: MULTIPLE POINT CUTTING

Process description: Milling is a cutting process in which material is removed by a rotating multiple-tooth cutter. Arbor milling is a form of milling in which the cutting takes place on a surface parallel to the axis of the tool rotation; for end milling the axis of the tool rotation is perpendicular to the cutting surface. The discontinuous chips that are formed during milling are, for the most part, swept away by the rotation of the cutter. Either the work piece may be fed into the cutter or the cutter may be fed into the work piece. Milling is most efficient when the work piece hardness is less than Rockwell C25. A large number of different milling cutters can be used to obtain different shapes.

Mechanical properties: Work piece properties change due to milling. A built-up edge on the cutter causes a rough work piece surface. Dull tools can cause severe surface damage and high residual stresses.

Material utilization: Material utilization in milling is extremely poor. Scrap is difficult and expensive to recycle due to contamination by the lubricant and changes in the microstructure of the scrap material.

---

[*] Bralla, 1986, ibid.
[†] G. Pahl, W. Beitz, J. Feldhusen, and K.-H. Grote, ibid.

Flexibility: The flexibility rating of milling is high. It is ideal for the production of individual pieces and small batches.

Cycle time: The cycle time is controlled by the relative hardness between the tool and the work piece, the lubrication, and the rate of cooling. The cycle time can be reduced with automation.

Operating costs: There is little dedicated tooling in milling. Machine costs depend on the degree of flexibility and automation. Operating costs are relatively low.

Dimensional accuracy: Milling can be used to improve surface quality and is only limited by the time and effort expended. Typical surface finishes range from 1.6 to 5.0 μm. Typical deviations from flatness are ±0.4 mm/m.

Shapes: Flat, convex, concave surfaces, and contoured surfaces can be milled. Typical tolerances are ±0.13 mm.

Products: Milling is used in the generation of simple to complex three-dimensional shapes over a very wide range of part dimensions.

Materials: Aluminum, brass, and plastics have good to excellent machinability; cast iron and mild steel have good machinability; stainless steel has fair machinability due to work hardening. Polymers, and some ceramics and composites can also be milled, although the tool wear rates can be high when machining ceramics and composites.

Design rules:
- Provide for raised flat surfaces; that is, mill only those portions of a surface that have to be milled.
- Arrange surfaces so that surfaces to be milled are on one level and parallel to the clamping surface.
- Avoid undercuts.
- Parts and tools must not deflect during operations.

For additional guidelines see Bralla,[*] Pahl et al.,[†] and Farag.[‡]

9.5.3 GRINDING

Process description—cylindrical grinding: Grinding is a process that removes material from a work piece using a tool made from abrasive particles of irregular geometry that are embedded on the surface of a rotating wheel. The process produces straight, tapered, and formed work pieces. To produce tapers, either the wheel or the table is swiveled. For cylindrical work pieces, the work piece is mounted between centers and the work piece and grinding wheel rotate in opposite directions. The grinding process produces highly accurate surfaces with smooth finishes. The geometry of the work piece is a mirror image of the grinding wheel. When the wheel is a formed wheel, the process is called *plunge grinding*. Wheels are usually 45% abrasive, 15% bonding material, and 40% porosity by volume. It is the porosity of the wheel that provides space for the chips and a path of delivery of the coolant. Fluids are needed to cool the wheel and work piece, to lubricate the wheel and work piece interface, and to aid in removing chips.

Process description—surface grinding: Surface grinding is similar to cylindrical grinding, except that in surface grinding a flat surface is formed as shown in Figure 9.11. It sometimes is used to finish a formed surface of steel. A grinding wheel, a reciprocating table, and work holding devices, such as a magnetic chuck for metallic work pieces

[*] Bralla, 1986, ibid.
[†] G. Pahl, W. Beitz, J. Feldhusen, and K.-H. Grote, ibid.
[‡] M. M. Farag, *Selection of Materials and Manufacturing Processes for Engineering Design*, 2nd ed., CRC Press, Boca Raton, FL, 2008.

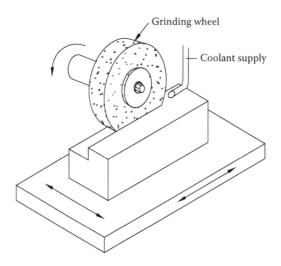

FIGURE 9.11 Typical surface grinding operation.

and adhesive materials or vacuum chuck for nonmetallic materials, constitute the basic equipment required. The wheel spindle can be either vertical or horizontal.

Mechanical properties: The grinding process creates residual surface stresses and a thin martensitic layer on the surface due to the high temperatures generated. Because of these two factors, fatigue strength of the work piece may be reduced. It can also cause a loss of the magnetic properties of ferromagnetic materials, and may increase a material's susceptibility to corrosion.

Material utilization: Material utilization is extremely poor in grinding. The scrap produced is difficult and expensive to recycle due to lubricant contamination and changes in its microstructure. In fact, the scrap produced gets into the grinding wheel and causes "loading" of the wheel, which greatly reduces the cutting efficiency. As soon as the wheel is loaded, the loaded surface layer must be removed and fresh cutting particles exposed.

Flexibility: Grinding is a highly flexible process.

Cycle time: The cycle time is relatively slow and is affected by the relative hardness of the work piece and the tool, the lubrication, and the amount of cooling. The production rate can be increased somewhat by increasing the degree of automation.

Operating costs: Some of the tooling in grinding is dedicated. The machine cost depends on the degree of flexibility and automation. Grinding is a process that consumes a lot of power for two reasons. First, each cutting particle is small and, therefore, cuts only a small area of the surface. The area around the groove cut by the abrasive particle forms a raised edge, which is removed by the following particles. This "ploughing" mechanism is an inefficient method of removing material. Second, the cutting particles are randomly oriented and have no uniform geometry; thus, there is no optimization of the equivalent of a tool's rake angle. Some particles will be shaped and oriented so that they do not cut at all and just rub the surface, which leads to higher power consumption.

Dimensional accuracy: Grinding is basically a finishing process, and the surface quality obtained depends on the amount of time and effort expended. Typical values of surface finish range from 0.2 to 0.8 μm and the typical tolerances are ±12.7 μm for diameters and ±2.5 μm for roundness. For precision applications, the tolerances can be as low as one-tenth these values. For surface grinding, the tolerances for flatness are typically ±0.05 mm and for parallelism ±0.075 mm. Deviations in flatness are typically ±0.08 mm/m.

Shape: Cylindrical grinding produces straight, tapered, and formed work pieces. Surface grinding is used for the grinding of flat and contoured surfaces and slots.

Products: Typical applications are to form shafts, pins, and axles over a range of sizes: 20 to 500 mm for diameters and 20 to 1900 mm for lengths.

Materials: Grinding can be used on all materials except those that become soft or gummy. The best results are obtained with cast iron and mild steel, the next best with aluminum, brass, and plastics, and the poorest with stainless steels, because they tend to work harden.

Advantages: Excellent surface finishes are obtained.

Disadvantages: High power consumption, low efficiency process with poor material utilization. The process creates residual surface stresses that reduce the fatigue strength of the work piece. Form grinding tooling can be expensive.

Design guidelines:
- Minimize surface to be ground and reduce part volume.
- Ensure that grinding wheel will be unimpeded.
- Keep grinding wheel edges from coming into contact with surfaces and provide runout for the grinding wheel.

For additional guidelines see Bralla[*] and Pahl et al.[†]

9.6   CUTTING—ELECTROMACHINING

9.6.1   ELECTRIC DISCHARGE MACHINING (EDM)

Process description—wire cutting: Cutting is achieved when thermal energy from electrical discharges between the wire and the work piece occur. At the small diameter, wire moves through the part to produce complex two-dimensional shapes as it advances from the supply reel to the take-up reel, as shown in Figure 9.12a. The work piece is fed continuously and slowly into the wire to create the desired shape. A dielectric fluid is used to flush the removed particles, to regulate the discharge, and to keep the tool and work piece cool. Typical diameters of the wires vary from 0.025 to 0.3 mm. The wire is inexpensive and is, therefore, not usually reused. Mirror image profile work and internal contours can be manufactured from a starting hole. Multiple work pieces can be stacked and cut, and the finished products do not contain burrs.

Process description—cavity type: Cavity-type EDM is a thermal-mass-reducing process that uses a shaped electrically conductive tool to remove electrically conductive material as shown in Figure 9.12b. This is done by means of thousands of controlled, repetitive spark discharges per second across a gap of approximately 0.025 mm. These discharges cause the work piece to vaporize and slowly produce the desired shape. Burr-free finished parts are obtained. A dielectric fluid is used to flush the removed particles, to regulate the discharge, and to keep the tool and work piece cool. A heat-affected zone of approximately 0.25 mm thickness is produced. In steel, this zone produces a thin carbide layer that lowers fatigue strength and creates microcracks. The finish on the work piece is affected by the gap voltage, discharge current, and supply frequency. The material removal rate is low.

Material utilization: The material utilization rating is extremely poor for this process and the scrap that is generated during the process cannot be recycled.

Flexibility: The tooling is dedicated in the EDM process. Set-up times can be short and, therefore, the process is a highly flexible one.

---

[*] Bralla, 1986, ibid.
[†] G. Pahl, W. Beitz, J. Feldhusen, and K.-H. Grote, ibid.

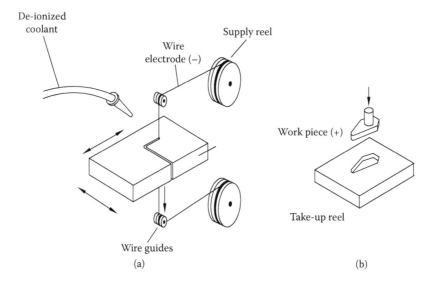

FIGURE 9.12 (a) Wire cutting EDM (b) cavity-type EDM.

Cycle time: The cycle time is normally long for EDM. The rate of material removal depends on various work piece properties, including its melting point and its latent heat.

Operating costs: The combination of the machines used and the setup required to obtain the operating conditions makes this process expensive.

Dimensional accuracy: The surface texture is inversely proportional to the material removal rate. Surface finishes are in the range 1.3 to 3.8 μm, and tolerances range from $\pm$ 4 to $\pm$13 μm at very slow material removal rates.

Shapes and products: The method is often used for making dies for extrusion, powder metallurgy, and injection molding. Frequently, punches and dies are manufactured using wire EDM.

Materials: All metals and all conducting non-metals can be used in this process.

Advantages: Used in cases where the materials that have to be machined are harder than typical tool materials. The process has a high repeatability.

Disadvantages: Care must be taken with some metals, as the surface has a heat affected zone, and in heat-treatable metals this layer will be very hard. Minute cracks can appear if very high amperage is used, which may result in premature fatigue failure.

9.7   FORMING—SHEET

9.7.1   BLOW MOLDING

Process description: Hollow products are manufactured by extruding a heated (softened) thermoplastic tube (called a *parison*) into a mold and, under air pressure, expanding the parison to match the inner contours of the mold, where it cools and hardens, as shown in Figure 9.13. When the mold is opened, the part is ejected. Parts are uniform in thickness. The process is limited to thin-walled hollow products. Parting lines and flash are present, but the flash is minimal. This process is suitable for the manufacture of thin-walled, hollow objects with internal volumes up to 200 liters. The parts can have metal inserts.[*]

---

[*] See http://www.pct.edu/prep/bm.htm for an animation of the process.

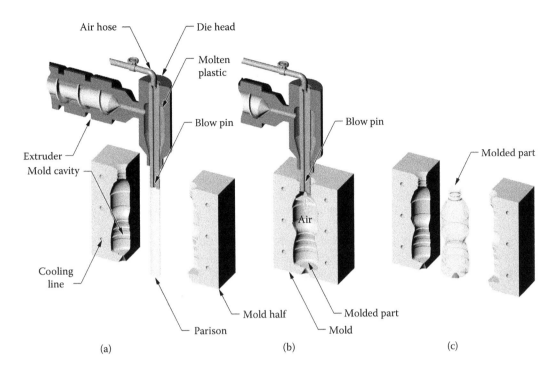

FIGURE 9.13 Blow molding system. (a) placement of parison, (b) inflation and cooling, (c) removal of part. (Copyright © 2008 CustomPartNet. Courtesy of CustomPartNet, Inc., www.custompart.net.)

Material utilization: The process is a near net shape one if the parison that is used is produced by injection molding. There is some scrap if the parison is produced by extrusion. Therefore, the material utilization rating of blow molding is high.

Flexibility: The tooling that is used in the process is dedicated, but set-up times are relatively short.

Cycle time: The cycle time is limited by the heating and the cooling of the polymer used, but is considered relatively low. This time can be reduced by automating the process. The production volume for this process varies from 1,000 to 10,000,000 pieces.

Operating costs: The tooling used in blow molding is relatively inexpensive. The machines are costly, particularly if the degree of automation is high. It is, however, the lowest-cost method of producing bottles and other closed containers at high production levels.

Dimensional accuracy: The surface texture that can be obtained by using this process is good. However, significant molecular orientation may be induced. Both the surface finish and the dimensional accuracy are fair. Tolerances for this process vary from ±0.5 to ±2.5 mm.

Shape: The process is limited to open and closed hollow products. Molded-in holes are not possible. However, products with intricate shapes and controlled wall thickness can be made.

Products: Typical products are bottles and other containers, hollow toys, and decorative objects.

Materials: The work piece materials are thermoplastics, principally polyalkenes and PET (polyethylene terephthalate). The use of thermosets is not possible.

Advantages: An economical process for rapid production of containers and other one-piece complex hollow objects.

Disadvantages: Limited to hollow objects and is not suitable for small quantities. Tolerances are relatively broad and wall thickness is difficult to control.

9.7.2 SHEET METAL WORKING

Process description—punching and blanking: Punching is a shearing process in which a scrap slug is separated from the work piece when the punch enters the die. This process is most economical for making holes in sheet metal at medium to high production rates. Blanking is a process in which the piece of sheet metal that is separated by the punch action is the required finished product, and the leftover sheet metal is the scrap. A punch mounted on a ram mates with a die mounted to the bolster plate. Multiple punches are often used to finish a part in a series of strokes, as indicated in Figure 9.14.

Material utilization: Material utilization can be increased by modifying the part's design to facilitate better nesting of the designs. There is always a minimal amount of scrap that is too expensive to recycle.

Flexibility: Tooling is dedicated, and the set-up time depends on the tooling complexity. Therefore, the process has a low flexibility.

Cycle time: Relatively fast cycle times can be obtained, but it depends on the number of punches used and the feed rate of the sheet metal strip.

Operating costs: Operating costs are determined by the complexity and the number of punches that are used. The amount of scrap will also determine the operating costs.

Dimensional accuracy: The process produces a burnished area, roll over, and die break on the side wall of the resulting hole. Typical dimensional accuracy is ±0.15 mm and typical surface finishes range from 0.8 to 1.6 μm.

Shapes: A large number of complicated shapes can be obtained. The only limitation is that the part must be of constant thickness.

Materials: Steel, aluminum, brass, and stainless steel are the most commonly used materials. Low carbon steels and medium carbon steels are the best for stampings, but they must be low temper steels. Stainless steel can be worked much more severely than ordinary steels (when annealed) but requires much higher press power. Nickel alloys are similar to stainless steels; however, they appreciably work harden during punching and may have to be annealed if additional operations are required. Brass and copper alloys form excellent stampings, but bronzes with lower ductility are not as desirable. The soft-temper aluminum alloys are the most easily worked and are preferred when their strength is adequate. The following materials can be punched or blanked and are ranked in decreasing order of blanking ease: lead, paper, leather, aluminum, zinc, cast iron, fiber, copper, German silver, brass, wrought iron, tin, low carbon steel, medium carbon steel, high carbon steel, untempered tool steel, nickel steels, and tempered tool steels.

Advantages: There is reduced material handling, and the process is easily automated. A large number of shapes and sizes can be obtained with good surface finishes.

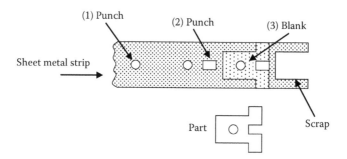

FIGURE 9.14 Sheet metal working: punching and blanking.

Disadvantages: Requires a die, has costly setup, produces scrap that cannot be recycled, and is restricted to thin sections.

Design guidelines:

- Design the part so that material utilization is maximized.
- When several different parts are involved, select their geometry so that their combination will maximize material utilization.
- Having one part dimension equal to the width of stock can improve material utilization.

For additional guidelines see Bralla[*] and Pahl et al.[†]

9.8   FORMING—BULK

9.8.1   FORGING

Process description: Drop forging is a metal shaping process in which a heated work piece is forced to conform to the shape of a die cavity by rapidly closing a punch and die as shown in Figure 9.15. The closing of die cavities may be either singular or repeated; typically, however, one blow is given in each die cavity. The drop hammer is powered by air, hydraulics, or mechanically. Striking forces from 50-1900 kN are obtained, depending on the mass of the ram and upper die, and on the drop height. As the forging forces increase, the dimensional tolerances improve. The product is usually manually loaded and removed. The resultant forging approximates the shape of the finished part, but secondary machining is usually required to obtain dimensional tolerances and good surface finish. The process produces a parting line and flash on the work piece, which must be removed. Parts can weigh from 1.4 to 340 kg and overall dimensions can vary from 0.08 to 1.3 m. If the preform mass is carefully controlled in closed die forging, it can be forged in totally enclosed dies with no flash gutters.

Open die forging is a process in which a solid work piece is placed between two flat dies and reduced in height by compressing it. It produces shapes like rings and shafts, and large components up to 270 metric tons can be forged economically this way.

Upset forging is a metal shaping process in which a heated work piece of uniform thickness is gripped between split female dies while a heading die (punch) is forced against the work piece, deforming and enlarging the end of the work piece. Upset forging is used to form bolts and to increase locally the area of any component.

Open die forging is used for manufacturing large parts, while closed die forging is used to manufacture smaller components.

Mechanical properties: Forging creates good to excellent mechanical properties, including improved fatigue and impact resistance. However, the process may create micro-cracks.

Material utilization: Hot forging in closed dies always produces scrap material in the form of flash. Cold forging can be a near net shape process. Cold forging has almost 100% material utilization; in hot forging it is lower.

Flexibility: Open die tooling is not dedicated, but it is in closed die forging. The length of the set-up period depends on the complexity of the tooling. The open die forging process has a good flexibility rating, while for the cold die forging process it is poor.

Cycle time: The cycle time is dictated by the rate of operation of the equipment. Open die forging has longer cycle times, because of the large size of the parts that are manufactured. The cycle time in closed die forging can be quite short, depending on the degree of automation.

---

[*] Bralla, 1986, ibid.
[†] G. Pahl, W. Beitz, J. Feldhusen, and K.-H. Grote, ibid.

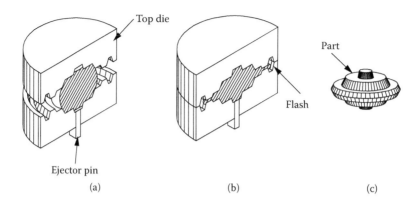

FIGURE 9.15 Closed die forging. (a) Billet in die, (b) die closure, and (c) part.

Operating costs: Tooling costs are moderate to high, depending on the part's complexity. There is a moderate material loss due to flash and secondary machining. Labor costs are moderate in closed die forging, while highly skilled operators are needed for open die forging. The degree of automation decides the cost in closed die forging. This process is economical for medium to high production levels.

Dimensional accuracy: The quality of the forged products depends primarily on the forging temperature. The surface texture and the dimensional tolerances deteriorate with increasing temperature. Therefore, hot forging results in products that have a poor surface finish and dimensional tolerance, and usually require further machining. Parts produced by open die forging tend to develop a barrel shape. Barreling is caused primarily by frictional forces at the die and work piece interface, where the outward flow of material occurs. Barreling can be minimized by using an effective lubricant. Typical dimensional tolerances range from ±0.5 to ±0.8 mm. Surface finishes range from 2 to 8 μm.

Shapes: The process can be used to produce solid shapes without undercuts or re-entrant angles. Stock work pieces are round, square, or flat with medium to high ductility. They are cut from a parent work piece to provide a favorable grain flow orientation.

Products: Forging is used to make mechanical parts that are subjected to high stresses such as aircraft engines and structures, land based vehicles components, connecting rods, crankshafts, valve bodies, and gear blanks.

Materials—cold forging: As the carbon and alloy content of the steels increase, their ability to be forged decreases. Most nonferrous metals such as copper, forging brass, naval brass, bronze, and copper alloys are easily forged. The materials are frequently selected on a compromise basis. First the required strength and the physical properties must be met. Corrosion resistance, size, toughness, fatigue resistance, heat resistance, and section thickness must then be balanced against the ability to be forged. Forgeability ratings for some commonly used metals are given in decreasing ease of forgeability based on the die life: forging brass, naval brass, alloy and low carbon steels (in increasing order of carbon/alloy content) and monel metal.

Materials—hot forging: All metals that can be cold formed can be hot formed. Also, metals unsuitable for cold working can be hot forged. Nonferrous materials have a narrow temperature range in which they flow easily. Danger of overheating these metals is high and the process, therefore, must be precise. However, their high forgeability, corrosion resistance, and color make them very useful. The relatively high forging temperature of copper, coupled with its tendency to form a black oxide, which has serious erosive action on dies, makes it better suited for cold forgings. Normalization of forged steel parts is necessary to obtain maximum grain refinement. It also improves machinability and relieves internal stresses created upon cooling. Normalization is a heat treatment process in which steel is heated to a temperature above its critical temperature, a temperature at which phase change occurs, and then cooled to below that range in air. Stainless steels and high carbon steels work similarly. At high temperatures, stainless steels are stronger and, therefore, are more difficult to forge.

Advantages: Controlled grain structure provides enhanced mechanical strength and results in parts with a high strength-to-weight ratio. There is a low loss of material and few internal flaws in the finished product.

Disadvantages: Machining is required to provide accurate finished dimensions; otherwise, tooling and processing costs become high.

Design guidelines:
- Choose parting lines such that the part lies entirely in one die half.
- Avoid nonplanar parting lines.
- Provide part with tapers.
- Locate parting line so that metal will flow parallel to the parting line.
- Design for the parting lines to be at half height.
- Avoid sharp changes in cross sections and excessively thin sections.

For additional guidelines see Bralla[*] and Pahl et al.[†]

9.8.2 ROLLING

Process description: A strip of ductile metal is fed continuously through a series of contoured rolls in tandem. As the stock passes through the rolls it is gradually formed into a shape with the desired cross section, which can vary in thickness from 0.1 to 3 mm and up to 0.5 m wide. Roll sets can also have side rolls and the size of the rolls depends on the work piece thickness and its formability. Rolling is a plane strain operation. By constraining the direction perpendicular to the direction of rolling, the displacement (deformation) of the length of the work piece is increased, not its width. Therefore, as the work piece elongates during rolling it speeds up; that is, the work piece speed is slower than the rolls while entering them and faster than the rolls when exiting them. The frictional force acts at a central point, where the work piece speed and roll speed are the same. Friction is necessary for rolling to occur, and the roll pressure depends on the coefficient of friction.

Material utilization: Scrap may be produced when the continuous product is cut to the proper length. The process has a high material utilization.

Flexibility: Most of the rolls are not dedicated. However, shaped rolls are dedicated and their set-up time is high. Therefore, the flexibility of rolling is fair.

Cycle time: The cycle time is dependent on the length of the product, but the production rate is high.

Operating costs: Production rates are high so that labor costs are low. Material utilization is excellent. Tooling and setup costs are high, so best economics occur with mass production.

Dimensional accuracy: The quality of the finished product depends primarily on the rolling temperature. The surface texture deteriorates with increasing temperature. Typical tolerances are ±0.4 mm for the width of the formed cross section and ±1.5 mm for its depth. The surface finish typically varies from ±0.25 to ±3.8 μm depending on the severity of the forming, and it typically varies from ±0.4 to ±1.3 μm for cold finished sheet stock.

---

[*] Bralla, 1986, ibid.
[†] G. Pahl, W. Beitz, J. Feldhusen, and K.-H. Grote, ibid.

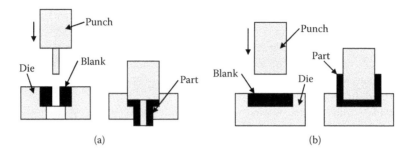

FIGURE 9.16 (a) Forward extrusion, (b) backward extrusion.

Shape: Finished part thickness ranges from 0.6 to 3 mm. Parts are typically long, with a constant cross section that can be complex. Short pieces can be obtained by cutting longer ones. Widths of the stock before forming normally range to 1 m and stock thickness to 5 mm. The resulting product from rolling is suitable for both decorative and structural applications.

Products: Products typically made by rolling are roof and siding panels for buildings, architectural trim, down spouts, window frames, stove and refrigerator panels and shelves, curtain rods, and metal picture frames.

Materials: Any metal that is ductile at the temperature of forming can be used. Aluminum, copper, and their alloys have excellent formability ratings. Nickel and magnesium have fair to good formability ratings and mild steel and stainless steel have fair to excellent formability ratings.

Advantages: The best applications are longer parts with complex cross sectional shapes. Production is rapid, and surface finish and dimensional consistency are good.

Disadvantage: Parts must have the same cross section for the entire length. Tooling and setup costs are high. The process may cause work hardening and microcracks.

9.8.3 EXTRUSION

Process description: Metal billets are forced by a mechanically or hydraulically actuated ram through a die hole of the desired shape or around a punch. The metal emerges from the die in solidified form and closely conforms in cross section to the shape and dimensions of the die opening. The punch controls the inside shape of the work piece. The die controls the outside shape of the work piece and may have more than one diameter. The work piece is ejected by either a counterpunch or an ejector. There are three types of extrusion processes: (1) forward extrusion in which the metal flows out (downward) through die; (2) backward extrusion in which the metal flows (upward) around the punch; and (3) combined forward/backward extrusion.* The forward extrusion and backward extrusion processes are shown in Figure 9.16. Forward extrusion can be used to produce complex shapes and reduces the cross section of the work piece drastically. Backward extrusion cannot produce very long shapes, but can produce hollow components with large length to diameter ratios. Wall thickness of the product is controlled by the clearance between the punch and the die. The final work pieces have excellent surface finish. Cylindrical cross sectional parts are

---

* For a visualization of these three types of extrusions, see http://www.jlometal.com/the_process/default.asp.

most common, but rectangular or odd cross sectional parts are also possible. The length of extruded part is limited by the column strength of the machine and by the guideline that it be less than 6 times the inside diameter of the part.

Material utilization: Extrusion is a near net shape process and the only scrap that is produced is during the cutting of the continuous product to its length. The material utilization of this process is high.

Flexibility: The tooling is dedicated and set-up times can be long; therefore, the process has only a fair flexibility.

Cycle time: The cycle time is dependent on the product's length, but the production rate is very high and, therefore, the cycle time is low.

Operating costs: Hot extrusions need protection of the work piece from air and oxidation. Therefore, hot extrusion is a more expensive process. Cold extrusion has high tooling and setup costs; therefore, this process is best for mass production.

Dimensional accuracy: The quality of the product is generally good for all metals, but depends mainly on the forming temperature. The surface texture deteriorates with increasing temperature. Typical tolerances are ±0.25 mm for diameters and ±0.4 mm for lengths. Surface roughness typically ranges from 0.5 to 3.2 μm. The tolerances and surface roughness depend on the press conditions: ram pressure, tool geometry, material size and shape, allowed length-to-diameter ratio, lubrication, and whether hot or cold extrusion is used.

Shapes—hot extrusions: Constant cross sections of any length up to 7.5 m are feasible. Cross sections can be large enough to occupy a circle 250 mm in diameter for aluminum and 150 mm in diameter for steel, and they can be very complex.

Shapes—cold extrusions: The parts can range from 13 to 160 mm in diameter, and the maximum length of the part is 2 m. This process is suitable for circular, hollow parts closed at one end. Cylindrical cross sectional parts are most common; however, rectangular or odd cross sectional parts are also possible. The work pieces can have combined shapes, stepped shapes, or cupped shapes. Typical work piece diameters are 13 to 500 mm and typical work piece lengths range from 13 to 760 mm.

Products: Typical products made by extrusion are building and automotive trim, window frame members, tubing, aircraft structural parts, railings, flashlight cases, aerosol cans, military projectiles, and fire extinguishers.

Materials—hot extrusion: Most commercially available shapes are extruded in copper, brass, steel, aluminum, zinc, and magnesium. A large number of copper-based alloys can also be extruded. Those with the highest zinc content are the most plastic and can produce the most complex shapes. Lightweight alloys, like those of aluminum and magnesium, are the most frequently used materials for extruded parts. Harder alloys are more difficult to extrude. Steels and stainless steels can also be extruded.

Materials—cold extrusion: Most cold extrusions use tin. The method has also been used on lead and aluminum parts. Zinc can be handled with a little preheating to get it to the ductile range (150°C).

Advantages: Intricate cross sectional shapes, including undercuts and hollows, can be obtained. The tooling cost is low. Extrusion increases the hardness and yield strength of the material. High material utilization reduces costs and, usually, further machining is not necessary.

Disadvantage: The process is limited to ductile materials and restricted to a maximum cross sectional size. Parts of nonuniform cross section require additional operations. Extrusion creates residual surface stresses and microcracks. Close tolerances are difficult to achieve.

Design rules:
- Ensure that parts are rotationally symmetrical and are without material protrusions.
- Avoid sharp changes in cross sections and sharp edges and fillets.
- Avoid tapers and sections with almost equal diameters.

For additional guidelines, see Bralla.[*]

9.9 POWDER PROCESSING

9.9.1 Powder Metallurgy[†]

Process description: Powder metallurgy is a net-shape manufacturing process in which a metal powder is compressed to a particular shape and then heated to bond the metal particles together. A powder is compacted into the required form and part density by the action of opposing punches, which move in a die as shown in Figure 9.17. The number of upper and lower punches depends on the complexity of the part. Compaction pressures may range anywhere up to 700 MPa. After compaction, a part is sintered by placing it in a furnace. Sintering, which takes place below the melting temperature of the primary constituent of the powder, causes the powder particles to bond together to produce the required properties of the part.

Material utilization: The process is near net shape and the material utilization is almost 100%.

Flexibility: The tooling expenses are moderate and the equipment is fairly flexible.

Cycle time: Due to the speed of the operations in producing parts from powder, economical quantities are high. Production volumes of at least 20,000 to 50,000 are necessary to obtain a cost advantage. Larger and complicated parts may prove economical at production volumes as low as 500 to 5,000 pieces. Even though the cycle time for compaction of a part is small, the sintering is a relatively slow and time-consuming process.

Operating costs: High costs of powders and normal die costs, which depend on the complexity of the part, preclude small production quantities. Labor costs are low and material utilization is excellent. Since this is a net-shape manufacturing process, secondary operations are minimal or absent, which also reduces the costs.

Dimensional accuracy: The dimensional accuracy of the parts produced by powder metallurgy is extremely high. Tolerances range from ±5 μm in small bores after repressing to ±0.13 mm in larger dimensions on parts that are not repressed. Cross sectional dimensions can be held to closer tolerances than those dimensions in the direction of pressing.

Shapes: Power-metal parts are normally small, less than 75 mm in the largest dimension. Complex shapes are feasible, but side walls are parallel, and undercuts and screw threads must be provided by secondary operations. Thin sections, feathered edges, and narrow and deep splines must be avoided.

Products: Typical products made using powder metallurgy are cams, clutches, brake linings, self-lubricating bearings, slide blocks, levers, gears, bushings, ratchets, guides, spacers, splined parts, connecting rods, sprockets, pawls, and other mechanical parts for business machines, sewing machines, firearms, and automobiles.

Materials: Materials used in powder form are iron, carbon steel, alloy steel, copper, brass, bronze, nickel, stainless steel, refractory metals such as tungsten, molybdenum, and tantalum, and precious metals. In addition, ceramics powders are used.

[*] Bralla, 1986, ibid.

[†] Additional information can be found at the Metal Powder Industries Federation Web site: http://www.mpif.org/index .asp and at the European Powdered Metallurgy Association (EPMA) site: http://www.epma.com/about_pm/web_pages/ nma_choosing__pm.htm.

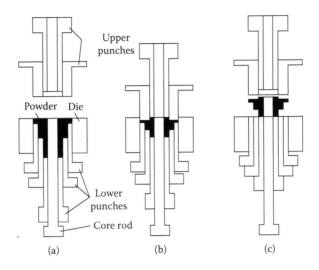

FIGURE 9.17 Powder metallurgy (a) fill, (b) compaction, and (c) ejection.

Advantages: Rapid production of parts with high dimensional accuracy, smooth surfaces, and excellent bearing properties. Parts can be somewhat intricate in shape. Scrap is low, since this is a near-net-shape process. Special properties such as self-lubrication in porous bearings are an important advantage of powder metallurgy. Powder metallurgy components have good damping characteristics that result in quieter operation of mating parts. Metallic brake linings and clutch plates that include a nonmetallic abrasive powder are manufactured using this process.

Disadvantages: Size of parts is limited, and not all shapes can be produced. Powder and tooling costs limit the process to high-production applications. Undercuts are not feasible. There are also strength limitations.

Design rules:
- Observe recommended minimum dimensions.
- Avoid sharp interior corners and feather edges.
- Avoid sharp corners and sharp reentry corners.
- Avoid undercuts.
- Avoid blind holes.
- Use small cross-sectional radii when possible.

For additional guidelines see Bralla[*] and Pahl et al.[†]

9.10 LAYERED MANUFACTURING

9.10.1 INTRODUCTION

Traditional manufacturing processes discussed in the previous sections typically require specific tooling, fixtures, and/or molds, and frequently long lead times. The complexity, special expertise needed, and cost of these requirements depend on the part shape, size, material, and the quantity manufactured. In the past two decades, new types of manufacturing processes have been

---

[*] Bralla, 1986, ibid.
[†] G. Pahl, W. Beitz, J. Feldhusen, and K.-H. Grote, ibid.

developed called layered manufacturing (LM). Layered manufacturing refers to an additive process that builds objects layer-by-layer. It has been made possible, in part, by the ability of LM to directly integrate the digital manipulation of 3D CAD models of the objects with the LM process itself. Layered manufacturing techniques are also known as *solid freeform fabrication*, *additive manufacturing*, and *rapid prototyping*. Layered manufacturing has been shown to reduce lead time, to require minimal expertise to use the process, and to be economically viable for low-volume production.

Layered manufacturing has the following advantages.

- **Does not require part-specific tooling or fixturing**, which means that after the object has been completed in a CAD system it is ready to be fabricated.
- **Very complex geometries can be built**, which enables one to build shapes with such features as deep undercuts and inaccessible cavities. These shapes would have not been realizable using traditional manufacturing processes.
- **Process planning is done automatically**, which enables one to use LM with minimal training.
- **Well-suited to small volume production**, since setup time and part-specific tooling have been eliminated.

Some of the disadvantages of LM are as follows.

- **Low part strength**, which is usually due to the weakness of the inter-layer interfaces.
- **Limited to relatively small parts**, due to the inherently slow nature of these processes.
- **High cost of the systems** and the need for highly specialized maintenance personnel.
- **Limited to small production runs**, because the speed of the process on a per part basis does not reduce significantly with the number of parts made.
- **Proprietary materials are usually required**, which can be expensive.

Major application areas for layered manufacturing are the following.

1. **Prototyping:** When 3D computer models of an object do not provide sufficient information, one is required to generate a prototype. LM has been used to create prototype cell phones, cameras, and household appliances such as mixers and screwdrivers, which are then used to evaluate their aesthetic appeal and ergonomic aspects. LM has also been used to evaluate assembly procedures for complex systems.
2. **Testing:** In such applications as flow studies around objects, LM has been used to generate scaled-down versions of objects to understand better their flow characteristics. LM is also used to produce test specimens on which one can perform load versus displacement tests.
3. **Rapid manufacturing of custom low-volume functional parts:** As LM processes become more capable of producing parts that have good strength characteristics for low stress applications and as the accuracy of the LM processes continually improve, objects can now be produced that function correctly in their intended applications. Some examples are LM parts for customized hearing aids and customized gas masks.
4. **Rapid tooling:** LM now creates tools that are used to produce other parts such as molds for injection molding (produced by selective laser sintering), sand casting patterns (produced by fused deposition modeling), and investment casting patterns (produced by stereolithography).

We shall now describe the following six-layered manufacturing processes: stereolithography, fused deposition modeling, solid ground curing, selective laser sintering, laminated object manufacturing, and 3D printing. These six LM processes employ the following six steps.

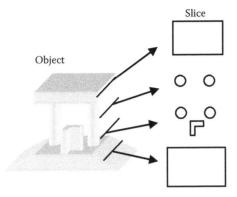

FIGURE 9.18 Layered manufactured object and several representative slices.

1. **Create a CAD model.** A computer-generated solid model of the object to be manufactured is created using 3D CAD software.
2. **Convert the CAD file into an STL file.** An STL file contains the coordinates of the vertices and direction normals of a set of triangles that are used to approximate the surface of the object to be built. The inherent error introduced by this approximation can be reduced by increasing the number of triangles; however, this tends to increase the subsequent computational effort.
3. **Select the orientation of the part with respect to how it is to be layered.** The choice of the build orientation affects the build time, the accuracy of the various surfaces of the part, and the part's strength. Hence, build orientation must be chosen carefully by accounting for the process' capabilities and the object's requirements.
4. **Slicing the part build orientation into thin layers.** After the part build orientation has been selected, it is digitally sliced into thin layers. The thickness of each layer depends on the LM process. A visualization of the slicing process is shown in Figure 9.18.
5. **Build the model layer-by-layer.** The coordinates of the approximation to the perimeter of each digitally-obtained layer are fed to the fabrication device that will build each layer sequentially, one on top of the other. The final part will be the totality of the layers.
6. **Post-build operations.** Depending on the process and the build orientation some post-build operations will have to be preformed, such as the removal of support structures that may have been added to prevent sagging of long overhangs and, for certain materials, additional cure time that must be provided.

After the six operations have been completed, the final object is ready to be used.

Elaboration of Step 5: In the fifth step given above, the four of the six LM processes to a large degree function as follows.[*] Each method uses a platform that moves in very small increments in the negative z-direction as shown in Figure 9.19. The area of the platform and the total vertical travel of the platform determine the maximum size of the object that can be made. Each of the four methods deposits in a manner to be discussed subsequently a thin layer of material. In some methods, the material is first deposited over the entire area of the platform and, then, where the material is desired it is hardened (solidified) in some manner (cured, fused). In other methods, the layer material is deposited at a specific location where it is allowed to harden. In either case, the specific locations where the hardening takes place are arrived at in the following manner. Referring to Figure 9.19, the process starts when the platform is at $z = z_1$. The system positions the hardening device at $x_1 = x_{1,min}$, which is one of two or more outermost points of the perimeter(s) at layer $z = z_1$. The hardening

---

[*] The two exceptions are laminated object manufacturing and solid ground curing.

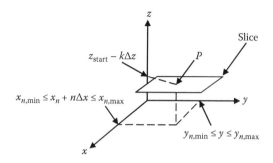

FIGURE 9.19 Coordinate system describing the location of a point on a layer.

device then moves in the y-direction from $y_{1,\min}$ to $y_{1,\max}$ and activates the hardening procedure at the required locations within the perimeters of the layer to be created. The locations $y_{1,\min}$ to $y_{1,\max}$ are the outermost points on the perimeter, or series of perimeters, in the y-direction at layer $z = z_1$. After all the points along the $x = x_1$ line have been selected, the system moves the hardening device to $x_2 = x_1 + \Delta x$, where the magnitude of Δx depends on the process, and the y-direction positioning proceeds as before except that y varies from $y_{2,\min}$ to $y_{2,\max}$. The procedure continues until all the material is hardened at all the desired points for that layer. Then the platform moves down an amount Δz, and the planar positioning process repeats. The magnitude of Δz depends on the process. This procedure is repeated until the last layer has been laid.

We shall now describe the six LM processes.

9.10.2 STEREOLITHOGRAPHY[*]

Process description: The model is built upon a platform that is situated in a vat of liquid photosensitive polymer. Referring to Figure 9.20, the platform is initially adjusted so that a thin layer of liquid of thickness d covers the platform surface. A highly focused laser beam traces out the cross-section of the layer as described previously in the elaboration of step 5. Upon exposure to the laser, the resin is polymerized and, hence, hardens. The region that is not part of the cross sections remains liquid. When a layer is completed, the platform is lowered by the thickness d. The magnitude of d determines the speed at which the laser beam traverses the cross section. When the platform is lowered, a fresh layer of liquid appears on the top of the part, again having a thickness d. If the part has any large overhanging features, then the additional areas are traced out to form support structures. The cross sections of these support structures are relatively small so that they can be removed easily. After all the layers have been completed, the part is removed from the liquid vat and placed in an ultraviolet oven to complete the curing process.

A wide variety of materials are available that can simulate the characteristics of polypropylene, ABS, polycarbonate, and nylon.

9.10.3 FUSED DEPOSITION MODELING[†]

Process description: Referring to Figure 9.21, a thermoplastic filament is extruded through a heated nozzle. The nozzle is positioned as described previously in the elaboration of step 5 to trace the cross section of the layer. The resistive heater in the nozzle keeps the filament at a temperature just above its melting point so that it easily flows through the nozzle during the extrusion process. The material is fed using the drive wheels and the flow is aided by gravity. The platform is maintained at

---

[*] http://www.3dsystems.com/.
[†] http://www.stratasys.com/.

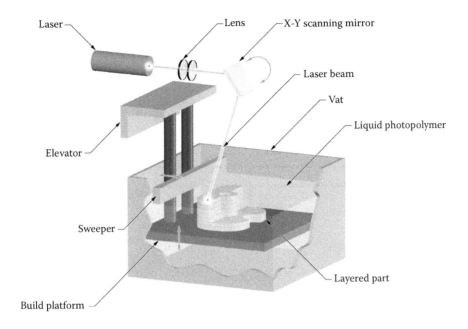

FIGURE 9.20 Stereolithography system. (Copyright © 2008 CustomPartNet. Courtesy of CustomPartNet, Inc., www.custompart.net.)

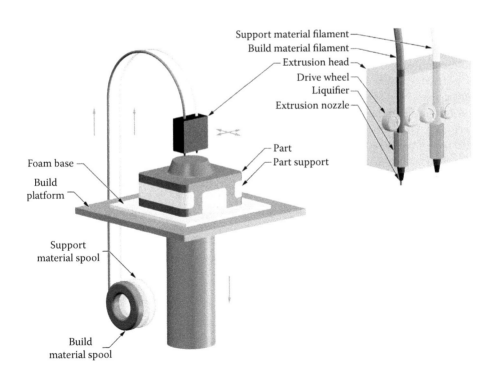

FIGURE 9.21 Fused deposition modeling system. (Copyright © 2008 CustomPartNet. Courtesy of CustomPartNet, Inc., www.custompart.net.)

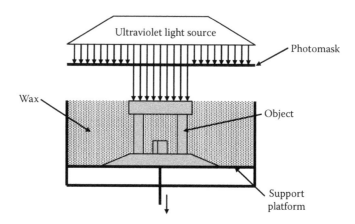

FIGURE 9.22 Solid ground curing system.

a lower temperature so that the material hardens quickly after the extruder places it at the intended location. When a layer has been completed, the platform is lowered by a depth equal to the layer thickness and a new layer is built on top of the already formed part. If there is a need for a support structure, then a second nozzle is used with a different filament material to build the support structures. The support material is removed by dissolving it in a solvent. The solvent is chosen such that it only dissolves the support material and does not dissolve the main material.

These systems use a variety of production-grade thermoplastics, including ABS, polycarbonates, and polyphenylsulfones.

9.10.4 SOLID GROUND CURING

Process description: As with the stereolithography process, the solid-ground curing process also uses photosensitive liquid resins. As shown in Figure 9.22, a layer of the object is built using a photomask to create the cross section of the object at that z location. The photomask is then transferred onto a glass plate. Next, a layer of the liquid resin is sprayed on the platform to a depth of the layer thickness being used to build the object. The layer of resin and the photomask are then exposed to ultraviolet light. Only those portions of the liquid resin that are exposed to the ultraviolet light are cured; the rest of the liquid is removed by a vacuum pump. The vacuum's nozzle moves across the entire surface and lifts the uncured polymer. The region that was occupied by the just-removed uncured fluid resin is filled with wax and cooled. Face milling is then used on the wax to create a uniform top surface on which the next layer will be built. Since the object is completely embedded in the wax during the fabrication, no additional support structures are needed. This process simultaneously cures the entire layer at once, whereas in the stereolithography process curing takes place as the laser beam traverses the cross section. After the build is complete, the wax is removed by immersing it in hot tap water or melting it by using a hot air gun to release the object. This process is also referred as the *solider process*.

9.10.5 SELECTIVE LASER SINTERING*

Process description: A focused laser beam is used to selectively fuse powered plastic or metal to form the object. Objects are built on a platform that is filled with the powder of the target material. This powder has a depth d and is rolled onto the platform from a bin before building the layer, as

---

* http://www.3dsystems.com/.

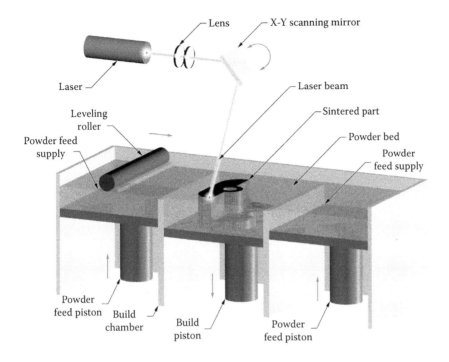

FIGURE 9.23 Selective laser sintering system. (Copyright © 2008 CustomPartNet. Courtesy of CustomPart-Net, Inc., www.custompart.net.)

shown in Figure 9.23. In every step, a layer of powder of thickness d is spread over the platform's region. This new layer is equal to the layer thickness. The powder is maintained at an elevated temperature so that it fuses easily upon exposure to the laser beam. The laser beam traces the cross section of the layer as described previously in the elaboration of step 5 and sinters the powder for that layer. After a layer is completed, the platform is lowered an amount d and a new layer of powder is applied. This process continues until the part is complete. Support structures are not needed in this process because the excess powder in each layer acts as a support to any overhanging sections.

This system uses nylon, glass-filled nylon, and steel powder as the raw material.

9.10.6 LAMINATED OBJECT MANUFACTURING[*]

Process description: An adhesive-coated thin sheet of paper, plastic, or composite is cut by a laser to form each layer of the object. As shown in Figure 9.24, a feeder/collector mechanism moves the sheet over the build platform. A roller applies both heat and pressure to bond the sheet to the previously built layers. The unused portion of the sheet is cut into small pieces by the laser beam using a hatched pattern so that this excess material can be disassembled from the object being fabricated easily during the post processing step. No special support structures are required because the excess material provides support for any overhangs and thin-walled sections. After a layer is completed, the platform is lowered by the height equal to the sheet thickness and the previous process step is repeated. Hence, blind cavities or cavities with limited access cannot be created using this process.

This system uses paper, plastic, and composite sheets as the raw material.

---

[*] http://www.cubictechnologies.com/.

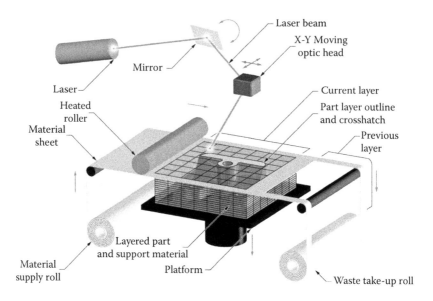

FIGURE 9.24 Laminated object manufacturing system. (Copyright © 2008 CustomPartNet. Courtesy of CustomPartNet, Inc., www.custompart.net.)

9.10.7   3D Printing[*,†,‡]

Process description: This process uses raw material in its powdered form. A thin layer of powder is spread over the platform and the powder is compressed with a roller as shown in Figure 9.25. A binder is then deposited on the layer of powder using an ink-jet printing head. The printing head is traversed back and forth to cover the entire cross section of the layer as described previously in the elaboration of step 5. After a layer is built, the platform is lowered and a new layer of powder is spread and the process is repeated until the part is complete. The loose powder surrounding the part—that is, the powder without the binder—acts a support for the part so no supports have to be added. After the part is completed, it is sintered to attain its strength.

These systems use ceramic, cellulose, and composites.

9.10.8   Comparisons of the LM Processes

In general, stereolithography offers much smaller feature sizes compared to other processes. Consequently, this process is a preferred alternative when very small feature sizes are needed. The fused deposition modeling and selective laser sintering processes employ materials that are commonly found in consumer products. Hence, if matching the product material to prototype materials is desired, then these processes are preferred alternatives. Selective laser sintering leads to prototypes with superior material characteristics; however, system costs are much higher than those of fused deposition modeling. Currently, selective laser sintering is the only viable LM process for making metal parts. Layered object manufacturing and 3D printing are relatively fast processes. Hence, when short build times are needed, these processes are preferred alternatives. Layered object

* http://www.zcorp.com/.
† http://www.cubictechnologies.com/.
‡ http://www.solid-scape.com/.

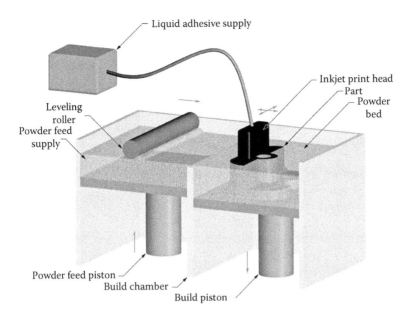

FIGURE 9.25 3D printing system. (Copyright © 2008 CustomPartNet. Courtesy of CustomPartNet, Inc., www.custompart.net.)

manufacturing is especially attractive for making solid parts, that is, parts without hollow interiors. On the other hand, 3D printing allows users to create multicolored parts.

To aid in determining which LM process is suitable for a particular project, it is recommended that the following factors be considered. When an LM process does not meet one of these requirements, it is eliminated as a viable process.

1. **Material** and material characteristics differ for each LM process; therefore, the LM process should produce an object with the desired material characteristics.
2. **Build envelope**, which determines the maximum size of the object that can be fabricated.
3. **Process accuracy** determines the geometric and surface variability, which are determined by such factors as the minimum layer thickness, warping and shrinkage of the material, and machine resolution.
4. **Speed of fabrication**, which includes the build time and the pre- and postprocessing time. Build time is the time that it takes the LM machine to build the object. Many processes can simultaneously build multiple parts. To decrease the total build time, the possibility of building several parts together should be considered. Pre- and postprocessing times include the time to translate the CAD file to the STL format* and the time it takes to set up the machine by loading raw materials and by performing calibration steps. The postprocessing time includes the time it takes to remove any support structures, the time it takes to perform any steps needed to increase part strength, and the time it takes to attain an adequate surface finish.

A summary of the basic specifications of LM systems is given in Table 9.3.

---

* Sometimes, the translation process to an STL file introduces geometric anomalies and, hence, manual editing is required.

TABLE 9.3
Representative Specifications of Several LM Processes

Technology	System Model	Build Envelop (mm)	Minimum Layer Thickness (mm)	Accuracy/Resolution
Fused deposition modeling	FDM 400 mc	$406 \times 356 \times 406$	0.127	0.127 mm or 0.0015 mm per mm, greater of
Selective laser sintering	Sinterstation Pro 230	$550 \times 550 \times 750$	0.1	0.125 mm + 0.025 × dimension
Stereolithography	Viper Pro	$1500 \times 750 \times 500$	0.05	0.05 mm + 0.025 × dimension
3D printing	Z Printer 650	$254 \times 381 \times 203$	0.089	600×540 dpi[a]
Laminated object manufacturing	LOM 1015	$381 \times 254 \times 356$	0.18	0.025 mm
Solid ground curing	SOLIDER 5600	$508 \times 508 \times 355$	0.1	0.1 % of size to maximum of 0.5 mm

[a] Dots per inch.

BIBLIOGRAPHY

M. F. Ashby, *Materials Selection in Mechanical Design*, 3rd ed., Butterworth-Heinemann, Oxford, 2005.

M. M. Farag, *Selection of Materials and Manufacturing Processes for Engineering Design*, 2nd ed., CRC Press, Boca Raton, FL, 2008.

M. P. Groover, *Fundamentals of Modern Manufacturing*, 2nd ed., John Wiley & Sons, New York, 2002.

C. Kai, L. Fai, and L. Chu-Sing, *Rapid Prototyping: Principles and Applications*, 2nd ed., World Scientific Publishing Company, Singapore, 2003.

S. Kalpakjain and S. R. Schmid, *Manufacturing Processes for Engineering Materials*, 5th ed., Prentice Hall, Upper Saddle River, NJ, 2008.

K. Lee, *Principles of CAD/CAM/CAE Systems*. Addison Wesley, Reading, MA, 1999.

P. K. Wright, *21st Century Manufacturing: Surveys of Products, Prototypes, Processes, and Production*. Prentice Hall, Upper Saddle River, NJ, 2001.

10 Design for "X"

Suggestions and guidelines are given for additional criteria that are frequently included in the concept evaluation, configuration, and embodiment phases of the product development process.

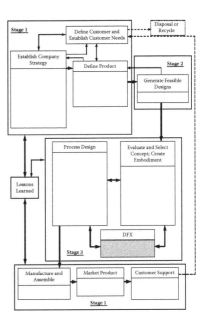

10.1 LIFE-CYCLE ENGINEERING

10.1.1 INTRODUCTION

In this chapter, we shall present several interrelated aspects of the design goals and constraints that appear in Table 2.1. Many of the topics discussed in this chapter fall into what is called *life-cycle engineering* or can influence what is called the *useful life* of the product. Life-cycle engineering is a process that emphasizes early in the product's development the following aspects of a product's journey from creation to disposal and/or recycling.

- Failure modes and reliability issues.
- Availability of spare parts.
- Diagnostic tools and personnel.
- Customer service and maintainability.
- Product operation and maintenance.
- Product installation requirements.
- Future upgrades and improvements.
- Logistic support to satisfy the preceding six aspects.
- Ease and safety of use.
- Product recycling and/or disposal.
- Costs associated with the environmental effects of any aspect of the above.

The useful life of a product is a measure of how long it will meet performance standards when maintained properly and when not subject to uses beyond its stated limits. The aspects of a product's design that can influence its life and should be addressed early in the product realization process are as follows.

- **Durability**, which is a measure of product life, that is, the amount of use one gets from a product before it breaks down and replacement is preferable to repair.
- **Adaptability**, which is the ability to incorporate continuous improvements of various functions; implies component and functional modularity.
- **Reliability**, which is the ability of a product neither to malfunction nor to fail within a specified time period and under the intended operating conditions.
- **Serviceability**, which is the ability to maintain performance for a specified period of time by one of the following methods:

Repair, which replaces nonfunctioning components to attain specified performance.
Remanufacture, which restores worn parts to like-new condition to attain specified performance.
Reuse, which finds additional uses for a product or its components after it has been retired from its original service.

- **Recyclability**, which is the reformation and reprocessing of products to recover some or all of their constituent materials; implicitly requires that the product can be disassembled cost-effectively, that the materials are identifiable, and that there has been a simplification and consolidation of its parts (recall Section 7.3).
- **Disposability**, which is the ease with which all non recyclable materials can be safely disposed and, when possible, energy can be recovered through incineration.

It has been suggested[*] for those products that have been designed for remanufacture, rather than for recycling, that the design process emphasize the ease of disassembly and reassembly, cleaning, inspection, and component replacement. Furthermore, individual components should have a high degree of standardization in its fasteners, interfaces, and modules. Recall Sections 6.3 and 3.4.

10.1.2 RELIABILITY

Reliability is the probability of a device performing properly for the period of time intended under the operating conditions encountered. The time period extends from one time (e.g., an air bag, electrical circuit fuse) to many millions of times (e.g., bearings, light switches, engines). One of the goals when designing for reliability is to do so while limiting the manufacturing costs and the life-cycle costs.

The determination of a component's or a system's reliability typically involves testing and analysis of such attributes as stress, temperature, and vibration. Testing should be performed to anticipate improper usage. A reliable design is one that has anticipated all that can go wrong. It is noted that testing for reliability is different from designing for testability, which results in a product that can have one or more of its performance attributes easily measured.

Reliability engineering addresses problems by adapting the laws of probability and a knowledge of the fundamental physics governing the mechanisms of failure to predict product failure. Reliability engineering measures can lead to the following:

- Techniques for reducing failure rates while products are still in the design stage.
- Failure mode and effect analysis, which systematically reviews how alternative designs could fail (see Section 10.1.3).
- Individual component analysis, which computes the failure probability of key components and aims to eliminate or strengthen the weakest links.
- De-rating, which requires that parts be used below their normally specified levels.
- Redundancy, which calls for a parallel system to back up an important component or subsystem in case it fails (e.g., dual brake systems on cars).

Failures are unexpected or unplanned occurrences that cause a component or system not to meet its normal or specified operating characteristics. Effective failure avoidance must deal with the entire spectrum of causal elements, which includes human errors, environmental effects, technical faults and malfunctions, inadequate design, and material-related deficiencies.

---

[*] T. Amezquita, R. Hammond, M. Salazar, and B. Bras, "Characterizing the Remanufacturability of Engineering Systems," ASME DE-Vol. 82, Vol. 1, Design Engineering Technical Conference, September 1995.

Failure of certain components of products can have a very large impact on the product's user, the manufacturer, or both. In late 2007, a quarter-million patients had been notified that a wire connecting their surgically-implanted electronic defibrillator to their heart may fracture.[*] It was found that 2.3% of these implants fracture within 30 months, meaning that more than 5,000 likely malfunctions could occur within this period of time. The question for the patients is whether or not to have the lead removed and replaced, which requires a procedure that costs around $12,000. The company is willing to provide new leads and pay only $800 toward the cost of this procedure. In addition to the cost, there is now a fairly high degree of uncertainty for the patients' continued use of this device.

On the positive side,[†] when the Boeing 777 first entered service around 1995, it was able to do so without having to go through a two-year test period, which would have restricted its airplanes from flying any route that was more than 1 hour from a landing point. Instead, it was able to fly routes that were 3 hours from a landing point. This was due to the improved reliability of their new jet engines, which have a probability of failure of only once every 25 to 30 years when used 8 to 10 hours per day.

It has been proposed[‡,§] that a design team can improve the reliability of a product by employing the following suggestions.

- Simplify the design and reduce the number and type of parts (recall Section 7.2.2).
- Standardize parts and materials. Establish preferred parts lists and their suppliers. The suppliers should have a documented and proven record of the product's reliability under the desired operational conditions.
- Design to counteract environmental factors and material degradation (recall Table 8.1).
- Design to minimize damage from mishaps in shipping, service, and repair. See Sections 10.3 and 10.4.
- When defects are found, their root causes must be determined and means devised to eliminate them. Root causes are the underlying reasons for the failure. This differs from a symptom, which is the result of a root cause.
- Use design techniques that lead to manufacturing methods that are insensitive to variations in materials, process variables, and environmental conditions. See Section 11.5.2.
- Use failure mode analysis techniques to minimize the risk of failures in the product. See Section 10.1.3.

10.1.3 FAILURE IDENTIFICATION TECHNIQUES[¶]

Reliability depends on the product's usage, which includes its method of transportation, its modes of operation including the number of on/off cycles, its maintenance and repair procedures, and its environmental conditions for both use and storage. To minimize the probability of failure, it is first necessary to identify all possible modes of failure and the mechanism by which these failures can occur. Although certain major considerations regarding reliability are taken into account in the product's concept generation and evaluation stages, the detailed examination with respect to reliability is performed in the embodiment stage. This usually requires combining the knowledge of the performance and experiences of similar products and materials, and a brainstorming session,

---

[*] B. J. Feder, "Patients Wonder Whether to Replace a Wire That Might Fail," *New York Times*, December 13, 2007, http://www.nytimes.com/2007/12/13/business/13defib.html?scp=11&sq=Feder&st=nyt.

[†] M. L. Wald, "F.A.A Allows Boeing 777 to Skip a Test Period," *New York Times*, May 31, 1995.

[‡] J. Doran and G. Hudak, "Parts Selection and Defect Control Reference Guide," Draft report for several of the templates defined in DoD 4245.7-M, August 1990.

[§] J. G. Bralla, 1996, ibid.

[¶] C. E. Witherall, *Mechanical Failure Avoidance: Strategies and Techniques*, McGraw-Hill, New York, 1994.

where the objective is to anticipate all possible actions and situations that can cause the product to fail to meet its performance criteria. All participants who were involved in bringing the product to the marketplace should help uncover the various factors that may cause problems.

To give an indication of the magnitude and complexity of the reliability problem, we introduce the following common environmental conditions and their possible effect on the reliability and performance of electronic components.[*]

- **High temperature**, where values of electrical components will vary; insulation may soften; devices will suffer thermal aging; oxidation and other chemical reactions are enhanced.
- **Low temperature**, where values of the electrical components will vary; ice forms when moisture is present; heat losses increase.
- **Thermal shock**, where electrical properties of components may become permanently altered due to cracks in the device or packaging.
- **Vibration**, where electrical signals may become modulated; component's materials may crack, displace, or loosen.
- **Humidity**, which penetrates porous materials and causes leakage paths between electrical conductors; causes oxidation, which leads to corrosion; causes swelling in certain materials.
- **Salt atmosphere and spray**, where salt water is a good conductor of electricity and can lower insulation resistance and cause corrosion.
- **Electromagnetic radiation**, which causes erroneous and spurious signals and may cause complete disruption of a system's functions.
- **Nuclear and cosmic radiation**, which causes heating and thermal aging; can alter chemical, physical, and electrical properties of materials; damages semiconductors.
- **Low pressure (high altitude)**, where air bubbles in materials may explode; electrical insulation may suffer breakdown; out gassing is more likely.

One of the major reasons that products and systems fail is material degradation, in which the material changes in some way from mechanical, thermal, or environmental effects, or some combination of them, or when they are accidentally damaged. Material integrity, therefore, is often a focus of life-cycle design. Many of the factors listed in Table 8.1 under the headings of temperature, environmental stability, damage tolerance, and application history are examples of the type of material effects with which one has to be concerned. Other factors often considered are wear and fatigue. These types of failures can occur singly or as a combination of those listed. However, not all materials are susceptible to all of the factors listed.

Several methods for failure identification will now be discussed. One method is called fault-tree analysis (FTA), which is a deductive logic model that depicts in a graphic form conditions and combinations of conditions that can produce the fault or failure under consideration. As usually constructed, the failure or other unwanted event is shown at the apex of its diagram. Beneath the apex are branched links, shown with the aid of logic symbols, to causal conditions or elements.

Another tool that is used in conjunction with FTA is failure modes and effects analysis (FMEA). Its objective is to predict what failures might occur, what the effect of such failures might be on the functional operation of the product, and what steps might be taken to prevent the failure or the effect on the operation. Its logic is the reverse of the FTA in that it starts with the most basic components and works its way up into more complex assemblies and subsystems. Its content is usually depicted in a spreadsheet format, as shown in Table 10.1, which is for the steel frame joining tool discussed in Section 6.4.

---

[*] M. Pecht, *Handbook of Electronic Package Design*, Marcel Dekker, New York, 1991. See also Table 4.6.

TABLE 10.1
FMEA for Some of the Components of the Steel Frame Joining Tool Described in Section 6.4

Part	Function	Mode	Cause	Result	Consequence	Severity[a]	Probability[b]	Minimization Method
Motor	Provide rotational energy	No function	Short circuit, Open circuit, Shaft jams, Brushes worn out, Overuse	No function, Fire	No function	2–3	C	Properly rated materials
		Overheat	Inadequate cooling	Fire, Motor failure	Loss of equipment	2–3	D	Provide airflow
		Noise	Bearings worn, Shaft worn	Mechanical failure	Loss of efficiency	3	C	Properly rated materials
Gears	Increase torque	Jam	Misalignment of teeth, Loose on shaft, Bearings worn	Broken gears	Loss of equipment, Loss of efficiency	2–3	C	Properly rated materials, Proper tolerances
Piston actuator	Rotary motion	Jams	Jams, Bent stroke arm, Does not reset, Center shaft cannot support forces	Improper cycling, Mechanical failure	Loss of equipment, Injuries	2–3	D	Properly rated materials, Proper tolerances
						2	D	
		Noise	Bearings worn			3	C	
Compressor	Create working pressure	No pressure	Seal fails, Structure leaks	No function	No function	3	A-B	Properly rated materials, Proper tolerances
		No movement	Seal binds motion			3	B-C	
Body	Provide structure	Loss of structural integrity	Cracks/breaks due to operator error, Cracks/breaks due to internal mechanical forces, Cracks/breaks due to age, Body-to-body fastening fails, High temperature	Sharp edges, Loss of some functionality, Separation of components	Loss of equipment, Injuries	3, 2	D, D	Properly rated material, Monitor injection molding process parameters
						2	D	

a 1 = Catastrophic (definite injury), 2 = critical (risk of injury, tool destroyed, external damage), 3 = marginal (repair required, external damage), 4 = Negligible (joint creation unreliable)

b Of failures occurring: A = frequent (>1 in 5), B = reasonably probable (between 1 in 5 and 1 in 10), C = occasional (>1 in 100), D = Remote (between 1 in 100 and 1 in 1,000), E = extremely improbable (<1 in 1,000)

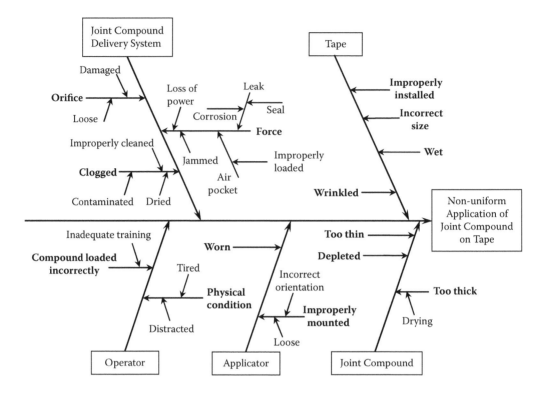

FIGURE 10.1 Cause-and-effect diagram for one function of a drywall taping system of Section 5.2.5.

Another tool that is used is called a cause-and-effect diagram, which resembles that created for FTA. This diagram classifies the various causes thought to affect the operation of a system, indicating with arrows the cause-and-effect relationship among them. The cause-and-effect diagram is sometimes called a *fishbone diagram*, an *Ishikawa diagram* (after the late Kaoru Ishikawa), or a *characteristics diagram*, which refers to its use in identifying the cause of degradation in various quality characteristics.

The diagram consists of the causes that can result in the undesirable effect, which is a particular problem resulting from one or more of the causes. That is, the causes are the factors that influence the stated effect. The diagram consists of arrows indicating the relationship between the effect and the causal factors. An example of a completed cause-and-effect diagram is shown in Figure 10.1, which was used to determine the possible causes of failure in one aspect of the operation of the drywall taping system given in Section 5.2.5. To a large degree, this diagram is independent of any specific embodiment. It was obtained as the result of a brainstorming session in which the IP^{2}D^2 team members came up with as many effects as they could think of that would cause nonuniform application of the joint compound onto the tape.

10.1.4 DESIGN FOR WEAR

An important aspect that can influence durability and reliability is wear, which occurs when two contacting parts have relative motion between them. Consequently, one may have to design for wear. There are several wear mechanisms. The first is adhesive wear, which is due to excessive loading between two surfaces. The second is abrasive wear, which is caused by particles that are harder than one or both of the rubbing surfaces. The third is lubricant wear, which is due to a change in the lubricant properties because of elevated temperature and/or the environment.

The following design guidelines have been developed to help implement design for wear.[*]

- Different classes of materials require different designs—attempting to adapt design rules for metals to plastics will not be correct.
- Minimize the presence of abrasive particles—use filters for the lubricant and shields and seals at the interfaces. If abrasives are present, make the wear surfaces harder than the abrasives.
- Use lubricants and dissimilar materials for the wear surfaces whenever possible, greatly reducing adhesive wear.
- Increasing wear surface hardness tends to reduce wear and prolong system life.
- Use rolling instead of sliding whenever possible—rolling requires that the slip between the rolling element and the contact surface be minimized.
- Avoid impacts in sliding and rolling contacts.
- Use sacrificial elements when available materials cannot achieve desired system life; choose materials such that only one of the wear members requires replacement.

10.2   POKA-YOKE

10.2.1   INTRODUCTION[†]

Poka-yoke is a technique for avoiding simple human errors at work.[‡] It was the Japanese manufacturing engineer Shigeo Shingo who developed the idea into a tool for achieving zero defects and eventually eliminating the need for quality control inspections. In effect, poka-yoke can be considered a technique of improving the reliability of a product. Shingo came up with the term *poka-yoke*, generally translated as "mistake-proofing" (to avoid [*yokeru*] inadvertent errors [*poka*]). The poka-yoke technique is applied during the embodiment stage, when the product's components are taking form and producibility issues are simultaneously being considered.

Many defects are caused by human errors. These errors and their safeguards are as follows.

- **Forgetfulness**—A required action is not performed.
  Safeguards: Alerting operator in advance or checking at regular intervals.

- **Misunderstanding**—A prohibited action is executed.
  Safeguards: Training, checking in advance, standardizing work practices.

- **Identification**—An error is made in selecting among alternatives.
  Safeguards: Training, attentiveness, vigilance.

- **Inexperience**—Information essential for an action is misinterpreted.
  Safeguards: Skill building, work standardization.

- **Inattentiveness**
  Safeguards: Discipline, work standardization, work instructions.

---

[*] R. G. Bayer, *Engineering Design for Wear*, Marcel Dekker, New York, 2004, Chapter 4.

[†] *Poka-Yoke: Improving Product Quality by Preventing Defects*, Edited by Nikkan Kogyo Shimbun, Ltd. Factory Magazine, Productivity Press, Portland, OR, 1988.

[‡] It is noted that intolerance for errors is now being addressed in the medical community. Medicare and several states and insurance companies will not pay hospitals the added cost of treating patients who were injured in their care if the injury was due to any one of ten reasonably preventable conditions appearing on a list that Medicare has created. See K. Sack, "Medicare Won't Pay for Medical Errors," *New York Times*, October 1, 2008: http://www.nytimes.com/2008/10/01/us/01mistakes.html?scp=7&sq=Kevin%20Sack&st=cse.

Although mistakes happen for many reasons, almost all of them can be prevented if one takes the time to identify when and why they happen. Then steps can be taken to prevent them by using poka-yoke methods and one or more of the safeguards listed above. One of the requirements of poka-yoke is that the safeguards employed to prevent defects should be inexpensive. Toyota uses 12 poka-yoke devices on the average at each manufacturing workstation and has a goal of limiting the cost of any one of the devices to less than $150.

If one breaks down daily activities in a factory, it will be found that there are five elements of production. These five elements are work instructions; acquisition of parts and materials; setting up parts and materials on machinery and equipment; establishing standard operating procedures; and workers implementing the standard operating procedures. These five elements determine whether a product is correctly manufactured or a defect is created. Defect-free products are assured by controls in each of these areas.

There are various types of defects. In order of importance, these defects can be as follows.

- Omitted processing.
- Processing errors.
- Errors in setting up work pieces.
- Missing parts.
- Wrong parts.
- Processing wrong work pieces.
- Incorrect operation.
- Adjustment error.
- Equipment not set up properly.
- Tools and jigs improperly prepared.

There is a strong correlation between the number of defects per unit and the complexity of the unit. It appears[*] that a good indicator of complexity is the product of the total number of operations required to produce a unit and the total time it takes to assemble it. The larger this product is, the larger the number of defects per unit irrespective of the functional aspects of the unit, that is, whether the unit is a computer disk storage device or an internal combustion engine.

10.2.2   THE BASIC FUNCTIONS OF POKA-YOKE

A defect exists in one of two states: it is either about to occur, or it has already occurred. Poka-yoke has three basic functions to use against defects—shutdown, control, and warning. Recognizing that a defect is about to occur is called *prediction*, and recognizing that a defect has already occurred is called *detection*. To detect defects, the poka-yoke system uses such devices and means as guide pins of different sizes, error detection and alarms, limit switches, counters, and checklists. The method involves the following actions.

- Analyzing the process for potential problems.
- Identifying items by their characteristics such as weight, dimension, and shape.
- Detecting deviations from procedures or omitted processes by examining the process sequences and the process-to-process sequences.
- Detecting deviation from fixed values by, for example, using a counter or by measuring a critical state of the artifact such as current or temperature.

---

[*] C. Martin Hinckley, *Make No Mistake: An Outcome-Based Approach to Mistake-Proofing*, Productivity Press, Portland, 2001.

Poka-yoke differs from statistical process control (SPC) methods in the following way: SPC essentially maintains the current level of defects, rather than seeking to eliminate them. In addition, the time it takes for the SPC method to provide feedback so that corrective action can be taken is slow when compared to the feedback obtained with poka-yoke. The poka-yoke technique essentially performs 100% inspection of certain attributes of each part at each critical manufacturing step. When an error is found, the feedback is immediate and the defect is prevented from occurring. In SPC, when an attribute is determined to be out of control, one usually has to determine its cause.

As indicated previously, there are several classes of devices that have proven useful in mistake-proofing assembly and manufacturing operations. The first set of devices uses mechanical objects such as guide pins of different sizes, blocks, and slots so that a part can only be oriented in a specific way before critical operations are performed. A second set of devices use electronic sensors such as limit switches, light, and proximity gauges to determine the presence or absence of a mistake. The third set of devices use special jigs and fixtures to detect upstream mistakes and to mistake-proof machine setup operations. Counters have been used to verify that the correct number of parts has been used or the correct number of operations has been employed. Numerous examples of how these devices can be used are available.[*]

Poka-yoke is also employed in numerous consumer products. It this case, the poka-yoke technique is used to prevent the user from incorrectly or unsafely using a product. Some common examples are[†]

- Electrical outlets for two-prong plugs, where each prong has a different shape so that it can mate in only one manner to ensure proper grounding.
- Cables that connect PCs and laptops to peripherals, each having a different shape so that, for example, the power cord will not fit into the printer port, and so on.
- Indoor garage entrances have a hanging bar to indicate the maximum height of the vehicle that can safely enter; this is effectively a go-no-go gage.
- Sinks and bath tubs have an additional drain near their top to prevent water from overflowing a fixture.
- Clothes dryers have a switch in their door that causes the drum to stop rotating when the door is opened.
- Lawnmowers have a lever on their push-bars that must be depressed for the mower to start and for the mower to remain on; this has dramatically reduced the number of self-inflicted finger and hand amputations.
- Cars that cannot be put in gear without the brake pedal being depressed.

10.3 DESIGN FOR MAINTAINABILITY (SERVICEABILITY)[‡,§,¶]

10.3.1 INTRODUCTION

Maintainability is composed of those elements of a product's design that assure that the product will perform satisfactorily throughout its intended life with a minimum expenditure of money and effort. In many cases, the best way to minimize maintenance is to increase the reliability of the

---

[*] C. Martin Hinckley, ibid.

[†] Many examples of poka-yoke applied to consumer products can be found at the following Web site: http://csob.berry.edu/faculty/jgrout/everyday.html.

[‡] J. G. Bralla, 1996, ibid.

[§] B. S. Blanchard, D. Verma, and E. L. Peterson, *Maintainability: A Key to Effective Serviceability and Maintenance Management,* John Wiley & Sons, New York, 1995.

[¶] M. A. Moss, *Designing for Minimal Maintenance Expense*, Marcel Dekker, New York, 1985.

components and the products. However, in most durable goods, some form of maintenance is required. An effective maintainability design minimizes the following.

- Down time for maintenance activities and maintenance-induced failures.
- User and technician time to accomplish maintenance tasks.
- Logistics requirements for parts, backup units, tools, and personnel.
- Equipment damage resulting from attempted maintenance.
- Personnel injury resulting from maintenance actions.
- Costs resulting from maintainability features and maintenance costs during operation of the product.

Experience indicates that those systems that have been designed for producibility are most likely to be maintainable. Each maintainability design feature is justified by its ability to support needed maintenance functions. These functions can serve one or more of the following three principal maintenance purposes.

- **Preventive:** To help keep the product in good condition.
- **Corrective:** To return a malfunctioning product to its usable condition.
- **Overhaul:** To return a product to its approximate original condition.

Maintainability features include the following: accesses; controls and displays; connectors, couplings, and lines; fasteners; handles; labels; appropriate mounting and positioning; and test and work points. The following are some guidelines for a design for maintainability:

- Provide easy and properly-sized access with common fasteners and attachments.
- Design for few serviceable items, each requiring simple procedures with minimum skills.
- Use modular construction.
- Place components/modules that need regular replacement or servicing in accessible locations.
- Require common hand tools and a minimum number of them.
- Provide built-in tests and indicators for maintenance diagnosis; provide accessible test points.
- Mistake-proof fasteners and connectors for easy reassembly.
- Require no, or minimal, adjustment; keep adjustable items easily accessible.
- Provide easy access to serviceable items; provide for their removal without having to remove unrelated components.
- Provide for visual inspection and for easy identification.
- Provide for easily replaced modules and parts; make them interchangeable and standard.
- Provide means to lift and manipulate heavy components.
- Provide safety interlocks, covers, guards, and switches.

10.3.2 STANDARDIZATION

Standardization is an important factor in product design, production, and life cycle costs. One of the areas in which standardization has important effects is in maintenance. The following suggestions illustrate some of the ways in which standardization impacts maintenance.

- Using a minimum variety of fastener sizes, threads, and head types minimizes supply problems, minimizes the number of tools required of maintenance technicians, and minimizes the likelihood that an improper tool will be used.
- Standardization provides cost-effective means of supplying parts and materials, especially in the field.

- Standardization can provide for interchangeability of components within and across equipment, thereby reducing supply and availability problems.
- Standardization provides consistency of nomenclature and labeling, which helps the technician to follow maintenance instructions and procedures.

10.4   DESIGN FOR PACKAGING

Packaging performs two distinct major functions. The first function is for it to provide protection during transit, and the second is for it to attract the consumer when placed in stores. To produce the best package for a product, the following considerations should be taken into account.[*]

General
- The size and physical characteristics of the product.
- The kind of package that will be most practical and economical.
- Whether the carton will be filled mechanically, by hand, or by a combination of both.
- How, when, where, and in what quantities the product will be sold.

Marketing
- The target group.
- The kind of construction, surface design, and finish that can be used to attract the attention of prospective purchasers.
- How competitive products are packaged.
- Whether or not there is a trademark that must appear along with the brand name on the package.
- The single advantage within the product that can be most effectively stressed in the illustration on the package to encourage its immediate purchase.
- The other features of the product that should be brought to the attention of the consumer.
- The style of package that will be most convenient for the ultimate consumer to examine and use.
- Whether an illustration will enhance the sales of the product or the visibility of the product is required.

Transit, Storage, and Display Environment
- The kind of material and construction that will offer the best protection against breakage, dirt, and deterioration.
- The kind of package that will meet the practical requirements of handling, storage, and display where the product will be sold.
- Whether the product requires a package with functional properties to inhibit rust, mold, insects, or other damage, or whether it requires waxing or other coating to make the package either moisture or grease proof.

10.4.1   Environmental Impact of Packaging

There is now a considerable effort being made by many major corporations to minimize the environmental impact of packaging materials in a cost-effective way, while still satisfying the considerations listed above.[†] Proctor and Gamble[‡] has introduced rigid tubes for its *Crest* toothpaste

---

[*] R. Bakerjian, ibid.

[†] It is mentioned that some biodegradable products, like plastics containing polylactic acid, cause difficulties when they are recycled and mixed with regular plastics, making them hard to reuse.

[‡] C. L. Deutsch, "Incredible Shrinking Packages," *New York Times*, May 12, 2007; http://www.nytimes.com/2007/05/12/business/12package.html?_r=1&oref=slogin.

so that the tube no longer has to be placed inside its own box. Owners of water bottled under the names Poland Springs and Deer Park have saved over 9,000,000 kg of paper in the last five years by simply using narrower labels on their bottles. Coke has recently designed its classic contour bottles to be lighter and more impact resistant. The bottle is smaller, but holds the same amount of soda. However, now they have to convince the customer that they are still getting the same amount of liquid. Many manufacturers of products that sell to Wal-Mart are being asked to get rid of excess packaging. Wal-Mart's goal is to become "packaging neutral" by the year 2025. This means that through recycling and reuse it will attempt to recover as much material that was used in the packaging that flows through its stores.

Wal-Mart and Costco have started to use a redesigned gallon (3.8 l) milk container so that the milk crates are gone.[*] Instead, the jugs are stacked with a cardboard sheet between layers and four such layers are shrink-wrapped. The initial estimates indicate that this kind of shipping has cut labor costs by half and water use by 60 to 70%. The water was used to wash the crates and typically used 100,000 gallons per day (380,000 l/day). More gallons of milk can be fit on a truck reducing the number of trips to each store to two a week, from five—resulting in a large fuel savings. Also, 224 gallons of milk can be stored in its coolers in the same space that used to hold 80.

HP also has embraced the principle of minimizing the impact of packaging on the environment and in doing so has arrived at the following guidelines to reduce packaging material while still protecting the product.[†]

- Eliminate the use of restricted materials such as lead, chromium, mercury, and cadmium in packaging, as well as ozone depleting materials.
- Design packaging for ease of disassembly by the end user.[‡]
- Maximize the use of postconsumer recycled content in packaging material.
- Use readily recyclable packaging materials such as paper and corrugated cardboard.
- Reduce packaging size and weight to improve transportation efficiency.

In addition, HP has stopped using polyvinyl chloride (PVC) in their new packaging designs and they are replacing molded polystyrene foam with molded recycled paper pulp. In all cases, HP determines its packaging design on the ability of different geographical regions to recycle certain materials, the total weight and size of the packaged product as it relates to the number of units that can be placed on a pallet, and the total costs including transportation and disposal.

10.5   DESIGN FOR THE ENVIRONMENT

The motivation for creating environmentally neutral products is coming from the recognition that sustainable economic growth can occur without necessarily increasing the amount of consumable resources. In order to create environmentally friendly products, a reevaluation of how the

---

[*] S. Rosenbloom, "Solution, or Mess? A Milk Jug for a Green Earth," *New York Times*, June 30, 2008.

[†] http://www.hp.com/hpinfo/globalcitizenship/environment/productdesign/packaging.html.

[‡] One type of packaging, called "clamshell" packaging, was originally introduced by electronics and toy manufacturers to attract shoppers to their products. The top surface of the package is clear plastic. In order to discourage shoplifting of the package's contents, the manufacturers decided to seal the hinges of the plastic with epoxy. The results have been the creation of an impregnable package that send about 6,000 people to the emergency room each year because of the various implements that have been used to open these packages. Several companies are now experimenting with "frustration-free" packaging that still provides security but can be straightforwardly opened. See B. Stone and M Richtel, "Packages you won't need a saw to open," *New York Times*, November 15, 2008.

environment should be considered within the product realization process has been taking place. In the emerging paradigm, the following three things are becoming clear.[*]

- In life-cycle design, the environment is being considered a customer.
- A product that negatively impacts the environment is, in a sense, a defective product.
- The product cost should reflect the product's total environmental impact.

In fact, businesses are realizing that reducing emissions, minimizing energy usage, using non-toxic processes and materials, and minimizing waste, scrap, and by-products improves a company's profitability.[†,‡]

A product's life is usually shortened for one or more of the following reasons.

- Technical obsolescence.
- Fashion obsolescence.
- Degraded performance and safety.
- Environmental or chemical degradation.
- Damage, either accidentally incurred or because of improper use.

To successfully reduce the environmental impact of a product's design, the IP^2D^2 team must evaluate its impact in terms of the materials selected and the processes used to manufacture its various components. With respect to materials, the IP^2D^2 team should explore the substitution of one material for another, provided that the functional requirements can be satisfied, and they should create designs and choose manufacturing processes that minimize material waste. Additionally, the IP^2D^2 team should consider both the environmental impact of the materials when the product is discarded and the environmental impact of producing the material. This includes the amount of energy used and the effects its manufacture has on the environment. Furthermore, with the proper production planning of the product and the use of just-in-time manufacturing methods, the product's manufacture can minimize potential waste if there is a sudden decrease in the sales of the product.

Proper selection of packaging materials (e.g., recyclable materials and biodegradable materials), the amount of material used, and the packaging of the product for the selected method of transportation can also minimize the environmental impact, the latter by permitting the carrier to transport the maximum amount of product possible for each load carried. Materials that are recyclable are[§]—

- **Metals:** iron, steel, copper, brass, aluminum, lead.
- **Thermoplastics:** polypropylene, ABS, polyethylene, nylon, acrylic, PVC, polycarbonate.
- **Other common materials:** nonlaminated glass, wood products, and paper, including cartons.

Materials that are not as economical to recycle are laminated materials, such as plastic and glass, plastic foam and vinyl, plastics, and metals and dissimilar materials; galvanized (zinc coated) steel; ceramics; thermosetting plastics such as phenolic and urea; and parts that are glued or riveted together.

---

[*] G. A. Keoleian et al., *Product Life Cycle Assessment to Reduce Health Risks and Environmental Impacts*, Noyes Publications, Park Ridge, NJ, 1994.

[†] "DuPont Adopts a Pollution-Free Industrial Culture," *Manufacturing News*, Monday, Vol. 3, No. 11, June 3, 1996.

[‡] "Japanese Throw Support Behind Zero Emissions," *Manufacturing News*, Monday, Vol. 3, No. 12, June 17, 1996.

[§] J. G. Bralla, 1999, ibid.

For the IP²D² team to be successful in designing for the environment, it is necessary that its members and its company understand their individual roles in the product's life cycle evaluation. Management must establish a corporate strategy toward the environment. Marketing and sales should provide the IP²D² with the customers' attitudes toward environmentally benign products. The legal department must be aware of and must interpret federal, state, and local statutes and ordinances and explain to the IP²D² team their role in the design, manufacture, and distribution of environmentally friendly products. The purchasing and procurement departments must know how to select suppliers who use manufacturing processes that minimally impact the environment.

The company's financial departments must know how to calculate the environmental costs and some of the less tangible costs associated with environmental considerations that enter the design and manufacture of a product. Some examples of these costs are monitoring equipment; operator training and protective equipment; reporting, inspections, and record keeping; and liability, such as fines, cleanup, injury, and damage. Some examples of intangible cost are consumer acceptance and loyalty, corporate image, and worker moral.

In addition to the suggestions given above, the following previously discussed guidelines are also environmentally sound design practices: the design guidelines given in Section 7.3.2 on the design for disassembly; many of the guidelines given in Section 10.3 on the design for maintainability; the HP guidelines for minimizing the impact of packaging on the environment given in Section 10.4; the JIT manufacturing method discussed in Section 1.2.2 and the use of poka-yoke discussed in Section 10.2 (because they minimize waste); and the net shape manufacturing methods discussed in Sections 9.2 to 9.4.

10.6 ERGONOMICS: USABILITY, HUMAN FACTORS, AND SAFETY[*,†,‡,§]

Ergonomics pertains to those aspects of the product that relate to human interfaces. Ergonomics are usually considered in the design process when it has been ascertained that one or more of the following issues are important.

- **The product's ease of use**

    The features of the product should effectively communicate their function, which results in a shorter time to learn the use of the product and the retention of operating skill over time. Ergonomic considerations help to minimize user dissatisfaction and discomfort and improve the chances of success in the marketplace.

- **The product's ease of maintenance**

    Maintenance is important in products that need service, both by the customer and the repairperson (e.g., an office copier). Ergonomic considerations often help to minimize poor product performance by the end user.

- **There are a large number of user interactions with the product**

    Ergonomic considerations help to minimize user discomfort, awkwardness, poor performance, and improper use.

- **There are safety and liability concerns**

    Ergonomic considerations help to minimize injury to the user and damage to the product.

[*] J. G. Bralla, 1999, ibid.
[†] B. M. Pulot and D. C. Alexander, Eds., *Industrial Ergonomics: Case Studies*, McGraw-Hill, New York, 1991.
[‡] C. H. Flurscheim, *Industrial Design in Engineering: A Marriage of Techniques*, Springer-Verlag, Berlin, 1983.
[§] K. T. Ulrich and S. D. Eppinger, *Product Design and Development*, 4th ed., McGraw-Hill, New York, 2008.

Regarding safety issues, ergonomics frequently play an important role.* Consider the child-resistant cap for medicine bottles that was introduced in 1972 because of actions taken by the U.S. Consumer Product Safety Commission (CPSC). It has been extremely successful in greatly reducing the number of poisonings of children. Unfortunately, many older people (>60 years of age) have difficulty opening them, and often do not replace the cap once it has been removed. It has been found, however, that about 20% of the childhood poisonings occurs at a grandparent's house. Because of this, the CPSC has required that a new type of cap be introduced. The present cap requires both strength and dexterity to remove it. One is required to both push down on the lid and, while pushing down, either turn it or line it up with an arrow, one on the lid and one on the bottle, and then push the cap off. The new caps require less strength and more coordination. One now lightly squeezes the opposite sides of the bottle with the fingers of one hand and turns the cap with the other hand.

Ergonomic design, coupled with creative engineering design, has been responsible for many safety-improved products. Some of these are (1) low kickback chain saws, which has greatly reduced severe face and head injuries; (2) ground fault current interrupters required on all hair dryers, which interrupts the flow of current and reduces it to a nonlethal level; and (3) the need to continually grasp a throttle control bar of a lawn mower in order for the engine to remain running, which has greatly reduced finger and hand amputations.

The goals of ergonomics are to create user-friendly products, which can be done by considering the following guidelines.

- Fit the product to the user's physical attributes and knowledge. Avoid awkward and extreme motion and forces from the user. Use standard conventions, arrangements, and systems.
- Simplify the tasks and the number of tasks that are required to operate the product. Make the controls and their functions obvious, and display the operating information clearly, visibly, and unambiguously.
- Anticipate human error, provide constraints to prevent incorrect actions by the user, and provide feedback to the user to indicate that a certain feature or operating mode has been selected.

Typically, ergonomic considerations take into account the end user's age, gender, reach and grasp, dexterity, strength, and vision.

Although ergonomic design of products is important, there are many ergonomic concerns that can be caused by the product's design that may affect the manufacturing system that produces the product. Some of the product characteristics that influence the ergonomics of the manufacturing environment are the following.

- Fragility and weight of the components.
- Size, quantity, and tightening torque of fasteners.
- Position of locating surfaces.
- Accessibility and clearances of components.
- Identification and differentiation of components.

Many of these characteristics are similar to those that concern the ease of assembly, which is discussed in Section 7.2.2, and design for maintainability, which is discussed in Section 10.3.

Usability does not mean that the product can't also be pleasurable to use.† Typically, the hierarchy of the needs of a product has four levels: the lowest level is concerned with safety and well-being; the second level is the product's functionality; the third level is the product's usability; and

---

* Ergonomics and safety issues in the workplace are monitored by the U.S. Department of Labor Occupational Safety and Health Administration (OSHA). They issue rules and regulations pertaining to several industries, which can be found at http://www.osha.gov/SLTC/index.html.
† W. S. Green and P. W. Jordan, Eds., *Pleasure with Products: Beyond Usability*, Taylor & Francis, London, 2002.

the fourth level is the pleasure one receives using the product. This pleasure can be derived from many sources.

- **Tactile**, which manifests itself by surface quality, softness, and grip of the product.
- **Prehensile**, which manifests itself by shape and size of the product.
- **Functional**, which manifests itself by how one uses, activates, and manipulates the product.
- **Thermal**, which manifests itself by the rate that heat is removed from or supplied to the product and the length of time that it retains its heat.
- **Acoustical**, which manifests itself by sound generated by the product when activated or for feedback.
- **Visual**, which manifests itself by the surface colors of the product and by its shape.

10.7 MATERIAL HANDLING

A material handling system is a system that moves, stores, and controls material. A systems view of material handling[*] is one of using the right method to provide the right amount of the right material at the right place at the right time, in the right sequence, in the right position, in the right condition, and at the right cost. The best material handling system is one in which virtually no material handling takes place, that is, one in which less material movement occurs, less storage is used, and less material control is required.

A good material handling system is one that has the following attributes.

- Is well planned.
- Combines handling with processing whenever possible.
- Uses mechanical handling whenever possible to minimize manual handling.
- Is safe.
- Provides protection for the material.
- Uses a minimum variation in the types of equipment.
- Maximizes the utilization of the equipment.
- Minimizes backtracking and transferring operations.
- Minimizes congestion and delay.
- Is economical.

The goal of greatly reducing all aspects of material handling is an integral part of JIT manufacturing, which was discussed in Section 1.2.2. In JIT, one produces what is needed only when it is required. JIT dramatically reduces the need for storage of raw materials, semifinished assemblies, and finished products. To an appreciable extent, however, the use of JIT is practical only for large manufacturers. The company must be able to predict relatively accurately its material requirements in advance so as to coordinate the activities of all its vendors. Material requirement-planning techniques can be used to reduce inventory and still meet a specific production schedule. However, manufacturers will always have some need to store material, partially finished products, and finished goods. In common usage, the term *storage* is generally associated with raw material and in-process goods, whereas warehousing refers to the storing of finished goods.

An important part of material handling is logistics, and a particular aspect of logistics is the distribution of the final product to the customer. It has been estimated that of the 300 billion dollars annual grocery sales in the United States an estimated 75 to 100 billion dollars worth of groceries are in the pipeline, tied up between the manufacturers' suppliers, the manufacturers, and the retailer. In other words, between one-quarter and one-third of the products are essentially in

---

[*] J. A. White, "Material Handling in Intelligent Manufacturing Systems," in *Design and Analysis of Integrated Manufacturing Systems*, W. D. Compton, Editor, National Academy Press, Washington, D.C., pp. 46–59, 1988.

inventory. To minimize this form of manufacturing inefficiency, it has been proposed that a manufacturer examine the requirements of each of its customers for all of its products using the following six characteristics.[*]

- Annual sales volume.
- Annual unit volume.
- Coordination requirements (simple, complex; i.e., feeding a JIT system).
- Order fulfillment requirements (<1 week, 1 to 3 weeks, etc.).
- Destination volume (low, high).
- Handling characteristics (cartons, pallets, etc.).

The analysis of the company's distribution system using these six characteristics can help determine how to best get the product into the hands of the customer with a minimum cost, delay, and inventory.

10.8   PRODUCT SAFETY, LIABILITY, AND DESIGN

The creation of safe products requires the elimination of injury-causing characteristics found from the operation and use of products. This usually involves knowledge of ergonomics, which, as mentioned in Section 10.6, deals with the interaction between human capabilities and product characteristics to determine the limits of safe, effective, and comfortable use. There are many types of injuries that can occur from both the proper and improper use of products. For example, mechanically-caused injuries can produce fractures, lacerations, amputations, and crushing. Electrically-caused injuries can shock and electrocute. Chemically-caused injuries are poisoning, both long term and immediate, and immunological reactions. As an examples of these types of injuries, it is noted that there have been cases where infants have been strangled by venetian blind cords, construction workers killed by their tools, and farmhands poisoned by pesticides. The U.S. Consumer Product Safety Commission maintains several databases[†] that give some indication of the types of injuries and their severity caused by consumer products. These data bases cover 15,000 types of consumer products and 4,000 product recalls. Associated with the occurrence of these types of injuries are the economic costs to the company who makes the product: restitution (voluntary and involuntary [loss of a lawsuit]), attorneys' fees, and insurance premiums. In addition, there can be a loss of reputation.

To effectively design a safe product the IP^2D^2 team has to analyze the product to determine the hazards associated with a product of its type, of its intended user, and of its use. In performing these analyses, with such tools as fault tree analysis discussed in Section 10.1.3, the IP^2D^2 team must anticipate the intended use as well as the unintended use.

Typically, the strategy to produce a safe product is to perform the following steps in the order given.

- Remove the hazard.
- Limit access to the hazard.
- Inform the user of the hazard using labeling and instruction manuals.
- Train the user to avoid the hazard.

When there is no satisfactory remedy to the elimination or minimization of the hazard, then the choice is to either learn to live with the hazard or to ban the product, which requires an action of the federal government.

---

[*] J. B. Fuller, J. O'Conner, and R. Rawlinson, "Tailored Logistics: The Next Advantage," *Harvard Business Review*, pp. 87–98, May–June 1993.

[†] http://www.cpsc.gov/cpscpub/prerel/prerel.html.

A warning label should accomplish three things:

- Get the user's attention.
- Describe the danger in vivid terms.
- Give specific instructions on how to avoid injury.

However, a study reported in the *New York Times*[*] of 3,500 articles about the effectiveness of warning labels had not found any reliable study that documented a reduction in accidents that resulted from warning labels. Despite this, there is an increased use of warning labels by manufacturers as many legal cases are being made around the idea that manufacturers have not sufficiently warned customers of the risks in using their products. On the other hand, at least one company has been successful in using labels to eliminate lawsuits resulting from the misuse of its product. In the early 1980s, spray cans containing household cleaners were being used by teenagers to get high, even though the label clearly warned of death or serious injury if the product was inhaled. The warning label was finally rewritten to state that inhalation could also cause hair loss and facial disfigurement. This false warning succeeded in discouraging the target group from using the product.

There are also products that are designed with safety as its primary function, especially for babies and toddlers, for example, heat-sensitive spoons that turn white when food is too hot, diaper-changing tables with seat belts, and nets that keep children in their cribs.

The hazardous characteristics of products and processes should be evaluated as part of the product development process. This evaluation may occur as part of a preliminary hazard analysis, a failure mode and effects analysis, or some other review of a product before its design is finalized. This evaluation is to determine the need for action to eliminate or control any hazards.

Hazardous characteristics of products and processes can be categorized as follows.[†]

- Acceleration.
- Chemical reactions.
- Electrical.
- Flammability and fires.
- Heat and temperature.
- Pressure.
- Vibration and noise.

Acceleration occurs when an object is set in motion or its speed is changed. This change, either an acceleration or deceleration, can impart additional forces that can cause damage.

Chemical reactions can be extremely rapid, such as an explosion or a fire, or very slow, such as corrosion, the effects of which may not be noticed until a failure occurs.

Several types of **electrical hazards** are listed below.

- Electrical shock.
- Ignition of combustible materials by electrical sparks and arcs or by elevated temperature of components due to electrical current.
- Heating and overheating of electrical equipment, which affects reliability and can cause the ignition of combustibles.
- Inadvertent activation of electrical equipment, such as when a piece of equipment starts up unexpectedly.

---

[*] J. M. Broder, "Warning: A Batman Cape Won't Help You Fly," *New York Times*, March 5, 1997.
[†] W. Hammer, *Occupational Safety Management and Engineering*, 4th ed., Prentice Hall, Englewood Cliffs, NJ, 1989.

- Failure of a device to operate when and as required.
- Electrical explosions in transformers and circuit breakers, and in batteries and capacitors if their polarity is reversed.

Fires require the presence of a fuel, an oxidizer, and an ignition source. Fuels come in the form of gases, liquids, and solids. Flaming combustion occurs in the gas state. Therefore, gaseous fuels are in a state where a flammable mixture can form and ignite. Liquid fuels must evaporate to a gas in order to burn. Solids vaporize and burn in a number of ways. Some solids sublime directly to a vapor; others melt to a liquid first and then evaporate to the gas state. Some solids may undergo a surface reaction known as smoldering.

Temperature and heat can cause damage through either high-temperature or low-temperature effects. Liquids may boil or freeze, causing undesirable stresses in a system. The mechanical properties of many solid materials are affected very strongly by temperature. Most materials tend to shrink and become more brittle at low temperature, and they tend to elongate and become more ductile at high temperature.

Pressure hazards are those related to the existence of a differential pressure across a boundary. Typically, pressure hazards are related to the magnitude of the differential forces of fluids or gases inside containers and the strength of the containers. A positive internal pressure differential can cause a rupture, whereas a vacuum may cause an implosion.

Excessive **vibration** levels can cause fatigue failures in structures, and can require shorter periods of operation of devices by personnel (e.g., a jack hammer). Excessive *noise* levels can cause temporary hearing loss or permanent hearing loss if one is exposed for long periods of time, can annoy, and can create dangerous environments by masking audible warnings.

10.8.1 PRODUCT LIABILITY LAW[*]

The goal of product liability laws is to help protect consumers from dangerous products by holding manufacturers, distributors, and retailers responsible for putting into the marketplace products that they knew or should have known were dangerous or defective. Product liability law is based on two principal legal theories: negligence and strict liability. Negligence theory places the burden of proof on consumers. If a defectively designed product caused an injury to a consumer, the consumer must prove that the manufacturer's conduct in designing the product was unreasonable. Reasonable conduct requires that reasonable people would foresee the consequences of their actions with the product and measure those consequences. Strict liability theory on the other hand disregards the conduct of the parties and instead focuses on the quality of the product that caused the injury. The consumer no longer has to prove the manufacturer's conduct in making or designing a product was unreasonable; it only has to be shown that the product itself is defective.

There are three types of product defects that incur liability in manufacturers and suppliers: design defects, manufacturing defects, and defects in marketing. Design defects are inherent; they exist before the product is manufactured. While the item might serve its purpose well, it can be unreasonably dangerous to use due to a design flaw. Manufacturing defects occur during the construction or production of the item. Only a few out of many products of the same type are flawed in this case. Defects in marketing deal with improper instructions and failures to warn consumers of latent dangers in the product.

To determine defect, the following seven factors are analyzed.

- The product's usefulness.
- The availability of safer products to meet the same need.
- The likelihood and probable seriousness of injury.

---

[*] J. D. Vargo, "Understanding Product Liability," *Mechanical Engineering*, October 1995, p. 46.

- The obviousness of the danger.
- The public expectation of the danger.
- The avoidance of injury by care in the use of the product, including the effect of instructions and warnings.
- The manufacturer's or seller's ability to eliminate the danger of the product without making it useless or unduly expensive.

The legal requirements of both negligence and strict liability are closely related to what most engineers consider to be good design. The major difference between these legal requirements and the design process used by engineers is that, in law, liability is assessed by looking backwards whereas in engineering, the design process looks forward. The IP^2D^2 team should, therefore, adapt the legal basis—hindsight—as a forward approach in product design. That is, the team should anticipate all the ways that the product can be improperly used and include in its design solutions to its potential liabilities.

The Internet may now have an important role in strict liability cases.[*] With the Internet, some products are "co-designed" with multiple user inputs. Still, a customer's involvement in the design of a product or system, whether in the consumer or industrial context, will not discharge the manufacturer of that product or system from responsibility for any injuries caused by its defects. In addition, the advent of blogs where consumers tell of their difficulties with products may make it harder for manufacturer's to claim, for example, that a certain defect was an isolated occurrence.

---

[*] S. L. Olson, "Net's Impact on Strict Product Liability Law," October 7, 2008: http://www.law.com/jsp/legaltechnology/pubArticleLT.jsp?id=1202425060704.

11 Product and Process Improvement

Several statistical experimental design methods are introduced as a means of reducing the variability of products and the processes that make them.

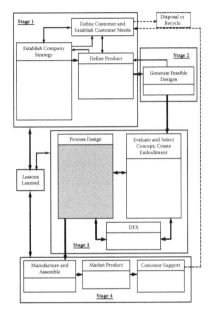

11.1 INTRODUCTION

Well-designed statistical experiments are powerful means to improve the performance of products and processes. When used correctly they can result in products that have fewer defects, reduced variability and closer conformance to target values, reduced development time, and reduced costs.

A general representation of a process is shown in Figure 11.1. A process has one or more inputs and one or more outputs. Affecting the outputs of the process are controllable and uncontrollable factors. When there is more than one output each output is considered independently. For example, consider a wave solder machine used to manufacture printed circuit boards. The input to the machine is the printed circuit board and its components, and its output is the printed circuit board with the components soldered to it. A defect occurs when the solder does not bond the component to the board. Some of the controllable variables are the solder temperature, solder flux type, solder depth, and conveyor speed. Some of the uncontrollable variables are the thickness of the board, layout of the components on the board, operator skill, and type of components used. The purpose of a designed experiment in this case would be to substantially reduce the number of defects by determining the settings (values) for the controllable variables, while at the same time making the process insensitive to the uncontrollable variables.

How one goes about designing such an experiment to understand and predict the output of the process is presented in this chapter. In particular, methods will be introduced that can be used to determine, with a certain degree of confidence, the process' output when variations occur in the controllable and uncontrollable variables. These methods are typically used when either there is no physics-based model of the process or such a model is too difficult to obtain.

There are several implicit assumptions made when discussing designed experiments. The words "designed experiment" mean that the system's parameters are under the control of the investigator, and that the investigator will determine what parameters will be varied, what quantities will be measured, and how the results will be analyzed. In the present context, it does not mean the details of how one is to physically construct the product or process to be tested, the selection of the transducers and sensors, and the test protocol, although these must exist.

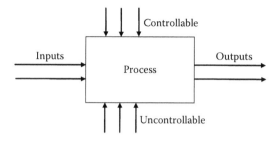

FIGURE 11.1 General representation of a process.

11.2 WHAT IS EXPERIMENTAL DESIGN?

Experiments are performed to discover something about a particular process or system. A designed experiment is a test, or a series of tests, in which purposeful changes are made to the inputs of a process or system so that one may observe and identify the reasons for changes in the outputs. The main reasons for designing an experiment are to obtain unambiguous results at a minimum cost, to learn about interactions among the variables, and to measure the experimental error in order to have an indication of the confidence that one can place in the conclusions. Designed experiments usually fall in one of four categories: (1) comparative and evaluative—make a statement about the effectiveness of a change in a process variable; (2) screening—determine the most important factors from a large number of factors that affect the output; (3) optimization—determine the settings for the various factors that either maximize or minimize some attribute of the output; and (4) regression—obtain a mathematical relationship among the factors and the output.

Before embarking on a test program the following questions should be answered:

- How do you know you have a problem?
- How do you know it affects the customer?
- Can the problem be corrected elsewhere?
- What happens if you do nothing?
- What benefit will your customer get?
- What is your competitor doing?

After these questions have been answered, the next question is, what is the goal of the planned experiment? The objectives of an experimental design usually include the following:

- Determining which variables are most influential on the output.
- Determining where to set values of the influential input variables so that—
  — Outputs are almost always near their desired values.
  — Variability of the output is small or minimal.
  — Effects on the output from changes in any uncontrollable variables are minimized.

Some facts about designed experiments are

- Not all experimental designs are equally good.
- Well-designed experiments enhance quality and ultimately save money.
- A well-planned experiment results in more information about more variables (factors) in fewer runs than either unplanned or one-factor-at-a-time experiments.
- To know about a process one must actively interfere with it.
- As the cost of experimentation increases, the more "developed" the process is.

We now introduce several terms that are associated with the design of experiments.

Response variable is a variable observed or measured in an experiment. An experiment will have one or more response variables. Regarding a response variable, one has to answer the following questions:

- Does this response variable directly affect the product's or process' performance?
- Is the response variable to be minimized, maximized, or targeted at a certain value?
- Does the attainment of the response variable affect other response variables? If so, are trade-offs required?

Factor is a quantity (variable) that is deliberately varied in order to observe its impact on the response variable (output). It is sometimes called a *primary variable*. All variables that could influence the response variable should be identified, and should be further identified as either primary or extraneous (see below), whether they are controllable or uncontrollable, and whether they are discrete or continuous. Such variables may be quantitative, such as temperature and pressure, or they may be qualitative, such as a preparation method or a batch of material. When using quantitative factors their units of measurements must be known, and an estimate of the accuracy of the settings (readings) must also be determined. Quantitative controllable variables are frequently related to the performance variable by some assumed statistical relationship or model. For example, if a linear relationship can be assumed, two levels (or conditions) may be sufficient (see below for the definition of level); for a quadratic relationship a minimum of three levels is required. Thus, the minimum number of conditions or levels per variable is determined by the form of the assumed model. The assumptions for the model also affect the range for the selected quantitative controllable variables. Assuming a linear relationship between the variables and the output, the wider the range of conditions or settings, the better are one's chances of detecting the effects of the variable. However, the wider the range, the less reasonable the assumption of a linear relationship may be. Also, one often would not want to go beyond the range of physically or practically useful conditions. However, the range chosen should include the current settings, if applicable. The selection of the range of the variables depends, in part, upon whether the ultimate purpose of the experiment is to learn about performance over a broad region or whether one is searching for an optimum condition; a wider range of experimentation would be more appropriate in the first case than in the second.

When there are two or more variables, there is the possibility that they might interact with one another; that is, the effect upon the response of one variable depends upon the condition of the other variable. This interaction is due to the process, not to the factors interacting with themselves; that is, the factors are independent of each other. Figure 11.2a shows a situation where two noninteracting variables, A and B, independently affect the output; that is, the effect of factor A on the output is the same for both levels of B, denoted B_1 and B_2. In contrast, Figure 11.2b shows an example of

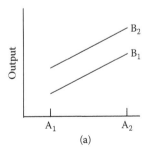
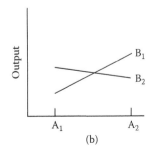

FIGURE 11.2 (a) Noninteracting factors, (b) interacting factors.

an interaction between A and B; that is, an increase in A increases the output for B_1, but decreases it for B_2. In some experiments, there may be one or more variables that one wishes to hold constant. Holding a variable constant limits the size and complexity of the experiment, but can also limit the scope of the resulting inferences.

QFD and the selection of factors: When one generates a QFD house of quality as described in Section 4.2.2, the characteristics that are important to the customer have been identified and have been related to the relevant engineering characteristics for each functional grouping or module. Recall, for example, the houses of quality in Figures 4.2 and 4.3. In addition, these engineering characteristics are measurable quantities, with each having a stated value as its goal—region 6 in the house of quality of Figure 4.1. Consequently, if one finds it necessary to design an experiment to optimize, in some sense, one or more modules, then these engineering characteristics are the factors to be used. If all of these factors can not be used, then the most important factors are used as indicated by the ranking of their absolute importance within each functional grouping—region 11 in the house of quality of Figure 4.1.

Level is the value or setting of a factor.

Extraneous variable is a variable that potentially can affect a response variable, but is not of interest as a factor. It is also called a noise, nuisance, or blocking variable (see below for the definition of blocking). Regarding an extraneous variable one should realize the following:

- In a well-designed experiment, one includes the extraneous variables and allows them to assume their natural range of variability.
- An effort should be made to identify the extraneous variables and their levels.
- In some experiments, the extraneous variables can be neutralized by either randomization or blocking.
- The extraneous variables may interact with the primary variables.
- When all extraneous variables are considered in the experiment, the number of observations may increase to an unmanageable amount.

Replication is the number of times that the basic experiment is repeated.

Randomization means that both the allocation of experimental materials and the order in which the individual runs of the experiment are performed are randomly determined. When planning an experimental design, the ease with which an experiment can be randomized should be determined. Randomization is carried out using random number tables. The advantages of randomization are that it neutralizes the effect of extraneous variables (influences), removes biases from the experiment (due to both the process and the experimenter), provides a more realistic estimate of the variance, and assures the validity of the probability statements of the resulting statistical tests. There are no disadvantages to randomization.

Blocking is a technique used to increase the sensitivity of the experiment by making a portion of the experimental data more homogeneous than the entire set of data. Blocking is implemented by fixing a level of an extraneous variable, and then running the various levels of a factor within each block. Its advantage is that it allows the effect of the extraneous variable to be controlled, estimated, and removed, thus gaining an increased sensitivity as to whether or not the factors are statistically significant.

As an example of blocking, consider an experiment to compare the wear of different types of automobile tires. Tire wear may vary from one automobile to the next, irrespective of the tire type, due to differences between automobiles, variability among drivers, and so forth. Suppose that one wishes to compare four tire types (A, B, C, and D), and four automobiles are available for the comparison. A poor way of doing this would be to use the same type of tire on each of the four wheels of an automobile and vary the tire type between automobiles, that is,

	Automobile		
1	2	3	4
A	B	C	D
A	B	C	D
A	B	C	D
A	B	C	D

Such an assignment would be undesirable because in the subsequent analysis the differences among tire types cannot be separated from the differences among automobiles. Separation of these effects can be obtained by treating automobiles as experimental blocks and assigning tires of each of the four types to each automobile as follows:

	Automobile		
1	2	3	4
A	A	A	A
B	B	B	B
C	C	C	C
D	D	D	D

When the symmetry exhibited in this example is not possible, so-called incomplete block designs are used.

In some situations, there may be more than one background variable whose possible contaminating effect must also be removed by blocking. In the automobile tire comparison, one may be concerned that the differences between wheel positions also affects the results, in addition to differences between automobiles. In this case, wheel position might be introduced into the experiment as a second blocking variable. If there are four tire types to be compared, this might, for example, be done by assigning the tires of each of the four types according to the following plan (known as a Latin square design):

Wheel Position	Automobile			
	1	2	3	4
1	A	D	C	B
2	B	A	D	C
3	C	B	A	D
4	D	C	B	A

Notice that each tire is in a different wheel position in each automobile.

11.3 GUIDELINES FOR DESIGNING EXPERIMENTS

The following guidelines are suggested for planning and conducting a set of statistical experiments:

- Generate a clear statement of the problem and a well-defined set of objectives. A well-planned experiment is often tailored to meet specific objectives and to satisfy practical constraints.
- Select the appropriate response variables (outputs). Know as much as you can about the process and the response variables.
- Choose the factors to be varied and their levels. Use screening experiments to determine the most important factors. (See Section 11.4.7.) Define the primary factors and the ranges (levels) that will meet the objectives. Unreasonable and dangerous conditions should be excluded from the experimental design. If there are any previously recorded benchmarks of performance, then include the benchmark conditions in the experiment, if possible, and use them to check the results. Know the process' nuisance and ambient factors and all, or suspected, interactions.
- Choose the appropriate experimental design, sample size, order of runs, and blocking, if necessary. Select the degree of precision (repeatability) of the final results, remembering that the greater the desired precision, the larger will be the required number of experimental runs. Select a model for the results; a more complex model usually requires more runs. Therefore, one must estimate how long it will take and how much it will cost to perform the complete experiment. Use randomization, replication, and blocking by blocking what you can and randomizing what you cannot. Realize that designing an experiment is often an iterative process, requiring rework as new information and preliminary data become available.
- Perform the experiment, which should be as simple as possible. Use a pilot study, if possible, and use sequential, or staged, experimental designs. A stage-wise approach can be used when units are made either in groups or one at a time and when searching for an optimum response, because it might allow one to move closer to the optimum from stage to stage. A single experiment may be desirable when there are either large startup costs at each stage or if there is a long waiting time between starting the experiment and measuring the performance, as when measuring product life. During the performance of an experiment, record the time sequence of events and the environmental factors.
- Analyze the data using the appropriate statistical techniques and display them graphically. Every model has assumptions; know them and test them (linearity, interactions, etc.).
- Draw conclusions and make recommendations. If an experiment has been properly planned, then the conclusions will be unambiguous and valid and provide insight into how things can be improved. Lastly, know the difference between practical significance and statistical significance.

11.3.1 DESIGNED EXPERIMENTS AND STATISTICAL PROCESS CONTROL

As previously mentioned, the role of designed experiments is to actively interfere with a process to gain knowledge about which controllable and extraneous variables affect the system's output and by how much. The results of such an experiment yield a means whereby a process can be controlled so that its outputs are predictable and their variability known. Statistical process control (SPC), on the other hand, is a passive statistical method to monitor a process either to verify that it is performing as desired, to provide information about the capability (variance) of the process, or to distinguish between random fluctuations in the process and abnormal variations. While SPC activities reduce the chance of shipping defective products, they do not decrease the occurrence of defective products. Using SPC without having a predictive model of the process is risky. Consider the situation when the process goes out of control and the usual remedies fail to bring it back in control. At this point, one is most likely to have to perform a designed experiment, one that should have been performed in the first place. In other words, SPC is not a diagnostic tool, only a means of identifying statistically meaningful changes in an attribute of the output of a process. Furthermore, if the designed experiment was performed properly, then the need for SPC to monitor the output is greatly reduced since the process' operating parameters could be selected to minimize the variability. However, in many manufacturing situations, it is still necessary to record statistically meaningful properties of the process, such as to meet ISO 9000 requirements.

11.4   FACTORIAL ANALYSIS

11.4.1   ANALYSIS OF VARIANCE (ANOVA)

If a population contains N elements, and n of them are selected, and if each of the $N!/[(N - n)!n!]$ possible samples has an equal probability of being selected, then the procedure is called *random sampling*. Suppose $x_1, x_2,..., x_n$ represent n sample observations, then the sample mean is

$$\bar{x} = \frac{1}{n} \sum_{i=1}^{n} x_i$$

and the sample variance is

$$s^2 = \frac{SS}{n - 1}$$

where

$$SS = \sum_{j=1}^{n} (x_i - \bar{x})^2 = \sum_{i=1}^{n} x_i^2 - n\bar{x}^2$$

is called the sum of squares. The sample's standard deviation is s. The quantity $n - 1$ is called the number of degrees of freedom and is equal to the number of independent samples in the computation of the variance.

Consider two random samples from two independent populations, one with a variance s_1^2 obtained from n_1 samples and the other with a variance s_2^2 obtained from n_2 samples. What we would like to know is how to determine if there is any statistically significant difference between s_1^2 and s_2^2. The phrase *statistically significant* refers to the ability to say whether any difference between s_1^2 and s_2^2 has to do with something that is different in how the samples themselves were created, or is different because of what naturally (randomly) occurs. To determine the statistical significance, we form the following ratio, called a *test statistic*,[*]

$$F_o = \frac{s_1^2}{s_2^2}$$

The measure of what is expected to occur naturally for this type of ratio is called the *f*-statistic, and its values are given in tables called *f*-tables. The *f*-statistic obtained from the *f*-tables is denoted f_{α,n_1,n_2}, where $(1 - \alpha) \times 100\%$ is the confidence level. Thus, if $\alpha = 0.05$, then the confidence level

---

[*] See, for example, G. E. P. Box, W. G. Hunter, and J. S. Hunter, *Statistics for Experimenters*, John Wiley & Sons, New York, 1978, for the formal conditions and procedures under which the ratio is valid.

is 95%, which means that there is a 5% chance that the conclusion reached after comparing F_o with f_{α,n_1,n_2}, is incorrect. The differences in the variances are not statistically significant when

$$F_o < f_{\alpha,n_1,n_2}$$

and the variances *are* statistically significant when

$$F_o > f_{\alpha,n_1,n_2}$$

at the stated confidence level.

The benefit of determining the statistical significance of a factor is that it prevents one from explaining or responding to results that are simply random effects.

11.4.2 SINGLE-FACTOR EXPERIMENT

Consider a single-factor experiment with the factor denoted A. We run an experiment varying A at a different levels, A_j, $j = 1, 2,..., a$, and we repeat the experiment n times, that is, obtain n replicates. The results are given in Table 11.1.

The results in the first column of the observations, x_{j1}, in Table 11.1 would be obtained by randomly ordering the levels A_j, $j = 1, 2,..., a$, and then running the experiment in this randomly selected order. Then the results in the second column of the observations, x_{j2}, $j = 1, 2,..., a$, would be obtained by generating a new random order for the levels A_j and running the experiment in this new random order. This procedure is repeated until the n replicates have been obtained. In running the experiment in this manner, one ensures that the values obtained for the x_{jk} have each been independently obtained. Thus, we can define two independent variances, using the quantities μ_i and s_i^2 defined in Table 11.1, as follows. The variance of the mean of factor A is

$$s_A^2 = \frac{n}{a-1}\sum_{i=1}^{a}(\mu_i - \bar{x})^2 = \frac{n}{a-1}\left(\sum_{i=1}^{a}\mu_i^2 - a\bar{x}^2\right) = \frac{SS_A}{a-1}$$

TABLE 11.1
Tabulations of the Results of a Single Factor Experiment with $n > 1$ Replicates

Level	Observations				Average	Variance
A_1	x_{11}	x_{12}	...	x_{1n}	$\mu_1 = \frac{1}{n}\sum_{j=1}^{n}x_{1j}$	$s_1^2 = \frac{1}{n-1}\sum_{j=1}^{n}(x_{1j} - \mu_1)^2$
A_2	x_{21}	x_{22}		x_{2n}	$\mu_2 = \frac{1}{n}\sum_{j=1}^{n}x_{2j}$	$s_2^2 = \frac{1}{n-1}\sum_{j=1}^{n}(x_{2j} - \mu_2)^2$
...			...	...	...	
A_a	x_{a1}	x_{a2}	...	x_{an}	$\mu_a = \frac{1}{n}\sum_{j=1}^{n}x_{aj}$	$s_a^2 = \frac{1}{n-1}\sum_{j=1}^{n}(x_{aj} - \mu_a)^2$

which has $a - 1$ degrees of freedom, and $\bar{x}$ is the grand mean given by

$$\bar{x} = \frac{1}{a}\sum_{i=1}^{a}\mu_i = \frac{1}{an}\sum_{i=1}^{a}\sum_{j=1}^{n}x_{ij}$$

The variance of the error is

$$s_{error}^2 = \frac{1}{a}\sum_{i=1}^{a}s_i^2 = \frac{1}{a(n-1)}\sum_{i=1}^{a}\sum_{j=1}^{n}(x_{ij}-\mu_i)^2 = \frac{1}{a(n-1)}\left(\sum_{i=1}^{a}\sum_{j=1}^{n}\mu_i^2 - an\bar{x}^2\right) = \frac{SS_{error}}{a(n-1)}$$

which has $a(n - 1)$ degrees of freedom.

These two variances, s_A^2 and s_{error}^2, are related by the following identity, called the sum of squares identity, which has $an - 1$ degrees of freedom:

$$SS_{total} = \sum_{i=1}^{a}\sum_{j=1}^{n}(x_{ij}-\bar{x})^2 = \sum_{i=1}^{a}\sum_{j=1}^{n}[(\mu_i-\bar{x})+(x_{ij}-\mu_i)]^2$$

The left hand side of the equation is called the total sum of squares. When the right hand side is expanded, it will be found that the cross product terms sum to zero, and we obtain

$$SS_{total} = \sum_{i=1}^{a}\sum_{j=1}^{n}(\mu_i-\bar{x})^2 + \sum_{i=1}^{a}\sum_{j=1}^{n}(x_{ij}-\mu_i)^2$$

$$= n\sum_{i=1}^{a}(\mu_i-\bar{x})^2 + \sum_{i=1}^{a}\sum_{j=1}^{n}(x_{ij}-\mu_i)^2$$

$$= SS_A + SS_{error}$$

$$SS_{total} = (a-1)s_A^2 + a(n-1)s_{error}^2$$

Thus, the identity has partitioned the total variance into two independent components: that due to the factor A and that due to the natural variation in the process.

In the analysis of variance, the convention is to define a quantity called the mean square, denoted MS, which is the sum of squares divided by the number of degrees of freedom. Thus, for the single factor experiment, we have that

$$MS_A = \frac{SS_A}{(a-1)} = s_A^2 \qquad a > 1$$

$$MS_{error} = \frac{SS_{error}}{a(n-1)} = s_{error}^2 \qquad n > 1$$

The objective of the experiment is to determine whether or not the various levels of A have any effect on the output x_{ij}. We now have the ability to determine this by forming the ratio of the mean square of the factor A with the independent mean square of the random error. This tells us whether or not the variance of A is a statistically significantly large portion of the total variance. Thus, the test statistic is

$$F_o = \frac{MS_A}{MS_{error}}$$

TABLE 11.2
ANOVA Table for a Single Factor Experiment with $n > 1$ Replicates

Factor	Sum of Squares	Degrees of Freedom	Mean Square	F_o	$f_{\alpha, a-1, a(n-1)}$
A	SS_A	$a-1$	MS_A	MS_A/MS_{error}	(From f-table)
Error	SS_{error}	$a(n-1)$	MS_{error}		
Total	SS_{total}	$an-1$			

and the analysis of variance test is

$$F_o < f_{\alpha, a-1, a(n-1)} \quad \text{or} \quad F_o > f_{\alpha, a-1, a(n-1)}$$

When the ratio F_o is sufficiently large (insufficiently large) at the $(1 - \alpha)100\%$ confidence level, then the values chosen for A cause (do not cause) a statistically significant change in the output. The results are usually presented in the form shown in Table 11.2.

11.4.3 FACTORIAL EXPERIMENTS

The results for a single factor experiment can be extended to experiments with several factors, which are called factorial experiments because the procedure requires that we run all combinations of all the levels of each factor for each replicate of the experiment. We will illustrate this for a two-factor experiment, which has the factor A at a levels and the factor B at b levels. The number of replicates is $n > 1$ and the output is x_{ijk}, where $i = 1, 2, \ldots, a$, $j = 1, 2, \ldots, b$, and $k = 1, 2, \ldots, n$. The intervals between each level of each factor do not have to be equal.

The starting point is the sum of squares identity. However, before proceeding with this identity, we introduce the following definitions for several different means

$$\bar{x}_{ijn} = \frac{1}{n} \sum_{k=1}^{n} x_{ijk}$$

$$\bar{x}_{ibn} = \frac{1}{b} \sum_{j=1}^{b} \bar{x}_{ijn} = \frac{1}{bn} \sum_{j=1}^{b} \sum_{k=1}^{n} x_{ijk}$$

$$\bar{x}_{ajn} = \frac{1}{a} \sum_{i=1}^{a} \bar{x}_{ijn} = \frac{1}{an} \sum_{i=1}^{a} \sum_{k=1}^{n} x_{ijk}$$

and the grand mean

$$\bar{x} = \frac{1}{abn} \sum_{i=1}^{a} \sum_{j=1}^{b} \sum_{k=1}^{n} x_{ijk}$$

The $\bar{x}_{ijn}$ are used in plotting the average output as a function of the various factors as shown in Figure 11.3.

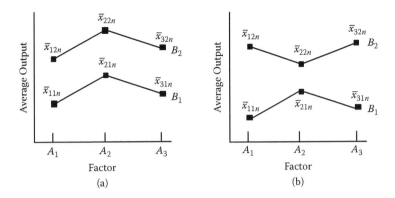

FIGURE 11.3 (a) Noninteracting nonlinear response; (b) interacting nonlinear response.

The sum of squares identity for a two factor analysis of variance is

$$SS_{total} = \sum_{i=1}^{a}\sum_{j=1}^{b}\sum_{k=1}^{c}(x_{ijk} - \bar{x})^2$$

$$= \sum_{i=1}^{a}\sum_{j=1}^{b}\sum_{k=1}^{n}\left[(\bar{x}_{ibn} - \bar{x}) + (\bar{x}_{ajn} - \bar{x}) + (x_{ijn} - \bar{x}_{ibn} - \bar{x}_{ajn} + \bar{x}) + (x_{ijk} - \bar{x}_{ijn})\right]^2$$

$$= SS_A + SS_B + SS_{AB} + SS_{error}$$

where

$$SS_A = \sum_{i=1}^{a}\sum_{j=1}^{b}\sum_{k=1}^{n}(\bar{x}_{ibn} - \bar{x})^2 = bn\sum_{i=1}^{a}\bar{x}_{ibn}^2 - abn\bar{x}^2$$

$$SS_B = \sum_{i=1}^{a}\sum_{j=1}^{b}\sum_{k=1}^{n}(\bar{x}_{ajn} - \bar{x})^2 = an\sum_{j=1}^{b}\bar{x}_{ajn}^2 - abn\bar{x}^2$$

$$SS_{AB} = n\sum_{i=1}^{a}\sum_{j=1}^{b}(\bar{x}_{ijn} - \bar{x}_{ibn} - \bar{x}_{ajn} + \bar{x})^2$$

and

$$SS_{error} = \sum_{i=1}^{a}\sum_{j=1}^{b}\sum_{k=1}^{c}(x_{ijk} - \bar{x}_{ijn})^2 = \sum_{i=1}^{a}\sum_{j=1}^{b}\sum_{k=1}^{c}x_{ijk}^2 - n\sum_{i=1}^{a}\sum_{j=1}^{b}\bar{x}_{ijn}^2$$

The quantities SS_A, SS_B, SS_{AB}, SS_{error}, and SS_{total} have $(a-1)$, $(b-1)$, $(a-1)(b-1)$, $ab(n-1)$, and $abn-1$ degrees of freedom, respectively. The sum of squares term SS_{AB} indicates the interaction of factors A and B. The ANOVA table for a two-factor experiment is given in Table 11.3.

It is seen that the analysis of variance isolates the interaction effects of the two factors, and provides a means of ascertaining, through the ratio MS_{AB}/MS_{error}, whether or not the interaction of the factors is statistically significant at a stated confidence level. This interaction effect is different from any nonlinear effects that may occur. To illustrate this, consider the two sets of curves shown in Figure 11.3. Figure 11.3a shows the case when the two factors do not interact, but exhibit a nonlinear response. Figure 11.3b shows a case where the two factors interact and exhibit a nonlinear response. It should be realized that the curves in Figure 11.3b do not mean very much unless the analysis of variance indicates that the interaction term is statistically significant. If the term proves statistically significant, then, when this series of tests is repeated, similar looking curves would be obtained. If they are not statistically significant, then these shapes and magnitudes may or may not be similar.

11.4.4 FACTORIAL EXPERIMENTS WITH ONE REPLICATE

There are instances when it is necessary to run an experiment with only a single replicate. In this situation, it is still possible to analyze the results, but with one important change. This change is brought about by the fact that a single replicate experiment does not permit one to estimate the mean square error, because the mean square value of the highest order interaction, that characterized by MS_{AB} in the two factor experiment, MS_{ABC} in a three factor experiment, and so on, remains part of the mean square error; that is, it cannot be separated from it. This inability to separate one or more mean square terms from other mean square terms is called *confounding*. It occurs when there is an insufficient number of degrees-of-freedom to calculate the various variances. Thus, when a one replicate factorial experiment is run, it is usually assumed that the highest order interaction mean square value is small compared to the mean square error. The computations for the two factor experiment simplify somewhat to

$$SS_{total} = SS_A + SS_B + SS_{error}$$

where SS_{error} now includes the includes SS_{AB} and has $(a-1)(b-1)$ degrees of freedom. In addition, SS_A and SS_B have $(a-1)$ and $(b-1)$ degrees of freedom, respectively, and SS_{total} has $ab-1$ degrees of freedom. The sum of squares quantities are calculated from the following relations:

$$SS_{error} = SS_{total} - SS_A - SS_B$$

where

$$SS_A = \sum_{i=1}^{a}\sum_{j=1}^{b}(\bar{x}_{ib} - \bar{x})^2 = b\sum_{i=1}^{a}\bar{x}_{ib}^2 - ab\bar{x}^2$$

$$SS_B = \sum_{i=1}^{a}\sum_{j=1}^{b}(\bar{x}_{aj} - \bar{x})^2 = a\sum_{j=1}^{b}\bar{x}_{aj}^2 - ab\bar{x}^2$$

$$SS_{total} = \sum_{i=1}^{a}\sum_{j=1}^{b}(x_{ij} - \bar{x})^2 = \sum_{i=1}^{a}\sum_{j=1}^{b}x_{ij} - ab\bar{x}^2$$

and

$$\bar{x}_{ib} = \frac{1}{b}\sum_{j=1}^{b}x_{ij} \qquad \bar{x}_{aj} = \frac{1}{a}\sum_{i=1}^{a}x_{ij} \qquad \bar{x} = \frac{1}{ab}\sum_{i=1}^{a}\sum_{j=1}^{b}x_{ij}$$

For the two factor experiment, the form of the ANOVA table is still that given by Table 11.3, except that the degrees of freedom change to those cited above and the row containing the factor AB is deleted.

TABLE 11.3
ANOVA Table for a Two Factor Experiment with $n > 1$ Replicates

Factor	Sum of Squares	Degrees of Freedom	Mean Square	F_o	$f_{\alpha, z, ab(n-1)}$
A	SS_A	$a - 1$	$MS_A = SS_A/(a-1)$	MS_A/MS_{error}	Value from f-table, $z = a - 1$
B	SS_B	$b - 1$	$MS_B = SS_B/(b-1)$	MS_B/MS_{error}	Value from f-table, $z = b - 1$
AB	SS_{AB}	$(a-1)(b-1)$	$MS_{AB} = SS_{AB}/(a-1)(b-1)$	MS_{AB}/MS_{error}	Value from f-table, $z = (a-1)(b-1))$
Error	SS_{error}	$ab(n-1)$	$MS_{error} = SS_{error}/ab(n-1)$		
Total	SS_{total}	$abn - 1$			

Although a full factorial design with replication is a near-ideal one and should be chosen if practical, the factorial design with one replicate is, in many cases, just as effective.

11.4.5 2^k FACTORIAL ANALYSIS

If the factorial experiments described in the previous section contain k factors, and each factor is considered at only two levels, then the experiment is called a 2^k factorial design. It implicitly assumes that there is a linear relationship between the two levels of each factor. This assumption leads to certain simplifications in how the tests are conducted and how the results are analyzed.

The convention is to denote the value of the high level of a factor with either "H" or "+," and the value of the low level "L" or "−". Then the 2^k combination of factors that comprise one run, which represents one replicate, is given in Table 11.4 for $k = 2, 3$, and 4. The table is used as follows. For the 2^2 ($k = 2$) factorial experiment, only the columns labeled A and B and the first four rows ($m = 1,\dots, 4$) are used. The four combinations of the factors are run in a random order. One such random order is shown in the column labeled 2^2. Thus, the combination in row 2 is run first, with A high (A_2) and B low (B_1). This yields the output value $y_{2,1}$. Then the combination shown in the fourth row is run, where both A and B are at their high levels (A_2 and B_2, respectively). This gives the output response $y_{4,1}$. After the remaining combinations have been run, one replicate of the experiment has been completed. A newly obtained random order for the run is obtained, one that is most likely different from the one shown in the column labeled 2^2, and the four combinations are run in the new order to get the output response for the second replicate. For $k = 3$ the factors are A, B, and C, and the first eight rows of the table are used; for $k = 4$ the factors are A, B, C, and D, and all 16 rows of the table are used. One set of a random run order is given for each of these cases in the columns labeled 2^3 and 2^4, respectively.

After the data have been collected, they are analyzed as follows, provided that the number of replicates is greater than one. Consider the tabulations in Table 11.5. The + and − signs in each column represent +1 and −1, respectively. The columns for the primary factors A, B, C, and D are the same as those given in Table 11.4, except that now the + and − signs stand for +1 and −1, respectively. The columns representing all the interaction terms are obtained by multiplying the corresponding signs in the columns of the primary factors. Thus, the signs in the columns designating the interaction ABC are obtained by multiplying the signs in the columns labeled A, B, and C. For example, in row seven ($m = 7$) $A = -1$, $B = +1$, and $C = +1$; therefore, the sign in the seventh row of the column labeled ABC is -1 [$= (-1)(+1)(+1)$]. Furthermore, the 2^2 experiment uses the first three columns and the rows $m = 1, 2,\dots, 4$; the 2^3 experiment uses the first seven columns and the rows $m = 1, 2,\dots, 8$; and the 2^4 experiment uses the first 15 columns and the rows $m = 1, 2,\dots, 16$.

TABLE 11.4
The Levels and Run Order of Each Factor for a 2^2, 2^3, and 2^4 Factorial Experiment

Run No.	Factors and Their Levels				Data ($y_{m,j}$)			Run Order*		
m	A	B	C	D	$j = 1$	$j = 2$	...	2^2	2^3	2^4
1	−	−	−	−	$y_{1,1}$	$y_{1,2}$		3	5	6
2	+	−	−	−	$y_{2,1}$	$y_{2,2}$		1	7	11
3	−	+	−	−	$y_{3,1}$	$y_{3,2}$		4	8	14
4	+	+	−	−	$y_{4,1}$	$y_{4,2}$		2	4	5
5	−	−	+	−	$Y_{5,1}$	$y_{5,2}$			2	13
6	+	−	+	−	$y_{6,1}$	$y_{6,2}$			1	2
7	−	+	+	−	$y_{7,1}$	$y_{7,2}$			3	16
8	+	+	+	−	$y_{8,1}$	$y_{8,2}$			6	15
9	−	−	−	+	$y_{9,1}$	$y_{9,2}$				9
10	+	−	−	+	$y_{10,1}$	$y_{10,2}$				7
11	−	+	−	+	$y_{11,1}$	$y_{11,2}$				10
12	+	+	−	+	$y_{12,1}$	$y_{12,2}$				3
13	−	−	+	+	$y_{13,1}$	$y_{13,2}$				8
14	+	−	+	+	$y_{14,1}$	$y_{14,2}$				4
15	−	+	+	+	$y_{15,1}$	$y_{15,2}$				1
16	+	+	+	+	$y_{16,1}$	$y_{16,2}$				12

* One set of randomly ordered runs for $j = 1$ only. For $j = 2$ a new randomly generated run order is used, and so on.

The sum of squares is obtained for $n > 1$ as follows:

$$SS_{total} = \sum_{j=1}^{n} \sum_{m=1}^{2^k} y_{m,j}^2 - 2^k n \bar{y}^2 \quad k = 2, 3, \ldots$$

$$SS_{error} = SS_{total} - \sum_{\lambda} SS_{\lambda}$$

$$SS_{\lambda} = \frac{C_{\lambda}^2}{n2^k} \quad \lambda = A, B, AB, \ldots; \quad k = 2, 3, \ldots$$

where

$$C_{\lambda} = \sum_{m=1}^{2^k} S_m \times (\text{sign in row } m \text{ of column } \lambda) \quad \lambda = A, B, AB, \ldots; \quad k = 2, 3, \ldots$$

$$\bar{y} = \frac{1}{n2^k} \sum_{m=1}^{2^k} S_m$$

and S_m is defined in Table 11.5.

The average value of the effect of the primary factors and their interactions is obtained from the relation

$$\text{Effect}_{\lambda} = \frac{C_{\lambda}}{n2^{k-1}} \quad \lambda = A, B, AB, \ldots; \quad k = 2, 3, \ldots$$

TABLE 11.5
Definitions of Various Terms That Are Used to Calculate the Sum of Squares and Mean Square Values for 2^2, 2^3, and 2^4 Factorial Experiments

| Factors and Their Interactions (λ)* | | | | | | | | | | | | | | | Data† | | | | | |
A	B	AB	C	AC	BC	ABC	D	AD	BD	ABD	CD	ACD	BCD	ABCD	j=1	j=2	...	j=n	$S_m = \sum\limits_{j=1}^{n} y_{m,j}$	m
−	−	+	−	+	+	−	−	+	+	−	+	−	−	+	$y_{1,1}$	$y_{1,2}$	...	$y_{1,n}$	S_1	1
+	−	−	−	−	+	+	−	−	+	+	+	+	−	−	$y_{2,1}$	$y_{2,2}$	...	$y_{2,n}$	S_2	2
−	+	−	−	+	−	+	−	+	−	+	+	−	+	−	$y_{3,1}$	$y_{3,2}$	...	$y_{3,n}$	S_3	3
+	+	+	−	−	−	−	−	−	−	−	+	+	+	+	$y_{4,1}$	$y_{4,2}$	...	$y_{4,n}$	S_4	4
−	−	+	+	−	−	+	−	+	+	−	−	+	+	−	$y_{5,1}$	$y_{5,2}$	...	$y_{5,n}$	S_5	5
+	−	−	+	+	−	−	−	−	+	+	−	−	+	+	$y_{6,1}$	$y_{6,2}$	...	$y_{6,n}$	S_6	6
−	+	−	+	−	+	−	−	+	−	+	−	+	−	+	$y_{7,1}$	$y_{7,2}$	...	$y_{7,n}$	S_7	7
+	+	+	+	+	+	+	−	−	−	−	−	−	−	−	$y_{8,1}$	$y_{8,2}$	...	$y_{8,n}$	S_8	8
−	−	+	−	+	+	−	+	−	−	+	−	+	+	−	$y_{9,1}$	$y_{9,2}$	...	$y_{9,n}$	S_9	9
+	−	−	−	−	+	+	+	+	−	−	−	−	+	+	$y_{10,1}$	$y_{10,2}$	...	$y_{10,n}$	S_{10}	10
−	+	−	−	+	−	+	+	−	+	−	−	+	−	+	$y_{11,1}$	$y_{11,2}$	...	$y_{11,n}$	S_{11}	11
+	+	+	−	−	−	−	+	+	+	+	−	−	−	−	$y_{12,1}$	$y_{12,2}$	...	$y_{12,n}$	S_{12}	12
−	−	+	+	−	−	+	+	−	−	+	+	−	−	+	$y_{13,1}$	$y_{13,2}$	...	$y_{13,n}$	S_{13}	13
+	−	−	+	+	−	−	+	+	−	−	+	+	−	−	$y_{14,1}$	$y_{14,2}$	...	$y_{14,n}$	S_{14}	14
−	+	−	+	−	+	−	+	−	+	−	+	−	+	−	$y_{15,1}$	$y_{15,2}$	...	$y_{15,n}$	S_{15}	15
+	+	+	+	+	+	+	+	+	+	+	+	+	+	+	$y_{16,1}$	$y_{16,2}$	...	$y_{16,n}$	S_{16}	16

* The "+" and "−" stand for +1 and −1, respectively, although they also indicate the high and low levels of the factors.
† The data are obtained as indicated in Table 11.4.

TABLE 11.6
ANOVA Table for a 2^k Factorial Experiment with $n > 1$ Replicates

Factor	Sum of Squares	Degrees of Freedom	Mean Square	F_o	$f_{\alpha,1,(n-1)2^k}$
A	SS_A	1	MS_A	MS_A/MS_{error}	Value from f-table
B	SS_B	1	MS_B	MS_B/MS_{error}	Value from f-table
C	SS_C	1	MS_C	MS_C/MS_{error}	Value from f-table
$\vdots$					
AB	SS_{AB}	1	MS_{AB}	MS_{AB}/MS_{error}	Value from f-table
AC	SS_{AC}	1	MS_{AC}	MS_{AC}/MS_{error}	Value from f-table
BC	SS_{BC}	1	MS_{BC}	MS_{BC}/MS_{error}	Value from f-table
$\vdots$					
ABC	SS_{ABC}	1	MS_{ABC}	MS_{ABC}/MS_{error}	Value from f-table
$\vdots$					
Error	SS_{error}	$2^k(n-1)$	MS_{error}		
Total	SS_{total}	$n2^k - 1$			

where Effect$_\lambda$ is called the effect of λ. As seen in Table 11.5, for $k = 2$ there are three λ's: A, B, and AB; for $k = 3$ there are seven λ's: A, B, C, AB, AC, BC, and ABC; and for $k = 4$ there are fifteen λ's: A, B, C, D, AB, AC, BC, AD, BD, CD, ABC, ABD, ACD, BCD, and $ABCD$.

The mean square values for the main effects and their interactions are simply

$$MS_\lambda = SS_\lambda$$

since the number of degrees of freedom for each factor is one. The mean square for the error is

$$MS_E = \frac{SS_E}{2^k(n-1)} \qquad n > 1; \quad k = 2, 3, \ldots$$

The test statistic is

$$F_o = \frac{MS_\lambda}{MS_{error}} = \frac{MS_\lambda}{SS_{error}/[2^k(n-1)]} \qquad \lambda = A, B, AB, \ldots; \quad k = 2, 3, \ldots$$

The ANOVA table for the 2^k factorial analysis is given in Table 11.6.

11.4.6 2^k FACTORIAL ANALYSIS WITH ONE REPLICATE

In some cases, it is more economical to perform a factorial experiment with only one replicate. When only one replicate is used there is no estimate of the mean square error. There is, however, a graphical method that permits one to identify the significant factors and interactions; but the error variance still remains unknown.

Before discussing how this procedure works, we introduce the plotting of data in which the ordinate (*y*-axis) is scaled using the cumulative normal distribution function. This is analogous to plotting data on which the ordinate has been scaled by the logarithm. On a graph in which the ordinate has been scaled logarithmically an exponential function will appear as a straight line. Similarly, on

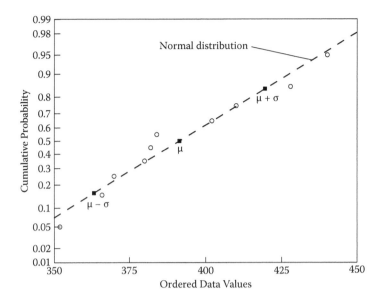

FIGURE 11.4 Graph of the data in Table 11.7 using a probability-scaled ordinate.

a graph in which the ordinate has been scaled with the normal cumulative probability function, a process that has its ordered values distributed normally will appear as a straight line. In other words, for a normal distribution the cumulative probability values $P(z)$ one standard deviation s on either side of the mean μ are $P(\mu + s) = 0.84$ and $P(\mu - s) = 0.16$, respectively, while that of the mean is $P(\mu) = 0.5$. Thus, on a probability transformed graph these three sets of coordinates $[\mu - s, 0.16]$, $[\mu, 0.5]$ and $[\mu + s, 0.84]$ specify three points that lie on a straight line as shown in Figure 11.4.

The procedure for plotting data with a probability-scaled y-axis is as follows. Consider a set of m data values y_i, $i = 1, 2,..., m$. Order the data from the smallest (most negative) to the largest (most positive) value and assign the lowest value the number 1, and the next lowest value the number 2, and so on, with the highest value having the number m. Call these ordered data values z_j, $j = 1$, $2,..., m$. Corresponding to each z_j, we assign a cumulative probability of $(j - 0.5)/m$, $j = 1, 2,..., m$; that is, the probability that $z \leq z_j$. The coordinates of each data value that are to be plotted on the probability scaled y-axis are $[z_j, (j - 0.5)/m]$. To illustrate this procedure, consider the ordered data shown in Table 11.7. Their mean value is $\mu = 391.4$ and the sample variance is $s^2 = 787.6$ ($s = 28.06$). These data are shown in Figure 11.4. If the data were truly normally distributed they would all lie on the straight line. However, based on the closeness and distribution of the points about this line it is reasonable to assume that these data are normally distributed.

We now proceed to use this technique to identify the statistically significant factors and interactions in a factorial design with one replicate, and for $k > 2$. The procedure is similar to the one used to obtain the results in Figure 11.4. The ordered data values are replaced by the ordered values of Effects$_\lambda$. The effects that are negligible will be normally distributed and will tend to fall on a straight line on this plot, whereas the effects that are significant will lie far from this straight line. Consider the data given in Table 11.8, which were obtained from a 2^4 design with a single replicate. Using the formulas of the previous section, we compute the sum of squares and the effects, which are tabulated in Table 11.8. These effects are ordered and the cumulative probabilities assigned as shown in the last columns of Table 11.8. The results are then plotted in Figure 11.5. It is seen from Table 11.8 that the sum of squares for A, C, D, AC, and AD accounts for 96.6% of the total sum. When the ordered effects are plotted using probability coordinates, it is seen that the effects due to A, C, D, AC, and AD fall far from the normal distribution line. This indicates that they are not normally distributed and, therefore, affect the process in a statistically significant manner.

TABLE 11.7
Ordered Data and Their Assigned
Probabilities

j	z_j	Cumulative Probability of z_j [$(j - 0.5)/10$]
1	352	0.05
2	366	0.15
3	370	0.25
4	380	0.35
5	382	0.45
6	384	0.55
7	402	0.65
8	410	0.75
9	428	0.85
10	440	0.95

TABLE 11.8
Analysis of a 2^4 Factorial Experiment with One Replicate[a]

Run No.	Factors and Their Levels				Data	Calculated Values						
m	A	B	C	D	$y_{m,1}$	λ	SS_λ	Effect$_\lambda$	Ordered Effect$_\lambda$	Ordered λ	Probability $(i\text{-}0.5)/15$	i
1	−	−	−	−	86	A	7482.25	43.25	43.25	A	0.9667	15
2	+	−	−	−	200	B	156.25	6.25	19.75	C	0.9000	14
3	−	+	−	−	90	C	1560.25	19.75	6.25	B	0.8333	13
4	+	+	−	−	208	D	3422.25	−29.25	5.25	BCD	0.7667	12
5	−	−	+	−	150	AB	0.25	0.25	4.75	BC	0.7000	11
6	+	−	+	−	172	BC	90.25	4.75	3.75	ABC	0.6333	10
7	−	+	+	−	140	AC	5256.25	−36.25	3.25	ACD	0.5667	9
8	+	+	+	−	192	CD	20.25	2.25	2.25	CD	0.5000	8
9	−	−	−	+	90	AD	4422.25	−33.25	0.75	BD	0.4333	7
10	+	−	−	+	142	BD	2.25	0.75	0.25	AB	0.3667	6
11	−	+	−	+	96	ABC	56.25	3.75	−2.75	ABCD	0.3000	5
12	+	+	−	+	130	BCD	110.25	5.25	−8.75	ABD	0.2333	4
13	−	−	+	+	136	ACD	42.25	3.25	−29.25	D	0.1667	3
14	+	−	+	+	120	ABD	272.25	−8.25	−33.25	AD	0.1000	2
15	−	+	+	+	160	ABCD	30.25	−2.75	−36.25	AC	0.0333	1
16	+	+	+	+	130	Total	22,923.75					
					$\bar{y} = 140.2$							

[a] The calculated values were obtained using a MATLAB program based on Example 14.13 from E. B. Magrab et al., 2005, *An Engineers Guide to MATLAB*, 2nd ed., Prentice Hall, Saddle River, NJ.

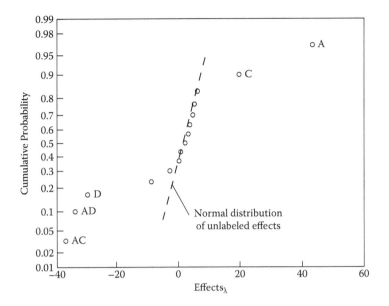

FIGURE 11.5 Statistically significant Effect$_\lambda$ for the data in Table 11.8.

11.4.7   REGRESSION MODEL OF THE OUTPUT

The results of ANOVA for the 2^k factorial or a 2^{k-p} fractional factorial design (see Section 11.4.7) can be used directly to obtain a relationship that estimates the output of the process as a function of the statistically significant primary factors and the statistically significant interactions. We first introduce the coded variable x_β

$$x_\beta = \frac{2\beta - \beta_{\text{low}} - \beta_{\text{high}}}{\beta_{\text{high}} - \beta_{\text{low}}} \tag{11.1}$$

where β is a primary variable; that is, $\beta = A, B, C,\dots$ If $\beta = A$, then $\beta_{\text{high}} = A_{\text{high}}$ and $\beta_{\text{low}} = A_{\text{low}}$ and when $\beta = A_{\text{high}}$, $x_A = +1$, and when $\beta = A_{\text{low}}$, $x_A = -1$. If $A_{\text{low}} \leq A \leq A_{\text{high}}$, then $-1 \leq x_A \leq +1$.

The estimate of the average output y_{avg} cannot be easily expressed in a general notation. Therefore, its generalization will have to be inferred from the following specific example. Consider again the data given in Table 11.8, and the results shown in Figure 11.5, which identified the statistically significant factors and their interactions. (When the number of replicates is greater than one, the statistically significant factors and their statistically significant interactions would be identified from its ANOVA table, Table 11.6.) The statistically significant primary factors and their statistically significant interactions were found to be A, C, D, AC, and AD. Then, an estimate of y_{avg} is given by

$$y_{\text{avg}} = \bar{y} + 0.5[\text{Effect}_A x_A + \text{Effect}_C x_C + \text{Effect}_D x_D + \text{Effect}_{AC} x_A x_C + \text{Effect}_{AD} x_A x_D]$$

$$= 140.13 + 21.63 x_A + 9.88 x_C - 14.63 x_D - 18.13 x_A x_C - 16.63 x_A x_D$$

where x_β takes on any value between -1 and $+1$ and $\beta = A, C,$ and D.

To verify that the equation for y_{avg} is reasonable, one compares its values to those obtained from the experiment, $y_{m,1}$, at each of the 16 combinations of levels appearing in Table 11.8. Thus, for example, for combination $m = 7$ in Table 11.8, we see that the measured value is 140 when $x_A = -1$, $x_C = +1$, and $x_D = -1$. Thus, $y_{\text{avg}} = 144.5$ and the difference between y_{measured} and y_{avg} is 4.50, which is called a residual. If this calculation is performed for each of the 16 combinations, and if the resulting differences (residuals) are ordered and plotted on probability scaled coordinates as shown in Figure 11.4, then we would have a basis on which to state whether or not y_{avg} is a reasonable representation of the process. In other words, if the residuals are normally distributed, then this equation is adequate.

11.4.8 2^k FRACTIONAL FACTORIAL ANALYSIS

Fractional factorial analyses are sometimes more useful than 2^k factorial analyses because they require fewer experimental runs. However, this decrease in the number of runs comes at the cost of introducing confounding. Recall that confounding is when one is unable to differentiate between either a main effect and its interaction or an interaction and other interactions. With confounding, it is not possible to get an estimate of the mean square error. Thus, fractional factorial designs are used most effectively when either the interactions and their magnitudes are known *a priori* or a screening set of experiments are to be run to determine which factors, if any, are important statistically. Fractional factorial designs are denoted by 2_R^{k-p}, where k is the number of factors, and $p = 1$ if the number of runs is to be decreased by a factor of two and $p = 2$ if the number of runs is to be decreased by a factor of four. The quantity R is the resolution of the experiment. The resolution indicates the severity of the confounding. When $R = III$, called a resolution three design, no primary factors are confounded with other primary factors, but they are confounded with other two factor interactions. An example of this is a 2^{3-1} design, which is a resolution three design denoted 2_{III}^{3-1}. When $R = IV$, a resolution four design, no primary factors are confounded with either other primary factors or with any two or more factor interactions. An example of this is a 2^{4-1} design, which under certain circumstances, is a resolution four design denoted 2_{IV}^{4-1}. When $R = V$, a resolution five design, no primary factors or two factor interactions are confounded with either other primary factors or two factor interactions. An example of this is a 2^{5-1} design, which, under certain circumstances, is a resolution five design denoted 2_V^{5-1}. The levels of the factors for the 2_{III}^{3-1}, 2_{IV}^{4-1}, and 2_V^{5-1} designs are given in Table 11.9.

TABLE 11.9
Levels of the Factors for Several Fractional Factorial Designs

Run No.	2_{III}^{3-1}			2_{IV}^{4-1}				2_V^{5-1}				
m	A	B	C	A	B	C	D	A	B	C	D	E
1	−	−	+	−	−	−	−	−	−	−	−	+
2	+	−	−	+	−	−	+	+	−	−	−	−
3	−	+	−	−	+	−	+	−	+	−	−	−
4	+	+	+	+	+	−	−	+	+	−	−	+
5				−	−	+	+	−	−	+	−	−
6				+	−	+	−	+	−	+	−	+
7				−	+	+	−	−	+	+	−	+
8				+	+	+	+	+	+	+	−	−
9								−	−	−	+	−
10								+	−	−	+	+
11								−	+	−	+	+
12								+	+	−	+	−
13								−	−	+	+	+
14								+	−	+	+	−
15								−	+	+	+	−
16								+	+	+	+	+

The effect of confounding, which is always present in fractional factorial designs, prevents one from estimating MS_{error}. Therefore, the data from a fractional factorial experiment are analyzed with the same graphical procedure that is used for the factorial experiment with one replicate.

11.5   EXAMPLES OF THE USE OF THE ANALYSIS OF VARIANCE

11.5.1   EXAMPLE 1—MANUFACTURE OF STIFF COMPOSITE BEAMS

The objective is to determine the manufacturing conditions that produce the stiffest fiberglass and epoxy composite beam when the beam is subjected to a three-point bending load. The manufacturing conditions are determined by running a three-factor, single replicate, full factorial experiment, which results in 27 different manufacturing combinations. The three factors are (1) the fiberglass fabric orientation; (2) epoxy resins from three manufacturers, with each resin having the same nominal strength characteristics; and (3) the amount of the hardener (curing agent) provided by each manufacturer and used with their epoxy resin. These combinations are summarized in Table 11.10. Twenty-seven beams are fabricated under 27 different combinations of these manufacturing conditions, which are tabulated in Table 11.11. The beams are cured at room temperature for at least 48 hours, and are initially in the form of thin plates 7.6 cm by 17.8 cm. These fiberglass plates are then trimmed to form 11.4 cm by 2.5 cm beams. Each of the 27 beams is then subjected to a three-point bending test to determine its stiffness. The three-point bending test supports the beam very close to its free edges while a load is applied at its center. The load is varied over a range of values, and at each value the displacement of the beam under the load is measured. The results are plotted and the best straight line through them is obtained. The slope of the line is the stiffness, when the x-axis is the beam's displacement.

The fiberglass fabric is nominally 0.7 mm thick. The composite has a 50% fiberglass content, which means that 50% of the volume is epoxy. The mold is filled with five layers of fiberglass cloth patches that are cut along different axes (bias) with respect to the nominal orientation of the fabric's weave. This gives a fabric with different weave angles as indicated in Table 11.10. The results of the tests are given in Table 11.12 and their analysis in Table 11.13. The average values of the statistically significant factors, A (fiber orientation) and C (resin system), are plotted in Figure 11.6. Based on the results shown in Figure 11.6, it is seen that in order to obtain the stiffest composite beams one should use a weave angle of 0° and a resin system from manufacturer #2. Since the hardener ratio is not a significant factor, one should use that recommended by the manufacturer.

11.5.2   EXAMPLE 2—OPTIMUM PERFORMANCE OF AN AIR-DRIVEN VACUUM CLEANER[*]

Pneumatic vacuum cleaners use compressed air to generate their suction. Figure 11.7 illustrates the major components of a pneumatic vacuum that produces the suction, which is called an ejector. It operates as follows. Compressed air, which is typically at 620 kPa, is discharged through the jet at the base of a nozzle. The exit velocity of the jet is choked (Mach number equals one), and the high velocity stream is decelerated in the mixing section of the nozzle by entraining the suction air. Further entrainment of the mixture and recovery of pressure occurs in the diffuser. The motive and suction air pass either directly to the outside or, if the exhaust is too loud, through an acoustic muffler.

The primary objective is to optimize the ejector by determining those dimensions that result in maximum suction flow, while having a design that would be insensitive to the ejector's dimensional variations. The pneumatic vacuum has to meet the following performance requirements: (1)

---

[*] T. E. Dissinger and E. B. Magrab, "Redesign of a pneumatic vacuum cleaner for improved manufacturability and performance," Paper no. 95-WA/DE-14, 1995 ASME International Mechanical Engineering Congress and Exposition, San Francisco, CA, November 1995.

TABLE 11.10
Manufacturing Conditions for the Coded Combinations Given in Table 11.11

Factor A: Lay-up and order and weave angle (E, F, G)

Lay-up layer order (1, 2, 3, 4, 5)

Lay-up E: (0°, 0°, 0°, 0°, 0°)

Lay-up F: (30°, 30°, 0°, −30°, 30°)

Lay-up G: (45°, −45°, 0°, −45°, 45°)

Factor B: Hardener (curing agent) ratio (L, M, N)

L = Recommended by manufacturer

M = Increase from that recommended by the manufacturer

N = Decrease from that recommended by the manufacturer

Factor C: Resin system (X, Y, Z)

X = Manufacturer #1

Y = Manufacturer #2

Z = Manufacturer #3

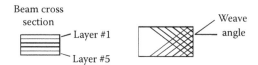

TABLE 11.11
Twenty-Seven Coded Manufacturing Combinations*

ELX(11)	EMX(4)	ENX(16)	ELY(24)	EMY(5)	ENY(12)	ELZ(17)	EMZ(25)	ENZ(20)
FLX(8)	FMX(22)	FNX(26)	FLY(3)	FMY(2)	FNY(18)	FLZ(23)	FMZ(15)	FNZ(1)
GLX(14)	GMX(6)	GNX(13)	GLY(9)	GMY(7)	GNY(27)	GLZ(19)	GMZ(10)	GNZ(21)

* Numbers in parentheses correspond to the order in which specimens are fabricated. The manufacturing conditions corresponding to the three-letter combinations are described in Table 11.10.

TABLE 11.12
Measured Stiffness (N/m) for the 27 Manufacturing Combinations Given in Table 11.11

Resin system	Fiber Orientation E Hardener Ratio			Fiber Orientation F Hardener Ratio			Fiber Orientation G Hardener Ratio		
	L	M	N	L	M	N	L	M	N
X	31.6×10^3	39.8×10^3	3.7×10^3	25.2×10^3	21.2×10^3	14.8×10^3	22.6×10^3	15.7×10^3	19.9×10^3
Y	48.2×10^3	48.0×10^3	44.1×10^3	31.0×10^3	39.7×10^3	33.8×10^3	22.1×10^3	36.9×10^3	25.4×10^3
Z	40.8×10^3	38.2×10^3	38.7×10^3	29.8×10^3	27.6×0^3	37.9×10^3	27.2×10^3	24.5×10^3	26.7×10^3

TABLE 11.13
ANOVA for the Data in Table 11.12[a]

Source of Variation	Sum of Squares	Degrees of Freedom	Mean Square	F_o	f at 95%
A (Fiber orientation)	7.1546×10^8	2	3.5773×10^8	7.97 ✓	4.46
B (Hardener)	1.2996×10^8	2	6.4980×10^7	1.45	4.46
C (Resin system)	1.0821×10^9	2	5.4105×10^8	12.05 ✓	4.46
AB	1.8146×10^8	4	4.5356×10^7	1.01	3.84
BC	3.5506×10^8	4	8.8764×10^7	1.98	3.84
AC	1.3010×10^8	4	3.2526×10^7	0.72	3.84
Error + ABC	3.5932×10^8	8	4.4915×10^7		
Total	2.9534×10^9	26			

✓ = Statistically significant at the 95% confidence level.
[a] The MATLAB function anovan was used to obtain these results.

maximum compressed air flow[*] of 1.13 sm³/min at 620 kPa; and (2) suction flow of 2.8 sm³/min at the end of the hose. The current design has a suction flow of 1.7 sm³/min.

Referring to Figure 11.7, it is seen that the design parameters are the internal diameter of the air jet D_j, the jet angle θ_j, the inlet diameter D_i, the inlet angle θ_i, the mixing section diameter D_{ms}, the mixing section length L_{ms}, the diffuser angle θ_d, and the diffuser length L_d. Initial analyses and testing showed that the nozzle and the jet were essentially uncoupled from each other, and, therefore, the jet could be designed independently. The inlet diameter was weakly coupled, and was selected as the smallest diameter that met the minimum flow requirements. This was found to be $D_i = 3.2$ cm. Preliminary tests also showed that the mixing section diameter primarily governed suction flow, and an analysis showed that a value of $D_{ms} = 2.0$ cm would work well. Additional tests showed that the flow rate increased as the diffuser length increased. Since there was a size limitation, L_d was chosen as large as practical, which turned out to be $L_d = 11.4$ cm.

The effects of the remaining geometric parameters, inlet angle θ_i, mixing section length L_{ms}, and diffuser angle θ_d, were determined using a three-factor, three-level experiment with a single replicate. Three levels were selected because a nonlinear response was expected for one or more of the factors. The levels selected and the corresponding measured flow rates are given in Table 11.14. To get these results, the nozzle's inlet section, mixing section, and diffuser were manufactured as separate segments, three versions of each, and then assembled into the 27 combinations as indicated in Table 11.14. The length of the inlet segment was 1.8 cm and the length of the diffuser was 11.4 cm. The axial position of the air-jet relative to the inlet of the mixing section also affected the performance. Tests were conducted to achieve maximum flow by adjusting the jet's axial position relative to the mixing section. This position remained constant for all 27 tests.

The statistical analysis of the results is given in Table 11.15, where it is seen that the main effects θ_i, L_{ms} and θ_d are significant, and that the two-factor interactions are not. The average responses for the statistically significant factors are shown in Figure 11.8, where it is seen that the maximum average flow rate is obtained when $\theta_i = 25°$, $L_{ms} = 7.6$ cm, and $\theta_d = 3°$. However, in practical terms θ_i can be as small as 10° and L_{ms} as small as 50.8 mm. To examine more closely the sensitivity of the flow with the diffuser angle, a set of tests were run with

---

[*] The number of standard cubic meters per minute (sm³/min) is obtained by multiplying the number of cubic meters per minute by the ratio of the pressure of the compressed air to atmospheric pressure, which, for this case, is 6.

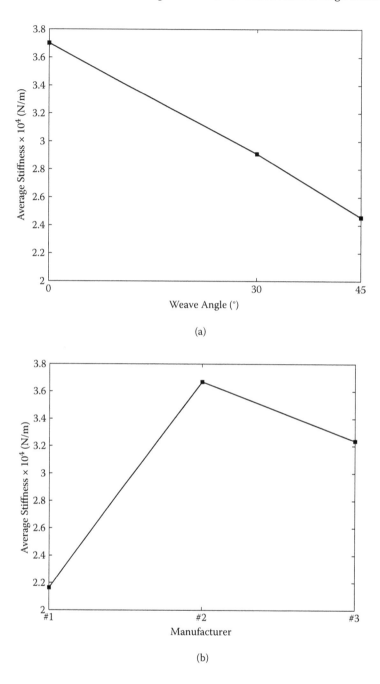

FIGURE 11.6 Average stiffness of the statistically significant primary factors: (a) weave angle and (b) manufacturer.

$\theta_i = 25°$, $L_{ms} = 7.6$ cm, $\theta_j = 10°$, $D_j = 3.2$ cm, and $L_d = 11.4$ cm, while θ_d was varied from $2°$ to $5°$ in one-half degree increments. The results, which are shown in Figure 11.9, indicate that over the range $2.5° < \theta_d < 3.5°$ the flow remains within 1% of its maximum average value. Hence, the final values for the three factors and their ranges of acceptable values are those given in Table 11.16. Thus, the design objective of wide manufacturing tolerances was met, and the minimum improved flow was exceeded by 10%.

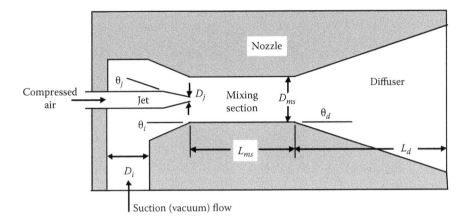

FIGURE 11.7 Cross sections of ejector components and their corresponding design parameters.

TABLE 11.14
Levels for the Three-Factor, Three-Level Experiment and the Measured Flow Rates (sm³/min)

θ_i	$L_{ms} = 50.8$ mm			$L_{ms} = 76.2$ mm			$L_{ms} = 101.6$ mm		
	$\theta_d = 1°$	$\theta_d = 3°$	$\theta_d = 5°$	$\theta_d = 1°$	$\theta_d = 3°$	$\theta_d = 5°$	$\theta_d = 1°$	$\theta_d = 3°$	$\theta_d = 5°$
10°	2.86	3.07	2.93	2.83	3.06	3.00	2.72	3.04	3.01
25°	2.88	3.00	2.89	2.85	3.14	3.03	2.73	3.02	2.89
40°	2.74	3.00	2.89	2.72	3.09	2.99	2.66	2.90	2.83

TABLE 11.15
ANOVA for the Three-Factor, Three-Level Ejector Design[a]

Source of Variation	Sum of Squares	Degrees of Freedom	Mean Square	F_o	f at 95%
A (Length, L_{ms})	0.0504	2	0.0252	19.79 ✓	4.46
B (Inlet angle, θ_i)	0.0440	2	0.0212	17.25 ✓	4.46
C (Diffuser angle, θ_d)	0.3507	2	0.1754	137.59 ✓	4.46
AB	0.0074	4	0.0018	1.44	3.84
BC	0.0030	4	0.0008	0.59	3.84
AC	0.0112	4	0.0028	2.19	3.84
Error + ABC	0.0102	8	0.0013		
Total	0.4769	26			

✓ = Statistically significant at the 95% confidence level.
[a] The MATLAB function anovan was used to obtain these results.

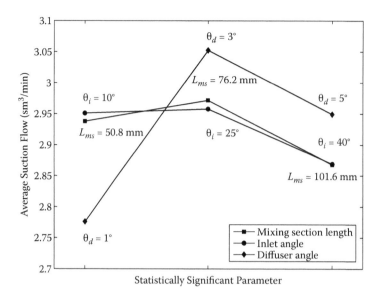

FIGURE 11.8 Average suction flow rates as a function of the inlet angle, diffuser angle, and mixing section length.

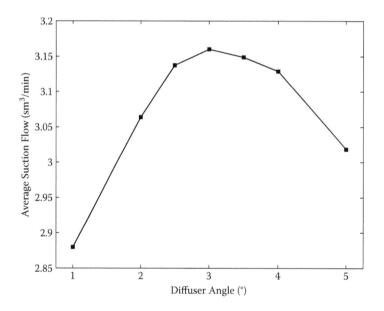

FIGURE 11.9 Suction flow as a function of diffuser angle.

TABLE 11.16
Dimensional Parameters of the Ejector and Their Acceptable Ranges

Parameter	Final Values and Ranges
Diffuser angle θ_d	$3° \pm 0.5°$
Inlet angle θ_i	$25° \pm 5°$
Mixing section length L_{ms}	64 ± 13 mm

11.6 THE TAGUCHI METHOD

Taguchi advocates a philosophy of quality engineering that employs experimental design as a formal part of the engineering design process. He considers three stages in a product's or process' development: system design, parameter design, and tolerance design. In system design, the engineer uses scientific and engineering principles to determine the basic configuration. For example, if one is to measure an unknown resistance, knowledge of electrical circuits indicates that the basic system should be configured as a Wheatstone bridge. On the other hand, if one is designing a process to assemble printed circuit boards, then one would specify the axial insertion machines, the surface-mount placement machines, the flow solder machines, and so forth. In parameter design, the specific values for the system parameters are determined. This would involve choosing the nominal resistor and power supply values for the Wheatstone bridge example, the number and type of component placement machines for the printed circuit board assembly process, and so forth. Usually, the objective is to specify these nominal parameter values such that the variability transmitted from uncontrollable (noise) variables is minimized. Tolerance design is used to determine the best tolerances for the parameters. In the Wheatstone bridge example, tolerance design methods would reveal which components in the design were most sensitive and where the tolerances should be set. If a component does not have much effect on the performance of the circuit, it would be specified with a wide tolerance.

Taguchi recommends that statistical experimental design methods be employed to assist in quality improvement, particularly during parameter design and tolerance design. Experimental design methods can be used to find a best product or process design, where "best" means a product or process that is robust to uncontrollable factors. A product or process is said to be robust when it is insensitive to the effects of sources of variability, even though the sources of variability have not been eliminated.

A key component of Taguchi's philosophy is the reduction of variability. The variation in a performance characteristic cannot be defined satisfactorily unless its ideal (target) value is known. Once the target value is determined, one can define variation in relation to it. A high-quality product performs near the target value consistently throughout the product's life and under its operating conditions.

In order to develop a process of designing products that provides on-target performance and maintains it in the face of variability, the nature of noise has to be recognized. There are three types of noise factors: (1) external noise factors, (2) unit-to-unit noise factors, and (3) deterioration noise factors. External noise factors are sources of variability that come from outside the product. Examples of external sources of variability are (a) environmental (temperature, relative humidity, dust, ultraviolet, electromagnetic interference) and (b) any unintended input of energy (heat, vibration, radiation) to which the system is sensitive. Unit-to-unit noise is a result of not being able to make any two items exactly alike. Manufacturing processes and materials are major sources of unit-to-unit variability in product components. Deterioration noise is often referred to as an internal noise factor, because something changes internally within the product or process. It is common for certain products to "age" during use or storage so that performance deteriorates—for example, the weathering of paint on a house.

The user's perception of the quality of a product is very closely related to its sensitivity to noise. Therefore, the effect of noise on the performance of the product or process has to be minimized. There are two ways to minimize this variability: (1) eliminate the source of noise, or (2) eliminate the product's sensitivity to the source of noise. For the latter approach, experimental design methods discussed in this chapter play a major role.

Taguchi considers any deviation of the product or process from its desired performance as creating a loss to the manufacturer, the customer, and society. Losses that may be incurred by the manufacturer could be inspection, scrap and rework, warranty costs, and returns. Losses to the customer could be time and effort taken to work around minor failures, lost profits due to a nonfunctioning machine, and service contract costs. Losses to society could be pollution and waste.

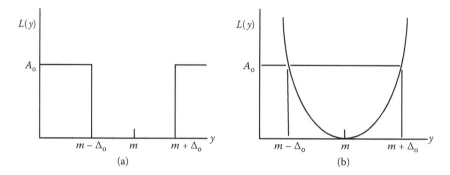

FIGURE 11.10 Loss function when (a) the loss is a minimum in the specification interval and (b) the loss varies nonlinearly in the specification interval.

11.6.1 QUALITY LOSS FUNCTION

All target specifications of continuous performance characteristics should be stated in terms of the nominal levels and the tolerances above the nominal levels. It is still a practice in some U.S. industries to state target values in terms of interval specifications only. This practice erroneously conveys the idea that the quality level is equally good for all values of the performance characteristic in the specification interval, and then suddenly deteriorates once the performance value exceeds the specification interval. This type of specification is illustrated in Figure 11.10a.

Taguchi proposed that the cost of deviation of the performance from its target value is quadratic in nature, instead of constant, such that the minimum cost is incurred at the target value as shown in Figure 11.10b. In other words, on-target performance is more important than conformance to a specification interval. The cost of the deviation from the target value is obtained from the quality loss function.

There are three types of cases that can be described by the quality loss function: (1) nominal-is-the-best; (2) smaller-is-the-better; and (3) larger-is-the-better. Examples of nominal-is-the-best are diameter of an engine cylinder and gain of an operational amplifier. Examples of smaller-is-the-better are microwave oven radiation leakage and automotive exhaust pollution. Examples of larger-is-the-better are the strength of an adhesive and the traction of a tire. The quality loss functions $L(y)$ for these three cases are as follows:

Nominal-is-the-best

$$L(y) = A_o (y - m)^2 / \Delta_o^2$$

Smaller-is-the-better

$$L(y) = A_o y^2 / \Delta_o^2$$

Larger-is-the-better

$$L(y) = A_o \Delta_o^2 / y^2$$

where Δ_o and A_o are defined in Figure 11.10b.

The implementation of the Taguchi method is similar to the analysis of variance technique. However, there are some important differences, which are discussed at some length in

Montgomery[*] and Fowlkes and Creveling.[†] It appears that the Taguchi method is an engineer's approach to factorial analysis, which places the speed at which results can be obtained ahead of some of the subtleties of statistical theory. Specifically, it does not use any statistical test for significance, and does not include a means to determine the interactions, which are frequently confounded with the main effects. In spite of this, the method is widely and successfully used.

11.7   SIX SIGMA

As mentioned in Section 1.4.3, Six Sigma derives its name from statistics, where sigma (σ) stands for the standard deviation of an attribute of a process, which we denote as X. The Six Sigma methodology assumes that the variation of the attribute of the process has a normal (Gaussian) probability distribution with mean μ and standard deviation σ. To quantify the meaning of Six Sigma, we introduce the cumulative distribution function for a normal distribution $\Phi_N(X < X_o; \mu, \sigma,)$. It indicates the probability that an attribute X is less than some specified value X_o when the population of X is normally distributed with a mean μ and standard deviation σ. Thus, the probability that a measured sample of the attribute X will have a value that is less then an upper specification limit (USL) is given by

$$p_U = \Phi_N(X < \text{USL}; \mu, \sigma)$$

and the probability that it will be less than its lower specification limit (LSL) is given by

$$p_L = \Phi_N(X < \text{LSL}; \mu, \sigma)$$

Then the probability that LSL $< X <$ USL is determined from

$$p_{U-L} = p_U - p_L = \Phi_N(X < \text{USL}; \mu, \sigma) - \Phi_N(X < \text{LSL}; \mu, \sigma)$$

Motorola, who developed and implemented the ideas behind Six Sigma, interprets this relationship as follows. They assume that the mean of the process can deviate from the desired mean μ by as much as $\pm 1.5\sigma$. If this is the case, then we have two limiting cases to consider: $\mu \to \mu - 1.5\sigma$ and $\mu \to \mu + 1.5\sigma$. The probabilities for these two cases are, respectively,

$$p_{(U-L)-} = \Phi_N(X < \text{USL}; \mu - 1.5\sigma, \sigma) - \Phi_N(X < \text{LSL}; \mu - 1.5\sigma, \sigma)$$

and

$$p_{(U-L)+} = \Phi_N(X < \text{USL}; \mu + 1.5\sigma, \sigma) - \Phi_N(X < \text{LSL}; \mu + 1.5\sigma, \sigma)$$

In addition, Motorola assumes that LSL $= \mu - 6\sigma$ and USL $= \mu + 6\sigma$; hence the name Six Sigma. When these upper and lower specification limits are used, it is found that

$$p_{(U-L)+} = p_{(U-L)-} = 0.9999966$$

In other words, only 3.4 defects per million will occur. If the mean value μ does not vary, then it is found that

$$p_{U-L} = 1.97 \times 10^{-9}$$

---

[*] D. C. Montgomery, *Design and Analysis of Experiments*, 3rd ed., John Wiley & Sons, New York, pp. 426–433, 1991.

[†] W. Y. Fowlkes and C. M. Creveling, *Engineering Methods for Robust Product Design*, Addison Wesley, Reading, MA, pp. 329–335, 1995.

that is, there are only 1.97 defects per billion. This is equivalent to having a wristwatch that has an error of 1 second every 16 years.

BIBLIOGRAPHY

T. B. Barker, *Quality by Experimental Design*, Marcel Dekker, New York, 1985.

A. Bendell, J. Disney, and W. A. Pridmore, Eds., *Taguchi Methods: Applications to World Industry*, IFS (Publications) Ltd, Bedford, UK, 1989.

G. E. P. Box, W. G. Hunter, and J. S. Hunter, *Statistics for Experimenters*, John Wiley & Sons, New York, 1978.

F. W. Breyfogle III, *Statistical Methods for Testing, Development and Manufacturing*, John Wiley & Sons, New York, 1992.

N. L. Frigon and D. Mathews, *Practical Guide to Experimental Design*, John Wiley & Sons, New York, 1997.

R. H. Lochner and J. E. Matar, *Designing for Quality*, Quality Resources, White Plains, NY, 1990.

R. D. Moen, T. W. Nolan, and L. P. Provost, *Improving Quality Through Planned Experimentation*, McGraw-Hill, New York, NY, 1991.

D. C. Montgomery, *Design and Analysis of Experiments*, 3rd ed., John Wiley & Sons, New York, 1991.

R. H. Myers and D. C. Montgomery, *Response Surface Methodology: Process and Product Optimization Using Designed Experiments*, John Wiley & Sons, New York, 1995.

M. S. Phadke, *Quality Engineering Using Robust Design*, Prentice Hall, Englewood Cliffs, NJ, 1989.

J. W. Priest, *Engineering Design for Producibility and Reliability*, Marcel Dekker, New York, 1988.

P. J. Ross, *Taguchi Techniques for Quality Engineering*, McGraw-Hill, New York, 1988.

G. Taguchi, *Introduction to Quality Engineering*, Asian Productivity Organization, Tokyo, 1986.

G. Taguchi, *System of Experimental Design: Engineering Methods to Optimize Quality and Minimize Costs*, Vols. 1 & 2, Kraus International Publications, White Plains, NY, 1987.

G. Taguchi, E. A. Alsayed, and T. Hsiang, *Quality Engineering in Production Systems*, McGraw-Hill, New York, 1989.

Appendix A: Material Properties and the Relative Cost of Raw Materials

TABLE A.1
Material Properties[a]

Material	Young's Modulus (GPa)	Poisson's Ratio	Coefficient of Thermal Expansion (μm/m/°K)	Thermal Conductivity (W/m/°K))	Yield Strength (MPa)
Plain carbon steel (<2% C)	170–180	0.275	11.5–13.7	15.22–32.3	276–621
Low alloy steel (<12% Cr)	76.5–223	0.280–0.300	10.1–14.9	25.3–51.9	180–2400
Stainless steels (>12% Cr)	68.9–240	0.220–0.346	7.02–21.1	1.35–37.2	15.0–2400
Cast irons	62.1–240	0.240–0.370	7.75–19.3	11.3–53.3	65.5–1450
Zinc alloys	63.5–97.0		19.4–39.9	105–125	125–386
Aluminum alloys	70–85	0.33	21–26	78–240	20–500
Magnesium alloys	45–50	0.35	25–30	45.0–135	70–400
Titanium alloys	85–120	1.33	8–11	4.9–12	350–1200
Copper alloys	10–170	0.181–0.375	5.80–26.3	2.00–401	0.250–2140
Brass	13.8–115	0.280–0.375	18.7–21.2	26.0–159	69–683
Bronze	41–125	0.280–0.346	16.0–21.6	33.0–208	69–793
Nickel alloys	28–235	0.230–0.339	0.630–27.3	3.50–225	35–4830
Tin alloys	30–53	0.330–0.400	14.0–36.0	19.0–73.0	11.9–448[b]
Cobalt alloys	100–235		1–15.7	6.50–200	379–1420
Molybdenum alloys	200–365	0.285	4.90–7.20	14.0–280	190–1100
Tungsten alloys	138–430	0.280–0.300	4.40–11.7	70.0–330	310–1240
Invar (Carpenter Invar 36® alloy, cold–drawn bars)	148		1.3 @ 93°C	8.5–11	483
Kovar (Carpenter Kovar® alloy)	138		4.9 @ 30–400°C	17.3	345
Super Invar (Carpenter super invar 32–5)	145	0.23	0.63 @ −55–95°C		276
Alnico (Alliance A–6S alnico magnetic material)	100–200		10–13 @ 20°C	10–15	80–300
Neodymium–iron–boron alloys (Alliance N–33 neodymium iron–boron magnetic material)	150–160		5 @ 20°C	9.0	80*
Copper–manganese–nickel (Sandvik Manganin® Manganin 43 resistance wire)			18.0 @ 20–100°C	22.0	180
Constantan (Thermocouple wire type E–standard)			8.30 @ 20°C	21.8	414[b]
Molybdenum (annealed)	330		6 @ 0–250°C	138	324[b]
Platinum (CP grade, annealed)	171	0.39	9.1 @ 20°C	69.1	125–165[b]

(continued)

TABLE A.1 (CONTINUED)
Material Properties[a]

Material	Young's Modulus (GPa)	Poisson's Ratio	Coefficient of Thermal Expansion (μm/m/°K)	Thermal Conductivity (W/m/°K))	Yield Strength (MPa)
Tantalum (annealed)	186	0.35	6.5 @ 20°C	54.4	450[b]
Tungsten	400	0.28	4.40 @ 20–100°C	163.3	750
Silicon carbide	350–480		4–5.2	110–205	400–650
Molybdenum disilicide ($MoSi_2$)	242		6.5 @ 20°C	66.2	
Graphite	4.80		0.60–4.30 @ 20°C	24.0	
ABS (molded)	1.52–6.10		80–200		20–65
Acrylic (Lycite, Perspex, Plexiglas) (PMMA, general purpose, molded)	0.950–4.50		54.0–150	0.128–0.190	50–70
Polypropylene (PP, molded)	0.85–1.5		18.0–185	0.187–0.216	21–38
Polystyrene (PS)	1.2–2.6	0.334–0.390	1.30–158		28–50
Epoxy (Cast, unreinforced)	2.3–3		61–110	0.0270–5.01	38–72
Fluorcarbon				0.167–0.795	
Phenolics (unreinforced; molded)	4.10–8.64				41.0–57.9[b]
Silicone (silicone rubber)	120–170	0.5	13.9–335		0.414–41.4[b]
Nylon				0.180–3.00	
Polyethylene (PE)	1.2–2.6		121–200		18–28
Acetal (copolymer, unreinforced)	1.38–3.20	0.35	11.0–234	0.231–0.310	37.0–120
Polyimide	1.36–47.0		2.50–60.0	0.042–0.488	23.3–230
Cellulose acetate (molded)	1.60–2.20		11.0–170	0.170–0.330	19.0–51.0
PVC (molded)	0.00159–3.24		50		17.0–52.0
Polyurethane (Elastomer, polyester grade)	0.002–0.02		25.2–171		1.10–42.1
Alkyds	0.0280–3.45		20.0–170		17.9–24.8
PBT (unreinforced, molded)	0.0280–3.45		20.0–170		17.9–24.8
PET (unreinforced)	1.83–3.70		25.0–92.0	0.190–0.290	47–90
Acrylonitrile–butadiene rubber (molded)	1.52–6.10		0.800–139	0.128–0.190	20–65
Neoprene			600–700	0.095–0.13	3.5–15
Butyl rubber	0.0015–0.004		110–300	0.08–0.95	2–3
Silicone rubber (silicone elastomers)	0.005–0.02	0.5	13.9–335	0.180–3.00	2.5–5.5
Aluminum nitride			6.2–7	80–310	
Alumina (99.9%, Al_2O_3)	290–360	0.22	7–12	20–30	360–650
Boron nitride (100%, Borazone)				20.0	
Carbon Fiber (CFRP)	78–170		1.1–3.2	1.2–2.8	560–1000
Silver	76	0.37		419	140[b]

[a] Compiled from the following sources: http://www.matweb.com/; http://www.custompartnet.com/materials/; and M. F. Ashby, *Material Selection in Mechanical Design*, Pergamon Press, Oxford, 1992.
[b] Ultimate strength.

TABLE A.2
Cost of Raw Materials Relative to Plain Carbon Steel Based on Their Weight

Material	Price/Weight Versus Steel	Source/Comments
Plain carbon steel (<2% C)	1	Hot rolled steel coil: http://www.steelonthenet.com/prices.html (10/31/08)
Low alloy steel (<12% Cr)	0.8–1.5	Ungureanu et al.[1]
Stainless steels (400 series)	3–20	http://www.chinamining.org (7/18/08)
Cast irons	0.55–1.1	Ashby[2]
Zinc	0.9–1.1	http://www.lme.co.uk/zinc.asp (10/31/08)
Zinc alloys	1.1–5.5	Ashby[2]
Aluminum	1.5–2.25	http://www.lme.co.uk/aluminium.asp (10/31/08)
Aluminum alloys	2.25–4.5	Ashby[2]
Magnesium	2.5–3.5	www.magnesium.com (10/31/08)
Magnesium alloys	8–10	Ashby[2]
Titanium alloys	40–140	Ashby[2]
Copper	3–4	http://www.lme.co.uk/copper.asp (10/31/08)
Copper alloys	2.5–9.0	Ashby[2]
Brass	2.5–9.0	Ashby[2]
Bronze	2.5–9.0	Ashby[2]
Nickel	9–12	http://www.lme.co.uk/nickel.asp (10/31/08)
Nickel alloys	8–18	Ashby[2]
Tin	11–14	http://www.lme.co.uk/tin.asp (10/31/08)
Cobalt	50–70	cobalt.bhpbilliton.com (10/31/08)
Molybdenum	40–70	www.infomine.com (10/31/08)
Tungsten alloys	25–40	Ashby[2]
Invar	35–45	National Electronic Alloys (9/19/2008)
Kovar	60–75	National Electronic Alloys (9/19/2008)
Platinum	22,500–27,500	www.infomine.com (10/31/08)
Silicon carbide	18–38	Ashby[2]
ABS	4–8	Ashby[2]
Acrylics (Lucite, Plexiglas)	3.5–6	Ashby[2] (values for poly(methyl methacrylate))
Polypropylene	1.7–2.6	Ashby[2]
Polystyrene	2.75–3.5	Ashby[2]
Epoxy	4.8–9.2	Ashby[2]
Silicone	5–12	Ashby[2]
Polyethylene	2.0–2.5	Ashby[2]
Acetal	5.5–10.5	Ashby[2]
Polyurethane	7.5–15	Ashby[2]
Silicone rubber	15–35	Ashby[2]
Aluminum nitride	150–250	Ashby[2]
Aluminum oxide (Al_2O_3)	6–8	Ashby[2]
Carbon fiber	80–160	Ashby[2]
Glass fiber	12–25	Ashby[2]
Silver	300–340	www.kitcosilver.com (10/31/08)

[1] C. A. Ungureanu, Das, S., and Jawahir, I.S., "Life–cycle Cost Analysis: Aluminum versus Steel in Passenger Cars," in *Aluminum Alloys for Transportation, Packaging, Aerospace, and Other Applications*, S. Das and W. Yin, Eds., The Minerals, Metals and Materials Society, 2007, pp. 11–24.

[2] M. Ashby, H. Shercliff, and D. Cebon, *Materials: Engineering, Science, Processing, and Design*, Butterworth–Heinemann (Elsevier), Burlington, MA, 2007.

Index

"f" indicates material in figures. "n" indicates material in footnotes. "t" indicates material in tables.

Milton Keynes UK
Ingram Content Group UK Ltd.
UKHW050307111024
449327UK00043B/2196